T0314293

River Restoration

River Restoration

Political, Social, and Economic Perspectives

Edited by

Bertrand Morandi
Université de Lyon, CNRS, France

Marylise Cottet
Université de Lyon, CNRS, France

Hervé Piégay
Université de Lyon, CNRS, France

ADVANCING RIVER RESTORATION AND MANAGEMENT

WILEY Blackwell

Registered Offices
John Wiley & Sons, Inc., 111 River Street, Hoboken, NJ 07030, USA
John Wiley & Sons Ltd, The Atrium, Southern Gate, Chichester, West Sussex, PO19 8SQ, UK

Editorial Office
9600 Garsington Road, Oxford, OX4 2DQ, UK

For details of our global editorial offices, customer services, and more information about Wiley products visit us at www.wiley.com.

Wiley also publishes its books in a variety of electronic formats and by print-on-demand. Some content that appears in standard print versions of this book may not be available in other formats.

Library of Congress Cataloging-in-Publication Data

Names: Morandi, Bertrand, editor. | Cottet, Marylise, editor. | Piégay,
    Hervé, editor.
Title: River restoration : political, social, and economic perspectives /
    edited by Bertrand Morandi, Marylise Cottet, Hervé Piégay.
Description: Hoboken, NJ : Wiley-Blackwell, 2022. | Includes
    bibliographical references and index.
Identifiers: LCCN 2021027145 (print) | LCCN 2021027146 (ebook) | ISBN
    9781119409984 (hardback) | ISBN 9781119409991 (adobe pdf) | ISBN
    9781119410003 (epub)
Subjects: LCSH: Stream restoration.
Classification: LCC QH75 .R584 2022 (print) | LCC QH75 (ebook) | DDC
    333.91/62153–dc23
LC record available at https://lccn.loc.gov/2021027145
LC ebook record available at https://lccn.loc.gov/2021027146

Cover Design: Wiley
Cover Image: © Marylise Cottet

Set in 9.5/12.5pt STIXTwoText by Straive, Pondicherry, India
Printed and bound by CPI Group (UK) Ltd, Croydon, CR0 4YY

C9781119409984_090921

Contents

Series Foreword

The field of river restoration and management has evolved enormously in recent decades, driven largely by increased recognition of the ecological values, river functions, and ecosystem services. Many conventional river management techniques, emphasizing hard structural controls, have proven difficult to maintain over time, resulting in sometimes spectacular failures, and often degraded river environments. More sustainable results are likely from a holistic framework, which requires viewing the "problem" at a larger catchment scale and involves the application of tools from diverse fields. Success often hinges on understanding the sometimes complex interactions among physical, ecological, and social processes.

Thus, effective river restoration and management requires nurturing the interdisciplinary conversation, testing and refining our scientific theories, reducing uncertainties, designing future scenarios for evaluating the best options, and better understanding the divide between nature and culture that conditions human actions. It also implies that scientists better communicate with managers and practitioners, so that new insights from research can guide management and results from implemented projects can, in turn, inform research directions.

The series provides a forum for "integrative sciences" to improve rivers. It highlights innovative approaches, from the underlying science, concepts, methodologies, new technologies, and new practices, to help managers and scientists alike improve our understanding of river processes, and to inform our efforts to better steward and restore our fluvial resources for more harmonious coexistence of humans with their fluvial environment.

G. Mathias Kondolf, *University of California, Berkeley, USA*
Hervé Piégay, *University of Lyon, CNRS, France*

Acknowledgments

Warmest thanks to all people who gracefully accepted to review the chapters of this book and for their useful comments.

Déborah Abhervé (Asca), Dr. Fanny Arnaud (CNRS), Dr. Régis Barraud (Université de Poitiers), Dr. Olivier Barreteau (INRAE), Dr. Brent M. Haddad (University of California, Santa Cruz), Dr. Gemma Carr (Centre for Water Resource Systems), Dr. Matthieu Couttenier (ENS de Lyon), Dr. Simon Dufour (Université Rennes 2), Dr. Coleen Fox (Dartmouth College), Julien Gauthey (Office Français de la Biodiversité), Dr. Ana González Besteiro (Université Lyon 3), Dr. Lesley Head (The University of Melbourne), Dr. Md Sayed Iftekhar (The University of Western Australia), Dr. Natacha Jacquin (Office International de l'Eau), Dr. Eleftheria Kampa (Ecologic Institute, Germany), Dr. G. Mathias Kondolf (University of California, Berkeley), Dr. Yves-François Le Lay (ENS de Lyon), Dr. Laurent Lespez (Université Paris-Est Créteil Val de Marne), Dr. Susanne Muhar (BOKU, University of Natural Resources and Life Sciences), Dr. Clare Palmer (Texas A&M University), Dr. Samantha Scholte (Sociaal en Cultureel Planbureau), Dr. Nick Schuelke (University of Wisconsin-Milwaukee), Dr. Kate Sherren (Dalhousie University), Dr. Marc Tadaki (Cawthron Institute), Dr. Amie O. West (University of Arkansas), Dr. Joanna Zawiejska (Pedagogical University of Cracow)

List of Contributors

John C. Bergstrom
University of Georgia
Athens
GA
USA

Kirsty L. Blackstock
Social, Economic and Geographical
Sciences Department
The James Hutton Institute
Aberdeen
Scotland

Brendon Blue
School of Environment
University of Auckland
Auckland
New Zealand

Kerstin Böck
Universität für Bodenkultur Wien
Institut für Hydrobiologie und
Gewässermanagement (IHG)
Vienna
Austria

Gabrielle Bouleau
Université Gustave Eiffel
Laboratoire Interdisciplinaire Sciences
Innovation Société (LISIS), CNRS, ESIEE,
INRAE, UGE
Marne-la-Vallée
France

Gary Brierley
School of Environment
University of Auckland
Auckland
New Zealand

Matthias Buchecker
Swiss Federal Research Institute WSL
Unit Economics and Social Sciences
Birmensdorf
Switzerland

Arjen Buijs
Forest and Nature Conservation
Policy Group
Wageningen University
Wageningen
The Netherlands

Nora S. Buletti
USYS TdLab
ETH, Zürich
Switzerland

John Cain
River Partners
Chico
CA
USA

Catherine Carré
Université Paris 1 Panthéon-Sorbonne
Paris
France

Caitriona Carter
UR ETBX INRAE
Centre de Nouvelle Aquitaine Bordeaux
Cestas Cedex
France

Emeline Comby
Université de Lyon
CNRS
ENS de Lyon
Environnement Ville Société
Lyon
France

Marylise Cottet
Université de Lyon
CNRS
ENS de Lyon
Environnement Ville Société
Lyon
France

Henry Dicks
Faculté de Philosophie
Université Jean Moulin Lyon
Lyon
France

Ludovic Drapier
Université Paris Est Créteil
UMR 8591 CNRS Laboratoire de
Géographie physique
Meudon
France

Olivier Ejderyan
USYS TdLab
ETH, Zürich
Switzerland

Silvia Flaminio
Institut de Géographie et Durabilité
Université de Lausanne
Géopolis
Lausanne
Suisse

Xavier Garcia
Barcelona Institute of Regional and
Metropolitan Studies
Cerdanyola del Vallès
Spain

Marie-Anne Germaine
Université Paris Nanterre
Nanterre
UMR 7218 CNRS LAVUE
France

Christopher Gibbins
School of Environmental and Geographical
Sciences
University of Nottingham Malaysia (UNM)
Selangor Darul Ehsan
Malaysia

Jean-Paul Haghe
Rouen University
Rouen
France

Dan Hikuroa
Te Wānanga o Waipapa
University of Auckland
Auckland
New Zealand

Alba Juárez-Bourke
Social, Economic and Geographical
Sciences Department
The James Hutton Institute
Aberdeen
Scotland

Caroline Le Calvez
Université d'Orléans
EA 1210 CEDETE
Orléans Cedex 2
France

Sophie Le Floch
UR ETBX INRAE
Centre de Nouvelle Aquitaine Bordeaux
Cestas Cedex
France

Yves-François Le Lay
Université de Lyon
CNRS
ENS de Lyon
Environnement Ville Société
Lyon
France

Laurent Lespez
Université Paris Est Créteil
UMR 8591 CNRS Laboratoire de
Géographie physique
Meudon
France

Jamie Linton
Geolab UMR 6042 CNRS
Université de Limoges
Limoges
France

Belinda Lip
WWF-Malaysia Sarawak Office
Kuching
Sarawak
Malaysia

John B. Loomis
Colorado State University
Fort Collins
CO
USA

McKenzie Augustine Martin
WWF-Malaysia Sarawak Office
Kuching
Sarawak
Malaysia

Sylvie Morardet
Université de Montpellier
AgroParisTech, BRGM, CIRAD
INRAE, Institut Agro, IRD
UMR G-EAU
Montpellier
France

Bertrand Morandi
Université de Lyon
CNRS
ENS de Lyon
Environnement Ville Société
Lyon
France

Stefanie Müller
Swiss Federal Research Institute WSL
Unit Economics and Social Sciences
Birmensdorf
Switzerland

Hervé Piégay
Université de Lyon
CNRS
ENS de Lyon
Environnement Ville Société
Lyon
France

Franziska E. Ruef
USYS TdLab
ETH, Zürich
Switzerland

Anne Salmond
Te Wānanga o Waipapa
University of Auckland
Auckland
New Zealand

Beth Styler-Barry
The Nature Conservancy – New Jersey
Field Office
Chester
NJ
USA

Marc Tadaki
Cawthron Institute
Nelson
New Zealand

Pere Vall-Casas
Universitat Internacional de Catalunya
School of Architecture
Barcelona
Spain

Riyan van den Born
Institute for Science in Society
Radboud Universiteit
Nijmegen
The Netherlands

Bernadette van Heel
Institute for Science in Society
Radboud Universiteit
Nijmegen
The Netherlands

Helena Zemp
Haute École d'Ingénierie – FHNW
Windisch
Aargau
Suisse

Part I

Introduction

1

What are the Political, Social, and Economic Issues in River Restoration? Genealogy and Current Research Issues

Marylise Cottet, Bertrand Morandi, and Hervé Piégay

Université de Lyon, CNRS, ENS de Lyon, Environnement Ville Société, Lyon, France

1.1 Introduction

1.1.1 River restoration at the heart of river management policies

Faced with the ever-increasing impact of human activities on the environment, the biologist E.O. Wilson (1992) announced, probably with much hope, the opening of an era of ecological restoration in the 21st century. Although its scope and consequences may be a matter of debate (Choi 2007; Sudding 2011), the realization of this hope seems to be currently confirmed. In response to the observed degradation of ecosystems (Palmer et al. 2004; Steffen et al. 2007; Cardinale et al. 2012), ecological restoration measures have become a structuring element of environmental management policies in both developed and developing countries (Aronson et al. 2006; Wortley et al. 2013). In the field of river management, they have been actively deployed since the 1970s (Gore 1985; Boon et al. 1992) because of particularly significant degradation resulting from the use of water and hydraulic installations, both old and increasingly numerous (e.g. Dudgeon et al. 2005; Vörösmarty et al. 2010; Grizzetti et al. 2017). Faced with significant water pollution and profound physical modifications of river ecosystems, many countries, particularly Western ones, have taken legislative and regulatory measures to preserve or restore a certain environmental quality to rivers. These measures consider, often in an integrated manner, physicochemical, biological, and hydromorphological issues (e.g. the US Clean Water Act, 1972; UK Water Act, 1973; French Water Laws, 1992, 2006; EU Water Framework Directive, 2000; Australian Water Act, 2007). These legislative frameworks have provided a fertile ground for the multiplication of restoration projects, as shown by several reviews conducted around the world (e.g. Bernhardt et al. 2005; Nakamura et al. 2006; Brooks and Lake 2007; Morandi et al. 2017; Szałkiewicz et al. 2018).

River restoration policies, whether at the legislative or technical level, are supported by substantial interdisciplinary research efforts. Scientific work focused on the dynamics of restoration research has shown a major increase in productivity, starting in the 1990s (Shields et al. 2003; Ormerod 2004; Bennett et al. 2011; Smith et al. 2014; Wohl et al. 2015).

This increase in scientific research accompanies an increase in the number of projects implemented (Bernhardt et al. 2005) and reflects the strength of science–management links in the field. Applied work in ecology, hydrology, and hydromorphology (Palmer and Bernhardt 2006; Vaughan et al. 2009; Wohl et al. 2015) has been carried out to better understand the functioning of river systems and to assess their responses to restoration actions. As these actions are often innovative, it is difficult to predict their effects, and adaptive approaches based on monitoring are therefore generally preferred (Downs and Kondolf 2002). The current thinking in the natural sciences, notably on the links between hydromorphological structures and processes (Kondolf 2000), or between habitats and biodiversity (Palmer et al. 2010), has been nourished by as much as it has nourished ecological restoration practices (Smith et al. 2014). Bradshaw (2002, p. 7) introduced "restoration as an acid test for ecology" and thus emphasized the links between ecological restoration (practice) and restoration ecology (science). While these special links have long led to the association of river restoration with natural sciences, they have not prevented calls to broaden the spectrum of disciplines involved in restoration approaches (Cairns 1995; Ormerod 2004; Palmer and Bernhardt 2006; Wohl et al. 2015).

1.1.2 An evolution in the positioning of societal issues in debates on river restoration

Societal[1] stakes have always structured thinking in the field of restoration. Back in the 1990s, Cairns (1995, p. 9) wrote that "ecological restoration must involve ecosocietal restoration. This is the process of reexamining human society's relationship with natural systems so that repair and destruction can be balanced and, perhaps, restoration practices ultimately exceed destruction practices." This thinking is part of a more general movement questioning the relationship between human and nature as it has been encompassed in modern Western thought. This questioning, which has permeated society since the 1970s, is sociocultural, anthropological, and philosophical (Gobster and Hull 2000). It is sometimes championed by natural science researchers who advocate a new ethics of nature (e.g. Jordan 2000; Clewell and Aronson 2013). These ethics are based on the recognition of nature as having an intrinsic value as much as they are based on observations of the degradation of this value by human societies. They are formed at the point of convergence between conservationist movements and scientific advances in the natural sciences. Human activities are characterized by the natural sciences, often in terms of pressures; it is a time of guilt. The ultimate goal of restoration projects is to return rivers to the good (pristine) state they were in before degradation by human societies.

Very quickly, the original idea of the river's return to a past state, what some authors have called the myth of the return to Eden (e.g. Dufour and Piégay 2009), was strongly criticized for reasons mainly related to scientific uncertainties and technical feasibility, but also to ethical positions. Restoration is certainly based on a new attitude toward rivers, but it must also deal with reality, including the social and economic reality of its implementation. The conceptualization of the Anthropocene is part of this evolution (Crutzen and Stoermer 2000). Studies highlight the importance of both biophysical and societal contexts in the implementation of actions. It is a time for pragmatism; leaders of restoration projects seek to do the best that they can ecologically within a given context. This realization is accompanied by a discussion of the concept of restoration and an evolution of its definitions (see Box 1.1). As Wohl et al. (2005, p. 2) point out: "because both technical and social constraints often preclude 'full' restoration of ecosystem structure and function, rehabilitation is sometimes distinguished from restoration." The concept of rehabilitation often carries the idea of a relaxation of the restoration objectives and a redefinition of the references used to define these objectives (Dufour and Piégay 2009). What is being restored? But above all, why is it being restored? Restoration, which until now was guided by the intrinsic value of the river, embodied by concepts such as ecological integrity, is opening up to more instrumental visions.

Box 1.1   River restoration: an "essentially contested concept"

The debates, and often polemics (e.g. Normile 2010), concerning what is and is not river restoration are numerous (e.g. Roni and Beechie 2013; Wohl et al. 2015). These debates on the definition of restoration have been fueled by the proliferation of concepts that are now widely used in the literature, such as "rehabilitation," "renaturation," "revitaliza- tion," "enhancement," and "improvement." If there is one certainty emerging from these debates, it is that attempts to provide an unequivocal and definitive answer to the question "What is restoration?" are doomed to failure. There has never been a consen- sus on definitions, and there certainly never will be. The concepts of restoration, reha- bilitation, or renaturation are "essentially contested concepts" (Gallie 1956, p. 169), that is to say, "concepts the proper use of which inevitably involves endless disputes about their proper uses on the part of their users." The interest in the debates lies not in their conclusions but in the debates themselves and the ideas that emerge from them.

   The definitions that have been used as references in the field of restoration show the evolution of these debates. Initially conceived as techniques, the definitions raised ethical questions as early as the 1990s and particularly questioned the hermetic sepa- ration between human and nature (Table 1.1). According to Westling et al. (2014, p. 2614), "This dichotomy is challenged, both by alternative theoretical frameworks arguing for the relevance of natural-cultural hybrid models for restoration, and by pragmatic perspectives that take restoration to be the balancing of ecological and human goals through rehabilitating or enhancing landscapes, rather than seeking

Table 1.1   Definitions of river restoration selected from the international scientific literature illustrating the place of societal issues in conceptual debates.

NRC 1992, p. 18	"Restoration is defined as the return of an ecosystem to a close approximation of its condition prior to disturbance. In restoration, ecological damage to the resource is repaired. Both the structure and the functions of the ecosystem are recreated."
Stanford et al. 1996, p. 393	"The goal of river restoration should be to minimize human-mediated constraints, thereby allowing natural re-expression of productive capacity. In some, if not most, intensely regulated rivers, human-mediated constraints may have progressed to the point that full re-expression of capacity is neither desired nor possible. Nonetheless, the implication is that basic ecological principles applied to rivers in a natural-cultural context can lead to restoration of biodiversity and bioproduction in space and time; but, the constraints must be removed, not mitigated."

(Continued)

Table 1.1 (Continued)

Downs and Thorne 2000, pp. 249–250	"It is now widely recognised that river restoration in the sense of Cairns (1991) – 'The complete structural and functional return to a pre-disturbance state' – is seldom feasible." "Practical 'river restoration' is, in fact, an historically-influenced exercise in environmental enhancement through morphological modification. It is probably more accurate to refer to the approach as river rehabilitation."
McIver and Starr 2001, p. 15, citing SER website	"Ecological restoration can be defined as 'the process of assisting the recovery and management of ecological integrity,' including a 'critical range of variability in biodiversity, ecological processes and structures, regional and historical context, and sustainable cultural practices.'"
Wohl et al. 2005, p. 2	"We define ecological river restoration as assisting the recovery of ecological integrity in a degraded watershed system by reestablishing the processes necessary to support the natural ecosystem within a watershed. Because both technical and social constraints often preclude 'full' restoration of ecosystem structure and function, rehabilitation is sometimes distinguished from restoration."
Palmer and Allan 2006, pp. 41–42	"River restoration means repairing waterways that can no longer perform essential ecological and social functions such as mitigating floods, providing clean drinking water, removing excessive levels of nutrients and sediments before they choke coastal zones, and supporting fisheries and wildlife. Healthy rivers and streams also enhance property values and are a hub for recreation."
Chou 2016, p. 2	"[R]iver restoration means different things to different people. In terms of scale and scope, it can be a complete structural and functional return to the pre-disturbance state, a recovery of the partly functional and/or structural conditions of rivers (i.e., rehabilitation), a recovery of the natural state of a river ecosystem without really aiming at the pristine, pre-disturbance state (i.e., renaturalization), or an improvement of the present state of rivers and their surrounding areas with the intention of enhancing their ecological, social, economic or aesthetic features (i.e., enhancement)."

return to a redundant, historical reference state." The efforts of scientists and managers are now aimed at clarifying the definitions of references and concepts such as "Lietbild" (Kern 1992) or "guiding image" (Palmer et al. 2005) to clarify the restoration objectives (Weber et al. 2018). In the definition of these objectives, the tension between an eco-centric approach that emphasizes the intrinsic value of the river and an anthropocentric approach carrying more utilitarian arguments is notable (Dufour and Piégay 2009).

Adhering to Chou's (2016, p. 2) assertion that "river restoration means different things to different people," we adopt a relatively broad definition of restoration in this book. We do not distinguish between some commonly used concepts such as restoration, rehabilitation, renaturation, or revitalization actions. We consider as restoration any human intervention on the river aimed at recovering a quality considered degraded or lost. This quality can be perceived in terms of biodiversity, hydromorphological dynamics, physicochemical parameters, landscape beauty, or even the possibility of recreational use.

According to Palmer et al. (2014), there has been a shift in ecological restoration from ecological theory to utilitarian concerns. This utilitarian approach to restoration is rooted in a more anthropocentric ethic, but paradoxically, it can also be instrumentalized to serve an eco-centric ethic. The articulation of intrinsic and instrumental values is at work in restoration approaches (Clewell and Aronson 2006). It is reinforced by the ever more pressing need to justify often costly policies in the fluvial domain (e.g. Bernhardt et al. 2005; Nakamura et al. 2006; Brooks and Lake 2007). The ecological or hydromorphological quality of a river is no longer justifiable as an end in itself, but must appear as a necessity with regard to the services that the good state and biophysical functioning of a river can provide, or the risks burdening societies in the case of nonaction. This utilitarian vision is strongly based on the notion of services that has developed since the 1990s (Costanza et al. 1997). It finds particular resonance in the context of urban river restoration projects, which are highly constrained socially, politically, and technically (Eden and Tunstall 2006; Bernhardt and Palmer 2007). Moreover, it is largely in the context of urban ecology that the concept of "novel ecosystems" emerges, which requires consideration of the hybrid character, both natural and artificial, of the rivers to be restored (Francis 2014). The specificity of restoration in relation to other management approaches can also be called into question, as Elliot (1982, p. 81) anticipated when he denounced "faking nature." Thus, there is a strong proximity between Martin's (2017) proposed definition of ecological restoration, considered as "the process of assisting the recovery of a degraded, damaged, or destroyed ecosystem to reflect values regarded as inherent in the ecosystem and to provide goods and services that people value" and Mitsch's definition of ecological engineering (2012, p. 5): "defined as the design of sustainable ecosystems that integrate human society with its natural environment for the benefit of both."

No matter how you look at this evolution, it has certainly increased the attention paid to societal issues in the field of restoration. Research in the humanities and social sciences has contributed to this evolution by proposing new methods to measure the socio-economic benefits that societies derive from restoration projects and also by analyzing the sociopolitical processes at work in the implementation of projects. If the objectives of restoration are no longer guided solely by a scientific conception of reference states, but also in response to political choices concerning what rivers should be, a new space for debate is opened up, particularly in respect to the questioning of decision-making. To the questions "Why are we restoring?" or "What are we restoring?" is added the question "Who is restoring for whom?" The work carried out in political ecology shows that new relationships of power are being established within the frameworks for implementing restoration projects. As much as the objectives of restoration, it is the political mechanisms leading to the definition of these objectives that are at the heart of debates. Work on participation is essential, and questions the place of scientific and technical expertise as much as that of other interests, be they economic or more simple public interests.

1.1.3 What do we know about research on societal issues in the field of river restoration?

Although several reviews highlight the increasing place of societal issues within the scientific work devoted to river restoration (Bennett et al. 2011; Smith et al. 2014; Wohl et al. 2015), none of them has focused on a specific analysis of this research field. This introductory chapter offers this analysis; without claiming to be exhaustive, it gives a broad overview of the dynamics at work on the societal stakes of river restoration.

1) What is the magnitude of these dynamics? How are they structured at the international level? The first part of the introduction looks at the way the scientific community is organized within the field. Particular attention is paid to the way in which the work carried out by this community fits within the dynamics of general research on river restoration, but also within work in human and social environmental sciences.
2) The second part then analyzes the main directions taken by researchers working on societal issues of restoration. What are the main problem areas that structure this field of research? In which disciplines or epistemological traditions are they rooted? What are the methodological approaches mobilized in the frameworks of the different projects?
3) Finally, the third part is devoted to the operational commitment of researchers working on societal themes. Do these researchers support the implementation of river restoration projects and policies and, if so, how?

The literature review that we present here is based on a qualitative and quantitative analysis of the body of international scientific articles devoted to river restoration. This body of literature was retrieved from a search of the titles of articles listed in the Web of Science (WoS) using the keywords river*, stream*, restor*, rehab*, renat*, and revit*. In total, this search identified 1677 articles published between 1971 and 2019 by authors from about fifty different countries. The content of these 1677 articles was analyzed and a subset of 121 publications specifically devoted to discussions of human and social sciences (HSS; e.g. historical, philosophical, sociological, economic, geographical, and political) on the theme of river restoration were identified. This collection of literature forms only a sample of the work published in the field, particularly because it prioritizes international articles written in English at the expense of national publications and books written in other languages. Nevertheless, it allows an overview of the relevant publications and, by extension, of the dynamics of HSS research in the field of river restoration. To this end, we base our analysis on scientometric methods (Mingers and Leydesdorff 2015), coupled with lexical analysis methods (Lebart et al. 1998) and content analysis (Berelson 1952) of the publications.

1.2 Genealogy of research on societal issues in river restoration

According to Palmer and Bernhardt (2006, p. 3), "the final research frontier is restoration science that is informed by social science scholarship." Addresses to the humanities and social sciences to invest more in the field of river restoration are recurrent (e.g. Ormerod 2004; Palmer and Bernhardt 2006; Wohl et al. 2015). However, it is some-times difficult to know which disciplines are expected to be involved, and what issues should be worked on. Bennett et al. (2011, p. 4) present the "recognition and promotion of human, societal, or cultural requirements for stream restoration" as a shift in restoration science, emphasizing the importance of participation. Palmer and Bernhardt (2006, p. 4) refer to "cultural anthropology, environmental education, landscape architecture and city planners" as "social sciences." Ormerod (2004, p. 548) refers to the "socio-economic sci-ences" with a more explicit focus on economic approaches. All these authors address dis-ciplinary contributions to the field of restoration according to the modern dichotomy between the "human sciences" and the "natural sciences" that is used to structure classi-cal classifications in the scientific field. Other distinctions could have been made, for example the Frascati Manual separates the "social sciences" (e.g. sociology, economics, psychology, geography) from the "humanities and the arts" (e.g. philosophy, history) (OECD 2015, p. 59). These divisions are sometimes inherited from schools of thought or institutional traditions that vary from one country to another. They are sometimes over-taken by fields of research that are structured in an interdisciplinary or even transdiscipli-nary manner. Many authors publishing on the societal issues of river restoration belong to environmental studies institutes, not to humanities and social sciences institutes. Epistemological positions are also rarely asserted in publications, and it is often difficult to identify the disciplinary tradition to which the authors adhere.

The tendency to break away from disciplinary divisions must be interpreted in the light of the epistemological evolutions that have marked the human and social-sciences-based work on environmental issues since the 1970s (e.g. Turner et al. 1994; Lester 1995; Hannigan 2006; Castree et al. 2016). These developments, which have led in particular to the structuring of the field of environmental humanities (Blanc et al. 2017; Emmet and Nye 2017; Heise et al. 2017), have made the disciplinary limits more labile. According to Emmett and Nye (2017, p. 4), "The environmental humanities have become a global intel-lectual movement that reconceives the relationship between scientific and technical disci-plines and the humanities, which are essential to understanding and resolving dilemmas that have been created by industrial society."

The WoS bibliometric analysis shows that the journals in which such research is pub-lished are very rarely humanities and social sciences journals (Figure 1.1). For the most part, studies on societal aspects of restoration are published in journals with an environ-mental editorial line, explicitly interdisciplinary and applied (e.g. *Environmental Management*; *Journal of Environmental Management, Ecology and Society*) or water man-agement (e.g. *Water Resources Research, Water Alternatives, River Research and Applications*) journals. Journals on environmental economics issues (e.g. *Ecological Economics, Water Resources and Economics*) are the ones with the strongest disciplinary roots on the social science side. Much work on economic or political issues is published in natural science journals (e.g. *Hydrobiologia, Journal of Hydrology*). Many societal issues have also been brought into the field of restoration by researchers in the natural sciences. For example, as early as the 1990s, it was the ecologist J. Cairns (1995) who proposed the notion of "ecoso-cietal restoration." Certain research approaches, initially undertaken by researchers in ecology, hydrology, or hydromorphology, have largely contributed to placing societal issues at the center of thinking (e.g. Wohl et al. 2005; Palmer and Bernhardt 2006; Dufour and

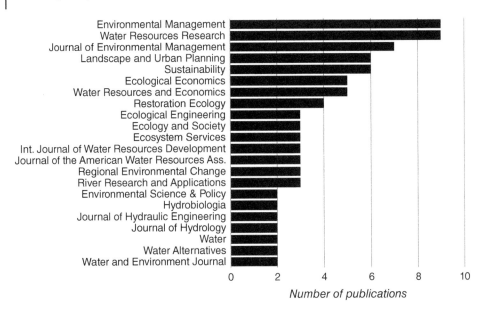

Figure 1.1 Main international scientific journals in which research on societal issues of river restoration is published (1992–2019).

Piégay 2009). This is the case with the Long-Term Ecological Research (LTER) network and then the Long-Term Socio-Ecological Research Network (Redman et al. 2004; Wells and Dougill 2019), which have been important steps in the emergence of more integrated approaches to restoration. This need to cross-disciplinary divides is expressed in another way in projects such as critical physical geography, which pays "critical attention to relations of social power with deep knowledge of a particular field of biophysical science or technology in the service of social and environmental transformation" (Lave et al. 2014, p. 2).

1.3 A scientific community organized regionally and occasionally around river restoration projects

Smith et al. (2014, p. 253) writes that "the integration of social science into restoration is relatively rare." Indeed, while research in the field of river restoration emerged in the 1970s and grew significantly from the early 1990s, studies specifically focused on societal issues only emerged in the 1990s and remained limited in number until the mid-2000s (Figure 1.2). Although societal approaches still represent a minority of the published work in the field of river restoration, it is a steadily growing minority. Over the decade 2010–2020, work on the social, economic, or political stakes of restoration represented 10% of all publications devoted to river restoration, whereas it represented only 5% in the 2000s and 2% in the 1990s. The results of this work seem to be increasingly mobilized by the interdisciplinary scientific community. According to WoS data, studies dedicated to societal issues are cited almost as much as studies addressing other topics in restoration (2.4 vs. 2.6 annual citations).

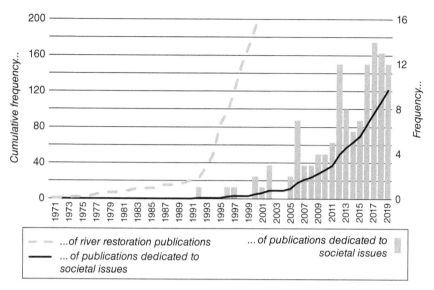

Figure 1.2 Chronology of international scientific publications on societal issues in river restoration.

The first works on societal issues listed in the WoS date from the 1990s, with these being published in Europe and the United States (e.g. Barendregt et al. 1992; Loomis 1996; Turner and Boyer 1997). Over the entire period, 72% of all publications dedicated to social, economic, and political issues were written by researchers from European and North American institutions, with 36% being written by US researchers alone. However, societal publications represent only 6% of all work in the field of river restoration published in the USA (Figure 1.3). Some countries, which publish less in the field of restoration, are proportionally more active on societal issues. This is the case for several European countries including Spain, Switzerland, Norway, and to a lesser extent Germany, which have been actively publishing since 2005. Studies published in Israel also give particular importance to work in economics (e.g. Axelrad and Feinerman 2009; Becker and Friedler 2013). Researchers from East Asian countries, whose first publications in the field of river restoration date back to the mid-2000s (Nakamura et al. 2006), very quickly became interested in societal issues (e.g. Tanaka 2006; Jia et al. 2010; Che et al. 2012). As a proportion of the total work they publish on river restoration, some countries, such as Taiwan, South Korea, Japan, and China, are among the most active on societal issues. In total, since the early 1990s, more than 400 authors from 31 countries have contributed to publications on societal issues in the field of river restoration.

Despite publications from a large number of countries, there does not seem to be an internationally structured scientific community around humanities and social sciences. While most of the publications listed in the WoS involve authors belonging to different institutions, a minority are the result of international institutional cooperation. Scientific communities do exist, but they are multifaceted and rather structured on national or even regional scales. Sometimes it is more accurate to speak of scientific teams than communities. While these teams contribute to the production of societal knowledge in the field of

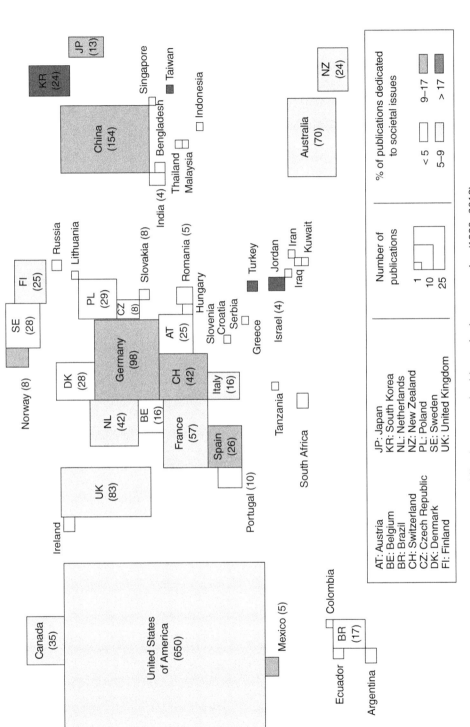

Figure 1.3 Geography of international scientific publications on societal issues in river restoration (1992–2019).

restoration, they are often structured around broader or related research issues. Thus, significant works on river restoration have been published by a network of researchers from several English universities working on flooding and public perception issues (e.g. Tunstall et al. 1999, 2000; Eden et al. 2000; Åberg and Tapsell 2013). Also, in Switzerland, in the canton of Zurich, a scientific community specializing in public participation issues has produced notable work in the field of river restoration (e.g. Schläpfer and Witzig 2006; Junker and Buchecker 2008; Seidl and Stauffacher 2013). The dynamics of regionalized research often seem temporary. Several researchers, whose work is now a reference in the field, have only had a momentary commitment to restoration issues (e.g. Pahl-Wostl 2006; Junker and Buchecker 2008; Buijs 2009). Out of more than 400 identified authors, less than one-tenth have participated in multiple publications on the topic of river restoration.

The ad hoc nature of scientific commitments can be explained by the fact that research is often conducted in connection with the implementation of fixed-duration restoration projects. The bibliometric analysis identified more than a hundred rivers for which research was undertaken on societal issues related to the implementation of restoration projects (Figure 1.4). Some of these projects are known mainly through these societal studies. For example, in South Korea, several research teams (notably attached to universities in Seoul province) have worked on the social, economic, and political evaluation of the restoration of the Cheonggyecheon and Anyancheon rivers, two rivers that are now emblematic of urban restoration (e.g. Lee and Jung 2016; Kim et al. 2017). In Israel and the Palestinian Territories, it is economists who have published scientific literature on restoration projects on the Yarqon, Jordan, and Kishon rivers (e.g. Becker et al. 2014; Garcia et al. 2016). Societal research has also focused on emblematic river restoration projects conducted on major rivers such as the Rhine (e.g. Buijs 2009), Danube (e.g. Bliem et al. 2012), Colorado (e.g. Bark et al. 2016), Sacramento (e.g. Golet et al. 2006), and Kissimmee (e.g. Chen et al. 2016). It is often in the context of such projects that an interdisciplinary culture is forged. In France, several studies on the social and political issues involved in the restoration of the Rhône (e.g. Barthélémy and Armani 2015; Guerrin 2015) were developed as part of a global approach to long-term socioecological research (Lamouroux et al. 2015; Thorel et al. 2018). Similarly, research on the Cole River and Skerne River (Tunstall et al. 1999; Eden and Tunstall 2006; Åberg and Tapsell 2013) in the United Kingdom was undertaken as part of an interdisciplinary European project (Holmes and Nielsen 1998).

1.4 A research field tackling several topics

Among the publications in the field of river restoration, those devoted to societal issues are distinguished by a specific lexicon (Figure 1.5). The qualitative analysis of these publications makes it possible to reduce the apparent thematic diversity and to schematically draw three main lines of research.

1) The first brings together research on human–river interactions. The public's environmental perceptions and preferences as well as social practices are often central to this

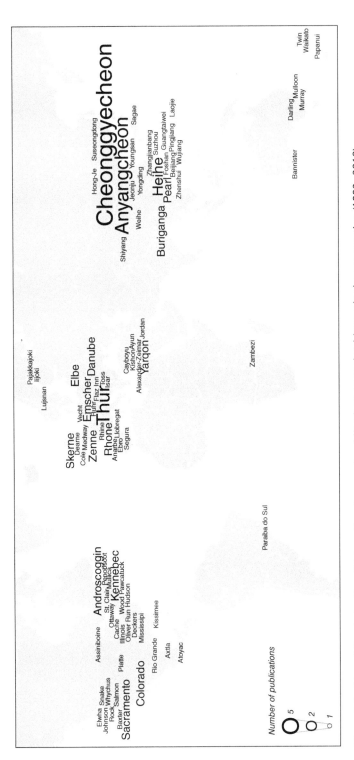

Figure 1.4 Map of the study sites of international research publications on societal issues in river restoration (1992–2019).

This wordcloud is produced from the calculation of lexical specificity scores. Based on a probability law, the specificity score assesses if some words occur more frequently in certain parts of a corpus (here, the words used in the titles of "societal" sub-corpus, n = 121) than in the whole corpus (here, the "river restoration" corpus, n = 1677) from which the parts are extracted.

Figure 1.5 A lexicon specific to international scientific publications on the societal stakes in river restoration.

first corpus of 53 publications. Analysis of the different links that individuals have with rivers, of their consideration within the framework of projects, or the way they are impacted by restoration measures, constitutes a first important challenge for researchers.

2) A second thematic corpus, composed of 38 publications, focuses more specifically on the political issues raised by the implementation of restoration projects. The analysis of governance and the roles played by the different actors involved in restoration projects structures the work in this corpus. In particular, the issue of public participation is at the heart of much of the work.

3) Finally, economic approaches structure a third problematic corpus consisting of 66 articles. Economic evaluations of restoration projects, sometimes based on cost–benefit analyses, are central to this work. Notably, strong methodological attention is paid to the evaluation of nonmarket ecosystem services restored by restoration projects.

These different thematic fields are not independent. Understanding environmental perceptions, for example, is often a first step in the political analysis of conflicts between actors. Similarly, the assessment of economic benefits, as well as the political analysis of opinions regarding restoration projects, often depends on the preferences of certain categories of stakeholders for different river states. Numerous publications therefore contribute to the advancement of knowledge within several thematic fields.

1.4.1 Understanding human–river interactions in the context of river restoration

Since the first works published on human–river interactions in the context of restoration in the early 2000s (e.g. Tunstall et al. 2000; Connelly et al. 2002; Piégay et al. 2005), a significant proportion of the publications in the corpus have dealt with this theme (Figure 1.6). These contributions are most often part of a constructivist conception of nature that posits that the humans are a stakeholder in the elaboration of the reality in which they intervene; reality is seen as a mentally constructed representation (Moscovici 2001; Dunlap et al. 2002). The implications of this conception are important in the field of environmental action. Since the river is no longer considered as an intangible reality, it is necessary to make room for the plurality of the modes of understanding in order to define the objectives and modalities of action. This is all the more true in the field of restoration, where determination of the reference – which can be defined as an approximation of the desirable state of the river, a standard chosen from several possible alternative states (Le Floc'h and Aronson 1995) – appears to be an eminently subjective value-laden activity (Hull and Robertson 2000); restoration is not only a scientific exercise based on rational criteria (Davis and Slobodkin 2004). Anchored in this epistemology, several research studies focused on the dynamics of human–river interaction within the framework of restoration projects.

This wordcloud is produced from the calculation of lexical specificity scores. Based on a probability law, the specificity score assesses if some words occur more frequently in certain parts of a corpus (here, the words used in the titles of "social" sub-corpus, n = 53) than in the whole corpus (here, the "societal" corpus, n = 121) from which the parts are extracted.

Figure 1.6 A lexicon specific to international scientific publications dealing with human–river interactions in the context of restoration.

1.4.1.1 A broad public at the heart of the debate to characterize society's support for river restoration projects

The majority of publications focus on the relationships that the "public" –"inhabitants," "residents," or simply "people," or "citizens" – have with restored or to be restored rivers (e.g. Larned et al. 2006; Weber and Stewart 2009; Hong et al. 2019), notably through survey approaches (see Box 1.2). These designations generally refer to all people making up society, regardless of the relationship they may have with the river. Within this population, local people (local residents) occupy a special place insofar as they live within the confines of the river (Tunstall et al. 2000; Seidl and Stauffacher 2013; Westling et al. 2014). Proximity to the river is also a subject of discussion in the definition of "local." From what distance is one concerned by a river restoration project? Are the expectations the same whether one lives near or far? These are questions raised by Soto-Montes de Oca and Ramirez-Fuentes (2019), whose objective was to evaluate the benefits of restoration perceived by inhabitants according to their proximity of residence to the Atoyac River (Mexico). This study shows that, more than the distance of residence, it is the frequency of use of the river that determines the interest in its restoration. River users, whether residents or nonresidents, are the focus of a number of scientific works (e.g. Polizzi et al. 2015; Deffner and Haase 2018; Zingraff-Hamed 2018). Generally speaking, publications dealing with human–river relationships often consider these different categories of stakeholders as forming public opinion and participating in political life and, through their positions, guiding the modalities of public action and, therefore, restoration projects. There are many publications linking the question of the relationship to the river to political issues (e.g. Junker et al. 2007; Davenport et al. 2010; Barthélémy and Armani 2015; Fox et al. 2017). The challenge underlying these publications is to understand how, in the context of projects, the relationships that local societies have with a river can give rise to support or opposition to its restoration. They sometimes position themselves ahead of projects to determine and anticipate, among the diversity of forms of relationships with the river, those that can help to ensure the adhesion of residents to the restoration project (e.g. Buijs 2009). They can also be located after projects to assess the degree of satisfaction or dissatisfaction with the changes that projects have brought about in terms of the links that residents have established with the river, or that they have established with each other around the river (e.g. Buijs 2009; Verbrugge and van den Born 2018). In particular, Buijs (2009) has shown, through the case of the restoration of the Rhine (Netherlands), that opposition to restoration projects generally responds to questioning of the foundations of people's attachment to the river.

1.4.1.2 Approaches focused on the values associated with rivers

To understand the relationship that the general public has with rivers, publications use different notions such as "perception," "preference," "attitude," or "attachment" to the environment. Each of these notions has its own definition, sometimes changing (Table 1.2); however, they all reflect the values and attachments that individuals have with rivers. Although such values have long been excluded from environmental action, their essential role "as rational motivators for the public to become involved and to voice their concern" (Vining et al. 2000, p. 145), is now well recognized. The assumption is that individuals are more attentive to environments they care about and are more willing to protect them.

Box 1.2   The field survey: a structuring method for many societal approaches in the field of river restoration

Surveys are the founding methods for HSS (see Bickman and Rog 2009; Gideon 2012). The first surveys in the field of river restoration date from the late 1990s and were conducted as part of economic studies (e.g. Loomis 1996; Loomis et al. 2000). They were quickly mobilized to develop studies on social and political issues (e.g. Tunstall et al. 2000; Purcell et al. 2002).

The majority of survey-based studies focus on the general public (Figure 1.7). They often target residents of a neighborhood or town near to a restored river site (e.g. Tunstall et al. 2000; Perni et al. 2012; Seidl and Stauffacher 2013; Hong et al. 2019). They are also interested in the users and visitors of the sites, whether or not they are residents (e.g. Loomis 2002; Becker and Friedler 2013; Kim et al. 2017; Deffner and Haase 2018). These surveys are particularly used to evaluate the economic, recreational, or landscape benefits of a restoration project. They are also deployed to provide information for restoration policies on the relationships that a regional or national population has with rivers or river management (e.g. Piégay et al. 2005; Junker and Buchecker 2008). The surveys also focus on specific categories of the population, whose role in the restoration process is considered specific. They are identified as project stakeholders, actors, participants, or interest groups. The definition of these categories is often debated, and survey work can help inform such discussions (e.g. Tanaka 2006; Junker et al. 2007). Political studies are mainly based on surveys targeting elected officials and staff of public institutions, leaders of environmental or citizen associations, economic players, and scientists. The objective of these studies is to understand the position of the various stakeholders with regard to restoration approaches, but also the interactions between the stakeholders, their power relationships, and the conflicts or agreements that may be generated by the implementation of restoration projects (e.g. Junker et al. 2007; Lave et al. 2010; Heldt et al. 2016; Druschke et al. 2017). Stakeholders are also surveyed as part of the assessment of restoration projects (which is not guided

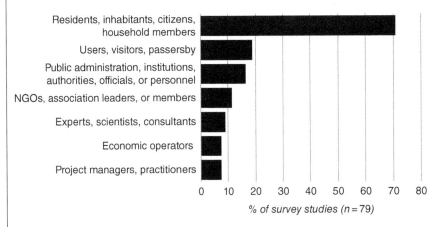

Figure 1.7   Main categories of respondents targeted by surveys on the societal issues of river restoration.

by HSS questions, although they can feed them). These surveys, carried out by environ-mental scientists, are intended to gather factual information on the restoration prac-tices implemented and on the evaluation of their effects from a biophysical point of view (e.g. Bash and Ryan 2002; Bernhardt et al. 2007; Kail et al. 2007).

This diversity of survey approaches is addressed by a variety of methods (Figure 1.8). Most of these are on-site surveys, interviews, and questionnaires conducted face to face with respondents. However, several studies used other survey methods such as tele-phone, mail, or Internet surveys. These different modes of survey administration, which are less developed, are used in questionnaire-based surveys because they often make it possible to reach a larger number of respondents (e.g. Loomis 1996; Buijs 2009; Bliem et al. 2012). For example, most economic studies use questionnaires for contin-gency evaluations that rely on the willingness to pay (e.g. Loomis 1996; Lee 2012; Kim et al. 2018), and then make quantitative analyses of their data. Studies on social prac-tices and perceptions also utilize questionnaires, sometimes coupled with interviews (e.g. Tunstall et al. 2000; Buijs 2009; Åberg and Tapsell 2013; Deffner and Haase 2018), and sometimes using specific survey methods such as photo-questionnaires (e.g. Piégay et al. 2005; Junker and Buchecker 2008; McCormick et al. 2015) or photo-based inter-views (e.g. Westling et al. 2014). Social approaches also give room for in-depth inter-views. The latter, and to a lesser extent observation methods, are the dominant methods supporting policy analysis (e.g. Tanaka 2006; Barthélémy and Armani 2015; Heldt et al. 2016; Druschke et al. 2017). Analyses of interview data are most often qualitative. They enable us to understand the complexity of the actors' roles and governance pro-cesses. Political and social studies often combine observations, interviews, and ques-tionnaires (e.g. Junker et al. 2007; Lave et al. 2010). Focus group interviews are rarely used as methods per se (e.g. Fox et al. 2017), but are often used to prepare the imple-mentation of a questionnaire, particularly in economic studies (e.g. Loomis 2002; Kenney et al. 2012; Kim et al. 2018).

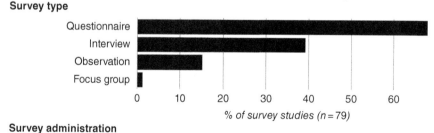

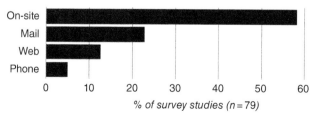

Figure 1.8 Main methods used in surveys of societal issues in river restoration.

Table 1.2 How are the notions of "perception," "attitude," and "place attachment" defined in the literature on societal issues in river restoration?

Perception	Although the notion of perception is often used in the field of river restoration, it is rarely defined in publications. Its epistemology is difficult to establish with certainty across different studies. However, the use made of it by authors leads to its inclusion in the field of "environmental perception" studies, which since the 1970s have been positioned at the intersection of different disciplines such as geography, psychology, sociology, or anthropology (Saarinen and Sell 1980). According to Zube (1999, p. 214), "Environmental perception has commonly been defined as awareness of, or feelings about, the environment." It is also "the act of apprehending the environment by the senses." It is on this apprehension of river landscapes or the elements that constitute them that several studies in the field of restoration have focused on (e.g. Piégay et al. 2005; Junker and Buchecker 2008; Seidl and Stauffacher 2013; Åberg and Tapsell 2013), with numerous works being specifically anchored in the field of "landscape perceptions" (Zube et al. 1982). Most are interested in "perceived landscape value" through different criteria such as aesthetics, naturalness, biodiversity, or ecosystem services. Other work is more detached from the landscape approach. Perception is then a mental construction as much as an act of sensory experience. This is the approach that seems to be defended in works focusing on the perception of environmental problems (e.g. Pahl-Wostl 2006; Alam 2011) or flood risk (e.g. Buijs 2009; Chou 2016). Behind the term "perception" is the idea of subjective evaluation, as Jähnig et al. (2011) explain when they compare the subjectivity of perception and the objectivity of scientific knowledge in the evaluation of restoration success. Although differences in the ways of perceiving reality may be certain, the hierarchy implied by the objective–subjective opposition is debatable. A number of works use the notion of perception in a more common sense, as being synonymous with opinion regarding project implementation (e.g. Davenport 2010; Feng et al. 2015). In some studies, a strong link is established between perception and attitude (e.g. Alam 2011; Åberg and Tapsell 2013; Deffner and Haase 2018). Knowing environmental perceptions would allow a better understanding of the support or opposition of certain categories of stakeholders to river restoration projects.
Attitude	The origin of the notion of attitude in the environmental field is to be found in psychological work (Kaiser et al. 1999). According to Gifford and Sussman (2012, p. 65), an "attitude is a latent construct mentally attached to a concrete or abstract object." Attitude would thus be distinguished from perception, which has a more sensory origin. It would also have a more direct link to behavior (Kaiser et al. 1999). It must be noted that, like the notion of perception, that of attitude is rarely defined or referenced in work carried out in the field of river restoration. It is often invoked in a generic sense synonymously with an opinion or perception of the restoration project and its conduct or effects (e.g. Tunstall et al. 2000; Purcell et al. 2002; Chin et al. 2005; Buijs 2009; Alam 2011). Support and opposition to restoration are sometimes defined as attitudes (e.g. Feng et al. 2015; Chou 2016). For example, Heldt et al. (2016, p. 5) speak of "positive," "neutral," or "negative" attitudes toward projects.
Place attachment	Born out of research on "people place relationships" (Lewicka 2011), the notion of "place attachment" is mobilized by various social science disciplines, including psychology, sociology, and human geography. Although ancient, this notion appears relatively recently in the field of river restoration (e.g. Buijs 2009; Alam 2011; Fox et al. 2016; Verbrugge and van den Born 2018). "In general, place attachment is defined as an affective bond or link between people and specific places" (Hidalgo and Hernández 2001). As Alam (2011, p. 637) reminds us, the definition of the notion of place attachment is not really stabilized and has many synonyms. For Buijs (2009, p. 2681), "place attachment" is, along with "place identity" and "place dependency," one of the dimensions of the "sense of place." Verbrugge and van den Born (2018, p. 242) propose a different perspective, presenting "place identity, place dependence, social bonding, and narrative bonding" as the four dimensions of place attachment. These authors also suggest that "place of attachment" is a dimension of the public's perception of restoration. This was not the case for Fox et al. (2016), who spoke of "attachment to landscapes" to define the place of attachment and established a stronger synonymy with the field of landscape perception, which also gives an important role to places. The specificity of the notion of place attachment would then be mainly attributable to the central place given to emotions and the affective dimension. Verbrugge and van den Born (2018, p. 241) speak of "emotional connections to place" in their definition of the concept.

In order to grasp human–river relationships, many publications give an important place to "landscape" or "riverscape." Landscape is defined as the sensitive side of the environment. Landscape perception is the mechanism by which individuals perceive ecological processes and the state of ecosystems. As such, landscape experience cannot be neglected in environmental action (Nassauer 1992; Gobster et al. 2007). Nevertheless, societal approaches to river restoration mobilize the landscape in a variety of ways. Some authors use it to highlight the cultural and historical anchoring of the relationship between residents and rivers that guides, or should guide, restoration projects (e.g. Fox et al. 2017; An and Lee 2019). Other researchers use the landscape to collect preferences toward different restoration scenarios (e.g. Junker and Buchecker 2008) or different river states according to their morphology or ecology (e.g. Piégay et al. 2005; McCormick et al. 2015). The landscape approach is also used to understand the impact of a restoration project on the relationship of riverside residents. In this respect, the work carried out in the United Kingdom by Åberg and Tapsell (2013) on the Skerne river, or by Westling et al. (2014) on the Dearne river, are particularly interesting; they are perfect examples of the impact that a restoration project can have on perceptions of river landscapes, both aesthetically and in terms of recreational practices. These works also present the original aspect of following up human—river relationships over the long term. In this respect, they stress the importance of understanding the temporal dynamics at play in the construction of such relationships. These relationships have a history, sometimes ancient, and evolve slowly, sometimes on the scale of several generations.

1.4.1.3 Practices supposed to guide the values associated with rivers

Although landscape perceptions are often at the heart of studies relating to human–river interactions, a significant part of the literature also focuses on the river-based activities that individuals engage in. These studies are often based on the idea that practices influence perceptions of the river, and therefore perceptions of restoration. In particular, a study on several urban rivers in the United Kingdom by Tunstall et al. (2000) showed that while most residents evaluate restoration projects very positively, particularly because of better access to the riverbanks and more recreational opportunities, some people are more nuanced. For example, anglers may denounce the degradation of fish resources following restoration, both in quantity and quality. Restoration projects may also create opportunities to renew practices between residents and their river, especially recreational ones, and increase the value they assign to it (e.g. Loomis 2002). Studies often show that the easier access provided by restoration projects contributes to intensifying or even creating new activities and attachments to rivers (e.g. Westling et al. 2014). These studies show that the diversity of links established between residents and rivers forges diverse expectations with regard to restoration actions. They also demonstrate that a restoration project will first be a political project, before being a technical one. This political project implies discussing both the diversity of links to the river and the diversity of expectations when defining restoration objectives (Wohl et al. 2005; Baker et al. 2014).

1.4.2 Studying the political stakes of river restoration

The first studies addressing the political dimension of river restoration were published in the 2000s, often in connection with analysis of the effects of restoration on human–river interactions (e.g. Tunstall et al. 2000; Connelly et al. 2002; Junker et al. 2007). They often involve both survey methods and literature reviews (see Box 1.3), and are based on the idea that environmental policies are the result of power plays between different interests (Baker and Eckerberg 2013). In fact, many policy approaches to river restoration are devoted to the issue of arbitration between different interests involved in decision-making. Which restoration objectives should be prioritized? How to intervene? On which sites should we prioritize action? The answers to these questions probably vary according to the actors questioned and the governance methods adopted for the project, but also more broadly according to the territorial context in which the project takes place. Thus, behind seemingly technical issues there are in fact significant and influential social and political processes that are at the heart of political approaches to restoration. For Light and Higgs (1996), the political stakes of

---

Box 1.3   Documentation sources: material that is little valued in river restoration research

While documentation sources are often used in societal studies, they are rarely mobilized as foundational materials for research work. They are most often complementary to survey methods (e.g. Buijs 2009; Barthélémy and Armani 2015; Heldt et al. 2016; Druschke et al. 2017), to provide elements of interpretation, confirmation, or discussion of the information obtained during interviews. The nature of the documents is often little discussed and the methods of their analysis are rarely explained; they are most often qualitative.

However, document sources are valuable for providing information on river restoration projects. The most used are policy documents or administrative documents produced by national, regional, or local administrations, and operational documents related to projects (planning documents, technical reports, and communication documents). These documents contain technical and scientific information that (for example) has contributed to the inventories of projects carried out by environmental scientists to evaluate restoration practices and their effects on river ecology since the 2000s (e.g. Bernhardt et al. 2007). The bibliographical study shows that these documents are also more mobilized to answer the societal questions raised by river restoration approaches. For example, operational documents are sources providing data on project costs for economic studies (e.g. Alam 2008; Carah et al. 2014; Langhans et al. 2014). Above all, they provide key material for understanding the political processes at work in restoration processes (e.g. Tanaka 2006; Gerlak et al. 2009; Lee and Choi 2012; Guerrin 2015). Their analysis makes it possible to trace the genealogy of a restoration project and to identify the role played by the various stakeholders in its implementation, their positions in a political sense, and their strategies for action. Some studies, less numerous, focus more specifically on how public opinion reacts to certain river restoration approaches (e.g. Bark et al. 2016; Heldt et al. 2016; Druschke et al. 2017). For this purpose, they mobilize other documentation sources, such as the news media and particularly the press.

restoration can be approached from two perspectives, often complementary. The first relates to the choices that are made, the reasons why they are made, and the manner in which they are made, and is referred to as "politics in restoration." In the field of river restoration, there are therefore many publications related to governance and stakeholder participation. The second approach is the "politics of restoration." It brings together work that looks at the way in which restoration approaches fit into further political, social, and economic mechanisms beyond the immediate realm. These two approaches examine the value of restoration policy (Light and Higgs 1996), questioning what is a "good" or "bad" restoration project, particularly from the perspective of legitimacy (who makes the decisions and in what capacity?) or equity (are the decisions fair and made in the name of the common good?).

1.4.2.1 A focus on stakeholders of river restoration: the participatory approach

According to Szałkiewicz et al. (2018), in Europe "52% of the projects analyzed have been designed and implemented without the participation of local stakeholders." Whether actively or passively concerned by a decision, the individuals or collectives (group or organization) that make up society nevertheless have interests that may be positively or negatively affected by its execution (or nonexecution). These interests often determine their relationship to the river and their stance on restoration. In policy approaches to river restoration, stakeholders are at the heart of the discussion (Figure 1.9), and the question is often

This wordcloud is produced from the calculation of lexical specificity scores. Based on a probability law, the specificity score assesses if some words occur more frequently in certain parts of a corpus (here, the words used in the titles of "political" sub-corpus, n = 38) than in the whole corpus (here, the "societal" corpus, n = 121) from which the parts are extracted.

Figure 1.9 Lexicon specific to international scientific publications dealing with the political stakes of river restoration.

to determine which river restoration stakeholders are or should be involved in decision-making (Junker et al. 2007). Political approaches to river restoration are often strongly biased toward broad participation. According to various authors, the river restoration decision, as well as its evaluation, is best made collectively (Jähnig et al. 2011; Deffner and Haase 2018). This positioning places river restoration at the center of a democratic debate. Indeed, it considers that a restoration project, insofar as it creates value (Light and Higgs 1996), must produce a value that is shared or at least collectively debated if the project is to have relevance and social legitimacy (Deffner and Haase 2018). The river is considered a common good around which a "community" coalesces. This community is made up of individuals who often have different or even divergent interests, but who come together around a shared project. By placing the democratic issue at the heart of restoration, political approaches make the river a tool for living together. In this sense, political action cannot be taken independently of the social and cultural contexts in which it takes place.

The form of the democratic debate is also at the center of considerations; representative democracy giving expert groups legitimacy to act is questioned. More and more authors are placing participatory approaches at the center of their work and considering the involvement of different stakeholders in the project; they approach the political dimension of river restoration from a governance perspective. According to Mansourian (2017, p. 402), "governance determines who takes decisions, and how these decisions are made and applied." Some of these studies are interested in the satisfaction, or dissatisfaction, of stakeholders regarding the degree and manner in which they were involved in the project (Junker et al. 2007; Heldt et al. 2016). Others propose a monitoring of governance, and analyze, often in a critical manner, the way in which the interplay of actors within the loop has been able to influence decisions (Tanaka 2006; Lee and Choi 2012; Hong and Chun 2018). Thus, within the framework of the restoration of the Anyangcheon river in Seoul (South Korea), Hong and Chun (2018) were able to highlight power asymmetries between the different stakeholders of the project that contributed to prioritizing, in the choice of restoration objectives, scientific values to the detriment of nonscientific values, such as cultural, aesthetic, social, or educational ones. The importance of the leadership of certain stakeholders, endowed with varied influence and capacity, for driving the concretization and orientation of projects is often mentioned (Lee and Choi 2012; Barthélémy and Armani 2015).

1.4.2.2 River restoration at the heart of power relationships: between conflict analysis and critical approaches

Several policy researches in the field of restoration offer analyses of power relations and conflicts related to project implementation. In particular, they seek to highlight the hidden tensions that result from the sociopolitical processes that animate and define the social situation. The political ecology field of research (e.g. Doyle et al. 2015; Sneddon et al. 2017; Drouineau et al. 2018), whose ambition is to study power relations (Benjaminsen and Svarstad 2019, p. 392), has particularly invested in these critical approaches, "questioning of the role and status of powerful actors as well as of what is taken for granted in leading discourses on environment and development." These approaches, politically committed to a fairer and more sustainable organization of society, bring a particular color to the thinking on river restoration, being rather driven by an environmentalist commitment. For example, a special place is given to local communities in the tradition of postcolonial studies (e.g. Fox et al. 2017; Woelfle-Erskine 2017). These communities arouse a strong interest because of the original and ancestral links they have forged with rivers and, paradoxically, because of the small weight they are generally able to have in the decision-making process in the face of more powerful institutional, political, or economic actors. Their involvement raises issues of cultural recognition and preservation of traditions and worldviews. Contributions in political economy have also made achievements in the field of river restoration (e.g. Lave et al. 2010; Lave 2016). They have shown, for example, how neoliberal logics have guided restoration practices in the United States by favoring the private sector in the production of expertise and promoting the creation of new markets related to ecosystem services (Lave et al. 2010).

Beyond the latent power balances, many publications also analyze restoration projects from the perspective of open conflicts, or at least the opposition they generate. These oppositions are particularly strong regarding certain restoration measures. Projects for restoring continuity, particularly dam removal, which generally result in major upheavals to landscapes and practices, appear to be the most controversial. These projects are therefore at the center of many political analyses (Figure 1.10) (e.g. Druschke et al. 2017; Sneddon et al. 2017; Drouineau et al. 2018).

While particular attention is paid to opposition to restoration, many of the works also focus on public support for the projects (Table 1.3). According to Junker et al. (2007), the

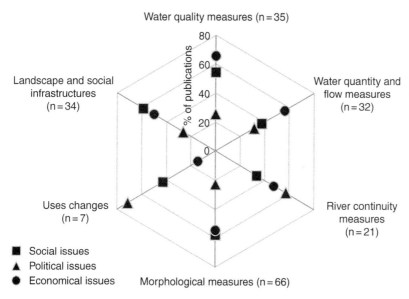

Figure 1.10 Restoration measures tackled in scientific publications on the social, political or economic issues of river restoration.

Table 1.3 How are the notions of "expectation," "support," and "acceptance" defined in the literature on societal issues in river restoration?

Expectation	The notion of expectation is rarely discussed and often used in a generic sense as "the action of waiting for something or someone; expectant waiting" (Trumble and Stevenson 2002). In the context of restoration, this foresight often concerns the way a project is conducted or the results it should have. Expectations are a priori positive. They are based on what people, practitioners (e.g. Chen et al. 2017), or the public (e.g. Tunstall et al. 2000; Junker and Buchecker 2008) imagine and want in relation to the river. Contrary to the notions of perception or attitude that most often refer to existing objects or phenomena, the notion of expectation requires projection, and is mobilized in the pre-restoration phase to nurture the restoration project (e.g. Åberg and Tapsell 2013). This does not prevent us from also looking at how the project has met these initial expectations, which is then more a matter of studying satisfaction. Satisfaction can be a criterion of success. Conversely, disappointment is a significant risk in the case of river restoration projects (e.g. Tunstall et al. 2000). This is all the more true since such a project can not only reveal latent expectations but can also generate new ones. Expectations can change over time and through the stages of a project.
Support	The notion of support is most often used in the scientific literature in a common sense. It is the support that certain categories of stakeholders bring to a restoration project. The perspective is political. It is generally the "public support" that is at the center of analyses (e.g. Connelly et al. 2002; Schläpfer and Witzig 2006; Buijs 2009). As with the notion of willingness, the intensity associated with the use of the notion of support can vary widely, from accepting an action to publicly encouraging and defending it. Beyond the evaluation of project support, the various studies in the field of restoration seek to understand its determinants. The notion of support is thus often backed up by notions of perception or attitude, to understand the reasons why certain stakeholders or certain categories of stakeholders support or oppose restoration (e.g. Tanaka 2006; Schläpfer and Witzig 2006; Buijs 2009). For some authors, this analysis can help target environmental education approaches and build public support (e.g. Chin et al. 2008; Chen and Cho 2019), whereas, for others, it is more the debating of different opinions in a participatory perspective that will help build support for restoration (e.g. Junker et al. 2007).
Acceptance	According to Depraz (2005), the notion of acceptance has its roots in German social psychology. This may explain why, in the scientific literature on river restoration, its use is highly standardized in Germany (e.g. Heldt et al. 2016; Zingraff-Hamed et al. 2018; Deffner and Haase 2018) and Switzerland (e.g. Junker and Buchecker 2008; Schläpfer and Witzig 2006; Seidl and Stauffacher 2013). A generic definition of acceptance is "the act or fact of accepting, whether as a pleasure, a satisfaction of claim, or a duty" (Trumble and Stevenson 2002), and the concept of acceptance is a form of agreement. According to Heldt et al. (2016, pp. 1052–1053), "It is highly important to understand that acceptance is not a stable property that can either be present or not." Acceptance appears to be a process that is built particularly within the framework of participatory approaches (e.g. Junker et al. 2007; Seidl and Stauffacher 2013; Heldt et al. 2016). In the field of river restoration, the concept obviously echoes those of support and willingness, with which it is sometimes synonymous (e.g. Junker and Buchecker 2008). The concept of acceptance is sometimes discussed because it can be interpreted as a prescriptive process in which adherence is sought more than participation. Heldt et al. (2016) distinguish between acceptance, which would be descriptive, and acceptability, which would be more normative.

primary objective of restoration – to increase the natural and ecological quality of rivers – would generally be a matter of consensus among the general public. Faced with the complexity of the socioeconomic issues generally raised by restoration projects, this positive view of restoration approaches deserves to be highlighted.

1.4.2.3 A political look at the place of science in river restoration practice

In political approaches to river restoration, scientists are considered as fully fledged protagonists of restoration policies and projects. Because of their knowledge, they are consulted in the development of projects and are often involved in their evaluation. Some authors critically analyze this expert stance (e.g. Lave et al. 2010; Lave 2016; Fox et al. 2017). For them, researchers are not immune to the influence of the political, economic, and social forces that influence the politics of restoration (Light and Higgs 1996). Scientific production is rooted in multiple values that have a particular weight in the practice of river restoration because of the credit generally given to scientific knowledge and the expert. Many publications interested in the political dimension of restoration aim to highlight these values and to discuss the ethical responsibility of researchers as they engage in dialogue with society. Among the avenues being explored, Drouineau et al. (2018) point to the importance of multidisciplinarity in this science–society interface. According to the authors, the crossing of disciplinary viewpoints is an essential approach to address any problem in an integrated manner, and to take into account its various facets. Other publications put scientific knowledge into perspective by comparing it with other more vernacular forms of knowledge (e.g. Woelfle-Erskine 2017; Hong and Chun 2018). They contribute to relativizing the weight of scientific knowledge in the decision-making process and make the scientist a stakeholder among others (Yun 2014). These approaches therefore reflect on processes for engaging local communities and citizens at large in decision-making processes (e.g. Heldt et al. 2016; Fox et al. 2017; Edwards et al. 2018). To this end, publications consider and evaluate the merits of alternative approaches for the production of knowledge in the field of river restoration. Pahl-Wostl (2006) insists, for example, on the principle of co-construction of knowledge to promote social learning in the field of river restoration. Restoration projects involving participatory science (e.g. Edwards et al. 2018) or collaborative research (e.g. Fox et al. 2017) are, however, rare. Fox et al. (2017, p. 532) showed that "spaces for the inclusion of new meanings, processes, and outcomes in restoration" were created through collaborations between researchers and indigenous communities. They have contributed to the recognition that the standards of indigenous communities are on a par with those of Western countries. According to the authors, however, this recognition is only one step on the long road to rebalancing power relationships in river restoration governance processes.

1.4.3 Economic evaluation of river restoration

Economic analyses were among the first societal approaches developed in the field of river restoration (e.g. Barendregt et al. 1992; Loomis 1996; Turner and Boyer 1997) (Figure 1.11). Several works are now widely referenced and utilized in the scientific community (Loomis

This wordcloud is produced from the calculation of lexical specificity scores. Based on a probability law, the specificity score assesses if some words occur more frequently in certain parts of a corpus (here, the words used in the titles of "economical" sub-corpus, n = 66) than in the whole corpus (here, the "societal" corpus, n =121) from which the parts are extracted.

Figure 1.11 A lexicon specific to international scientific publications on the economic stakes in river restoration.

et al. 2000; Chen et al. 2017; Acuña et al. 2013; Vermaat et al. 2016; Bliem et al. 2012; Gerner et al. 2018). Economic approaches focus primarily on producing an assessment of river restoration policies; however, in contrast to budgetary approaches to public policy evaluation, which are placed from the point of view of funders and primarily consider the financial costs of projects, they adopt a more socioeconomic perspective for this evaluation (Brouwer and Sheremet 2017). Thus, the majority of economic publications integrate discussions on social practices and environmental perceptions (Table 1.4). According to Brouwer and Sheremet (2017, p. 1), the economic value of a river restoration project is defined as "the calculation of all costs and benefits, including (second-order) indirect effects on sectors and (non-priced) environmental effects, often referred to as the broader social costs and benefits."

1.4.3.1 Contributions focused on the evaluation of benefits of river restoration
The financial amounts committed to river restoration are an important issue, highlighted by several reviews on river restoration (e.g. Bernhardt et al. 2005; Szałkiewicz et al. 2018).

Some economic publications are more focused on project costs alone. Cost reduction is very often the objective of these studies, which take note of the strong budgetary constraints imposed on the implementation of restoration actions. This search for cost optimization leads some authors to question and prioritize the restoration techniques that are adopted in projects (e.g. Hill et al. 2011; Kristensen et al. 2012; Carah et al. 2014). Other authors look at costs from a different perspective, assessing costs as losses to the stakeholders in the restoration. At the heart of these studies is the value of the assets that certain stakeholders must give up if the restoration project is to be carried out; these are opportunity costs. For example, Gómez et al. (2014) consider the opportunity costs associated with the reduced power generation resulting from the implementation of flushing floods which are known to restore basic environmental functions in highly engineered rivers. In particular, they demonstrate that this cost is lower than the willingness to pay for such restoration measures. Losses are also sometimes considered in terms of the compensation to be paid to the stakeholders affected by the restoration project (e.g. Cheng et al. 2019). Such cost-only approaches are not dominant in the literature.

The vast majority of the literature with an economic approach to river restoration compares costs against benefits or focuses on assessing benefits alone. In the cost–benefit analyses, it is above all the evaluation of benefits that is at the heart of the research effort. The costs are considered as fixed data that are not very open to discussion. Behind this interest in the benefits, there exists a form of engagement of the scientific community that is in favor of river restoration policies. By making visible the benefits induced by the projects and highlighting their effectiveness from the point of view of public investment, these publications help to justify and promote river restoration to decision-makers and the general public. This focus on benefits can also be explained by the methodological challenges faced by researchers. While it is indeed relatively easy to access project cost data – at least for technical implementation – the measurement of benefits is more challenging (Brouwer and Sheremet 2017). This is all the more true because environmental benefits are most often of a nonmarket value that is difficult to measure. To determine this value, which some authors describe as a "shadow price" (e.g. Grossmann 2012; Becker et al. 2014), economics relies on a substantial methodological arsenal (see Box 1.4).

1.4.3.2 What river restoration benefits are we talking about? The ecosystem services approach

River restoration is likely to produce multiple benefits, whether ecological (e.g. restoration of certain ecological functions), economic (e.g. creation of a market and thus of companies or consulting firms specializing in river restoration), or social (e.g. development of recreational opportunities associated with rivers). To capture these benefits, economic approaches to river restoration are essentially based on the concept of ecosystem services. They are thus part of an anthropocentric paradigm of interactions between societies and rivers. Ecosystem services are defined as the benefits that society derives from ecosystems and that therefore contribute to human well-being (Costanza et al. 1997; Daily 1997; MEA 2005)[2]. Ecological structures and processes perform ecological functions that are themselves sources of services for humans. However, these services are only regarded as benefits when society attributes a value to them (Haines-Young and Potschin 2010).

Box 1.4   Economic methods: assessing the benefits of river restoration

The evaluation of the benefits produced by river restoration in relation to the costs it generates is at the center of many economic studies. The bibliographical work makes it possible to distinguish two main approaches in the way this evaluation is conducted.

1) The first is cost-effectiveness analysis. Several publications measure the benefits of restoration according to ecological (e.g. Barendregt et al. 1992; Langhans et al. 2014) or socio-ecological (e.g. Golet et al. 2006; Jia et al. 2010; Kendy et al. 2018) indicators, which they compare against monetary project costs.
2) The second and more common approach compares costs to economic benefits. While the measurement of costs is not described in the literature as posing any particular problems, the measurement of benefits raises important methodological issues. The benefits of a project are multifaceted, and in practice correspond to very different units of measurement (e.g. the quantity of fish available for angling, the beauty of the landscape, the number of visitors). To alleviate this problem, authors often conduct a monetary evaluation of the benefits of restoration. The monetary amount resulting from this evaluation is more readily comparable to costs. Money is thus used as the "common denominator" in cost–benefit analyses (Brouwer and Sheremet 2017). Several meta-analyses of the scientific literature dealing with non-market valuation of benefits have been undertaken within the field of river restoration (e.g. Bergstrom and Loomis 2017; Brouwer and Sheremet 2017).

It is possible to distinguish, within the literature, two main approaches to the monetary valuation of the benefits of restoration (Figure 1.12). The first is the benefit transfer approach, which is based on existing data on the value of certain ecosystem services.

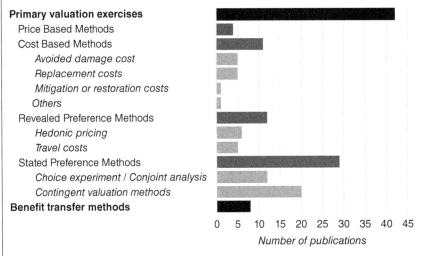

Figure 1.12   Economic methods used in international scientific publications to assess the benefits of river restoration.

These data, measured, for example, on a previous restoration project, are used to deduce the value of the same services on another project at another site. This approach is generally used when study resources do not permit the production of primary data. For example, Alam (2008) used the work of Haque et al. (1997) to estimate the benefits produced by the restoration of the Buriganga River (Bangladesh) in terms of health and housing and land values. Several authors call for great caution in the use of this method because of the significant biases that exist when extrapolating data from one area to another, particularly because of the heterogeneity of their characteristics (e.g. Brouwer et al. 2016; Lewis and Landry 2017).

The second approach to assessing benefits, utilized in the majority of publications, is through primary evaluation exercises. This includes all methods that produce data specific to one or a set of restoration projects. Among them, price-based methods are currently seldom applied to restoration projects. These methods use the amounts of actual transactions for certain ecosystem services. For example, Vermaat et al. (2016) accounted for the benefits of several European restoration projects using, among other things, the real market prices of certain supplies (e.g. agricultural commodities, drinking water), regulations (e.g. fertilizers), and cultural services (e.g. fishing and hunting licenses, kayak rentals).

Cost-based approaches are more represented in the scientific literature. These use the costs incurred in the absence of the ecosystem services as a proxy for ecosystem services values. For example, this method is used by Kenney et al. (2012, p. 608) to "estimate the cost-effectiveness of restoration for infrastructure protection by comparing it to the avoided cost of riprap bank protection."

Revealed preference approaches are used in about a quarter of primary assessments. These consist of the analysis of individual choices in marketable goods transactions. Individual preferences reveal the market value of these goods. For example, some studies conduct hedonic pricing approaches and examine changes in real estate prices in the context of river restoration (e.g. Chen 2017; Lewis and Landry 2017; Netusil et al. 2019). Others conduct travel cost approaches and measure the amount people pay to access a service (e.g. Loomis 2002; Becker and Friedler 2013; Akron et al. 2017).

Finally, stated preference methods are the most used in the field of river restoration. These methods simulate a market for ecosystem services through surveys of hypothetical changes in the provision of ecosystem services. People are asked to indicate explicitly or implicitly their willingness to pay for positive change (e.g. Loomis 1996; Chen et al. 2017; Vásquez and de Rezende 2019) or their willingness to accept negative change (e.g. Zhang et al. 2019). The extensive use of stated preference methods echoes the recommendations of some authors (e.g. Honey-Rosés et al. 2013) who prefer, for operational reasons, to use approaches that rely on the demand for services to qualify their value. Generally speaking, these are more directly anchored in the reality and social expectations of the studied area.

Finally, it should be noted that these different approaches are not exclusive and that many studies use several of them to address the various benefits of restoration in a holistic manner. In this perspective, many authors refer to the conceptual framework of "total economic value" (TEV) (Randall 1987).

Quantification of this value is an issue mobilizing a significant proportion of the economic science community working on restoration issues.

The perspectives adopted in the publications are, however, heterogeneous. Some authors focus their attention on a single type of service for which they wish to estimate the value. For example, Grossmann (2012) aimed to assess the economic benefits of retention of river-carried nutrients by floodplain wetlands. This evaluation of a so-called regulating ecosystem service (MEA 2005) had a demonstrative value, aiming to highlight the shadow price of the alluvial plains of the Elbe basin (Germany) according to evaluation of their capacity to retain pollutants. Following more pragmatic and operational approaches, other authors have attempted to estimate the value of ecosystem services based on the social demand for them, not only on their level of availability/supply (Honey-Rosés et al. 2013). These approaches generally lead the authors to anchor their thinking to a specific restoration project. It is then a question of determining all the services that could be affected and measuring the overall benefit to society. These approaches based on the assessment of social demand for ecosystem services sometimes mobilize different scenarios corresponding to different restoration measures. By evaluating the fluctuation of assigned value according to different scenarios, these studies make it possible to prioritize the measures to be implemented within the project. For example, several cost–benefit analyses of restoration projects in the field of pollution treatment carried out on the Alexander-Zeimar and Kishon rivers (Israel) show that, while the most ambitious restoration measures may present the best cost–benefit ratios (Becker and Friedler 2013), more-targeted restoration measures may sometimes be more economically profitable (Becker et al. 2019).

Among the publications reviewed, few presented a negative economic balance of river restoration projects (e.g. Lee and Jung 2016). Most report excess cost–benefit analyses (e.g. Lee 2012; Polizzi et al. 2015; Kim et al. 2018; Logar et al. 2019). In other words, river restoration projects are presented as being economically profitable overall in relation to the benefits they produce. However, these publications show that not all services are equal when it comes to understanding the overall benefit of restoration projects. Some projects do not stand up to cost–benefit analyses if they do not take into account cultural services, particularly aesthetic or recreational services (e.g. Kenney et al. 2012; Garcia et al. 2016). The authors of these studies also underline the importance of these specific services for supporting river restoration policies, especially in urban areas. Cultural objectives, particularly landscape and recreational objectives, can also be added to traditional ecological objectives to strengthen the social amenities of rivers, and these are known to greatly increase the value of projects (e.g. Bae 2011).

A number of publications also discuss the fact that the assessment of restoration benefits is highly dependent on the spatial and temporal scale at which one places oneself (e.g. Turner and Boyer 1997; Becker et al. 2014). For example, there may be imbalances between the territories that pay for projects and those that benefit from them. Thus, Logar et al. (2019) highlight the tensions that exist in Switzerland between the scales at which projects are financed (national scale) and the scales at which projects are implemented (local scale), and they stress the interest in securing additional local funds to guarantee territorial equity. From a temporal

Table 1.4 How are the notions of "preference" and "willingness" defined in the literature on the societal stakes of river restoration?

Preference	The notion of preference is used recurrently in publications dealing with human–river interactions in the context of restoration, but according to different epistemological anchors. A number of these uses are part of a tradition of research in environmental psychology that focuses on the affective and cognitive relationships that individuals have with the environment (Kaplan 1987). In this perspective, the notion of environmental preference is closely linked to that of environmental perception. According to van den Berg et al. (2003, p. 136), "environmental preference is determined by environmental properties that possess a potential functional significance for the perceiver." The notion of environmental preference is, however, more in line with a logic of prioritization and comparative judgment. It is therefore often mobilized for the aesthetic or landscape evaluation of different restoration scenarios (e.g. Chin et al. 2008; Junker and Buchecker 2008; McCormick et al. 2015; Hong et al. 2019). Preference is also a benchmarking process that is very often invoked in economic evaluations (Hausman 2011). The concept refers particularly to revealed and stated preferences methods (Adamowicz et al. 1994; Boxal et al. 1996). These methods are widely used in the framework of restoration benefit assessment (Brouwer and Sheremet 2017; Bergstrom and Loomis 2017). Preference is synonymous with choice regarding the services that restored rivers can provide; the individual is positioned as a consumer of these services. Preference is used as a founding notion to determine the value of the river.
Willingness	The notion of willingness is difficult to define because it can be understood as being synonymous with wishing in a proactive sense, as much as a more passive "without compulsion, voluntarily" (Trumble and Stevenson 2002). In any case, the notion seems to have its origin in economics work using willingness to pay (WTP) or willingness to accept (WTA) as the structuring concept for the method of evaluation; it is within this methodological framework that it is used in publications on river restoration (e.g. Loomis 1996; Bliem et al. 2012; Kim et al. 2018). The notion of willingness is closely associated with that of preference. It is mainly discussed methodologically, between an approach that measures the amount that an individual is willing to pay to access or retain a service and an approach that measures the price they are willing to accept for losing that service (Zhao et al. 2013; Polizzi et al. 2015). Beyond economic valuations, the notion of willingness is also used to evaluate support or opposition to restoration actions. Willingness is less a matter of consent than of commitment (Connelly et al. 2002; Alam 2011). Alam (2011, p. 636) uses the term "willingness to contribute" (WTC), understood in terms of money or time invested. Some approaches are even demonetized, such as the work of Januchowski-Hartley et al. (2012, p. 24), who are interested in "landholders' willingness to engage in restoration on their land."

point of view, Polizzi et al. (2015) showed that the benefits of a restoration project undertaken in Finland offset the costs in only 3–10 years, whereas Garcia et al. (2016) estimated the net benefit of an Israeli project at 30 years. Any evaluation of river restoration cannot be separated from consideration of the timescales of the project and its benefits.

1.5   A diversity of researchers' positions with regard to operational action

Research in the field of river restoration is essentially an applied research field; it is concerned with and involved in the transformation of river environments. Studies interested in restoration often express, among their objectives, a desire to contribute to the improvement of policies and to assist in the implementation of projects. The increasing number of restoration assessment procedures performed to enable appropriate management is illustrative of this desire to support environmental action. By studying restoration policies and projects and analyzing their objectives, the way they are implemented, and their socioeconomic effects, societal approaches obviously contribute to the critical work that is carried out on restoration approaches. Rarely does societal research maintain a distant position with respect to action. Most authors make recommendations or more generally suggest ways to improve actions. If the commitment of researchers is real, it obviously takes different forms, depending on the research issues, the traditions specific to different disciplines, the individual positions of researchers, and the expectations of practitioners.

If the objective of many works carried out on societal issues is to improve river restoration policies and projects, the manner in which these improvements are considered is obviously based on different ethical positions. The primary challenge may be to contribute to achieving the environmental objectives defined within the framework of restoration projects, most often on the basis of ecological criteria. Analysis of environmental perceptions or expectations of a project is then seen as a means of removing opposition to ecological restoration approaches. The production of knowledge is thus often articulated with environmental education approaches. For example, the work conducted by Chin et al. (2008, p. 894) "is advocated to bridge the gap between scientific knowledge and public perception for effective management and restoration of river systems with [in-channel] wood." Research thus focuses on the socioeconomic benefits of restoration in order to provide arguments to justify and support ecological action.

There may, however, be some mistrust within the community of researchers working on societal issues when faced with work aimed essentially at fostering social acceptance, which could be conceived as a simple recording of decisions taken. Many studies show a strong commitment to a participatory conception of public action in the field of restoration, both for political reasons – a vision of the democratic process – and for pragmatic reasons – involvement in the decision-making process reinforces public support for restoration practices (e.g. Junker et al. 2007). Improvement in restoration is also improvement in the governance processes. The question is then political, about who makes the decision and who participates in the decision-making. It is often a question of encouraging a more shared and therefore more consensual approach to restoration. Improvement is not about facilitating the implementation of decisions but about enhancing equity in decision-making. In a frequently referenced work, Pahl-Wostl (2006, p. 12) recommends "to implement a participatory process that facilitates social learning and institutional change, and leads to an adaptive management strategy for the restoration." Participation should thus facilitate a broad acceptance of environmental issues, but also enable projects to evolve by taking into account and bringing together a wide range of expectations. Restoration is thus approached from a more integrated perspective. Ecological restoration then becomes a means of environmental and territorial development leading to the long-term well-being of the inhabitants. While the ecological dimension occupies an important place, the social question, and in particular social justice within the constraints and benefits of restoration, is also at the heart of the debate and must not be neglected.

Some studies also question the objectives of restoration without guiding the participants in their choice. Social science researchers often have a particular relationship to the commitment and expertise. The deconstruction of the figure of the expert, discussions around the hierarchy of knowledge (e.g. Fox et al. 2017), and, more generally, the study of power relations linked to knowledge (e.g. Lave 2016) can lead to paradoxically opposed positions. Some researchers, in political ecology for example, adopt a resolutely committed stance and affirm it as a postulate of their research approach. Others may be cautious in their links to action. It is therefore necessary for those involved in restoration to think differently about the science–management relationship, which in the field of water-related issues has long been dominated by an engineering logic based precisely on the logic of solution. Work in the humanities and social sciences often carries with it a tradition of the question rather than the answer. This is also related to the spatial and temporal scales on which such work is conducted, which sometimes differ from the scales of river management. For example, the work of Lave et al. (2010) on the relationship between science and management in the United States does not have an immediate application in the framework of projects, but serves to accommodate the thinking of restoration stakeholders in the scientific and technical models on which their actions are based.

1.6   A book to share a diversity of societal approaches in the field of river restoration

1.6.1   What will you find in this edited book?

This book is the result of a long collaborative process. It stems from the observation that many researchers have worked for many years on the societal issues of river restoration. However, these scientific commitments remain relatively short term and are most often carried out within the framework of interdisciplinary projects led primarily by the natural sciences. Researchers addressing the social, political, or economic issues raised by restoration are relatively isolated when it comes to their ability to consider common reflexive approaches and share their work with the broader community working in river restoration. For example, there are few major dedicated events where they can come together to exchange and discuss the approaches, scientific results, and operational perspectives opened up by their research. This book was conceived to open up a space for sharing such perspectives, and to highlight the rich and dynamic interdisciplinary community.

This book forms a compendium of the diversity of scientific work carried out over the last ten years on societal challenges to river restoration. It proposes 14 contributions by researchers from different research fields, such as environmental economics, critical physical geography, human or social geography, hydromorphology, ecology and biology, sustainable management studies, philosophy and ethics, political ecology, political science, architecture and urban studies, and anthropology. Each of these disciplines enriches the book with concepts and methods from different epistemological traditions. Some contributions propose syntheses of already published works, whereas others present original studies. Most of them base their analysis on case studies and empirical data. To account for the diversity of the environmental, sociopolitical, and cultural contexts in which river restoration takes place, the contributions of this book are international. They come from research teams in Austria, France, the Netherlands, New Zealand, Spain, Switzerland, the United Kingdom, and the United States of America. Collecting these contributions together makes it possible to emphasize their complementary nature. This compendium is intended to be a showcase of research achievements: questions, methods, and scientific results. It provides stimulating perspectives for understanding restoration in its social, economic, and political dimensions.

1.6.2 Who is this book aimed at?

This book is intended for scientists working on the theme of river restoration, and for those working on broader environmental issues. It is, of course, intended for researchers in the humanities and social sciences who wish to learn about the latest work carried out in this field by their peers. It is also, and more generally, addressed to the interdisciplinary scientific community working in the field of river research and restoration, which has strong expectations regarding societal approaches.

It is also intended for students in environmental sciences who intend to go into research or river management professions. This book will help them to understand the social, political, and economic dimensions of river restoration, and their linkages with ecological issues; it will therefore help to forge an interdisciplinary culture to better respond to the issues raised by river restoration and environmental management in general.

Finally, this book is aimed at operational actors in river restoration, be they project leaders, technical partners, or financiers. It will be of interest to public bodies such as governmental, regional, and local agencies with expertise in the field of river restoration. It will also be of interest to NGOs, whether they are specialized in the field of river restoration or more broadly interested in river management. The book may also be of interest to design offices that advise such parties. These various professionals in the midst of the action are often the first to be confronted with the social, economic, and political stakes of restoration. This book will provide them with keys to better understand the mechanisms in action and to better integrate them into the implementation or support of river restoration policies and projects.

Notes

1 In this chapter, the adjective "societal" is used to refer to all work addressing the social, political, and economic issues involved in river restoration.

2 According to the IPBES (2020), ecosystem services are "the benefits people obtain from ecosystems. In the Millennium Ecosystem Assessment, ecosystem services can be divided into supporting, regulating, provisioning and cultural. This classification, however, is superseded in IPBES assessments by the system used under 'nature's contributions to people'. This is because IPBES recognises that many services fit into more than one of the four categories. For example, food is both a provisioning service and also, emphatically, a cultural service, in many cultures."

References

Åberg, E.U. and Tapsell, S. (2013). Revisiting the River Skerne: the long-term social benefits of river rehabilitation. *Landscape and Urban Planning* 113: 94–103.

Acuña, V., Díez, J.R., Flores, L., et al. (2013). Does it make economic sense to restore rivers for their ecosystem services? *Journal of Applied Ecology* 50(4): 988–997.

Adamowicz, W., Louviere, J., and Williams, M. (1994). Combining revealed and stated preference methods for valuing environmental amenities. *Journal of Environmental Economics and Management* 26(3): 271–292.

Akron, A., Ghermandi, A., Dayan, T., et al. (2017). Interbasin water transfer for the rehabilitation of a transboundary Mediterranean stream: an economic analysis. *Journal of Environmental Management* 202: 276–286.

Alam, K. (2008). Cost–benefit analysis of restoring Buriganga River, Bangladesh. *International Journal of Water Resources Development* 24(4): 593–607.

Alam, K. (2011). Public attitudes toward restoration of impaired river ecosystems: does residents' attachment to place matter? *Urban Ecosystems* 14(4): 635–653.

An, D.W. and Lee, J.Y. (2019). Influence and sustainability of the concept of landscape seen in Cheonggye stream and Suseongdong Valley restoration projects. *Sustainability* 11(4): 1126.

Aronson, J., Clewell, A.F., Blignaut, J.N., et al. (2006). Ecological restoration: a new frontier for nature conservation and economics. *Journal for Nature Conservation* 14: 135–139.

Axelrad, G. and Feinerman, E. (2009). Regional planning of wastewater reuse for irrigation and river rehabilitation. *Journal of Agricultural Economics* 60(1): 105–131.

Bae, H. (2011). Urban stream restoration in Korea: design considerations and residents' willingness to pay. *Urban Forestry & Urban Greening* 10(2): 119–126.

Baker, S. and Eckerberg, K. (2013). A policy analysis perspective on ecological restoration. *Ecology and Society* 18(2): 17.

Baker, S., Eckerberg, K., and Zachrisson, A. (2014). Political science and ecological restoration. *Environmental Politics* 23(3): 509–524.

Barendregt, A., Stam, S.M.E., and Wassen, M.J. (1992). Restoration of fen ecosystems in the Vecht River plain: cost–benefit analysis of hydrological alternatives. *Hydrobiologia* 233(1–3): 247–259.

Bark, R.H., Robinson, C.J., and Flessa, K.W. (2016). Tracking cultural ecosystem services: water chasing the Colorado River restoration pulse flow. *Ecological Economics* 127: 165–172.

Barthélémy, C. and Armani, G. (2015). A comparison of social processes at three sites of the French Rhône River subjected to ecological restoration. *Freshwater Biology* 60(6): 1208–1220.

Bash, J.S. and Ryan, C.M. (2002). Stream restoration and enhancement projects: is anyone monitoring? *Environmental Management* 29(6): 877–885.

Becker, N., Helgeson, J., and Katz, D. (2014). Once there was a river: a benefit–cost analysis of rehabilitation of the Jordan River. *Regional Environmental Change* 14(4): 1303–1314.

Becker, N. and Friedler, E. (2013). Integrated hydro-economic assessment of restoration of the Alexander-Zeimar River (Israel–Palestinian Authority). *Regional Environmental Change* 13(1): 103–114.

Becker, N., Greenfeld, A., and Zemah Shamir, S. (2019). Cost–benefit analysis of full and partial river restoration: the Kishon River in Israel. *International Journal of Water Resources Development* 35(5): 871–890.

Benjaminsen, T. A. and Svarstad, H. (2019). Political ecology. In: *Encyclopedia of Ecology* (ed. B.D. Fath), 391–396, Amsterdam: Elsevier.

Bennett, S.J., Simon, A., Castro, J.M., et al. (2011). The evolving science of stream restoration. In: *Stream Restoration in Dynamic Fluvial Systems: Scientific Approaches, Analyses, and Tools* (eds A. Simon, S.J. Bennett, and J.M. Castro), 1–8. Washington DC: American Geophysical Union.

Berelson, B. (1952). *Content Analysis in Communication Research*. New York: Free Press.

Bergstrom, J.C. and Loomis, J.B. (2017). Economic valuation of river restoration: an analysis of the valuation literature and its uses in decision-making. *Water Resources and Economics* 17: 9–19.

Bernhardt, E.S. and Palmer, M.A. (2007). Restoring streams in an urbanizing world. *Freshwater Biology* 52(4): 738–751.

Bernhardt, E.S., Palmer, M.A., Allan, J.D., et al. (2005). Synthesizing U.S. river restoration efforts. *Science* 308: 636–637.

Bernhardt, E.S., Sudduth, E.B., Palmer, M.A., et al. (2007). Restoring rivers one reach at a time: results from a survey of US river restoration practitioners. *Restoration Ecology* 15(3): 482–493.

Bickman, L. and Rog, D.J. (eds) (2009). *The SAGE Handbook of Applied Social Research Methods*. Thousand Oaks, CA: SAGE Publications.

Blanc, G., Demeulenaere, E., and Feuerhahn, W. (2017). *Humanités environnementales. Enquêtes et contre-enquêtes*. Paris: Éditions de la Sorbonne.

Bliem, M., Getzner, M., and Rodiga-Laßnig, P. (2012). Temporal stability of individual preferences for river restoration in Austria using a choice experiment. *Journal of Environmental Management* 103: 65–73.

Boon, P.J., Calow, P., and Petts, G.E. (eds) (1992). *River Conservation and Management*. Chichester: Wiley.

Boxall, P.C., Adamowicz, W.L., Swait, J., et al. (1996). A comparison of stated preference methods for environmental valuation. *Ecological Economics* 18(3): 243–253.

Bradshaw, A.D. (2002). Introduction and philosophy. In: *Handbook of Ecological Restoration. Principles of Restoration* (eds M.R. Perrow and A.J. Davy), 3–9. Cambridge: Cambridge University Press.

Brooks, S.S. and Lake, P.S. (2007). River restoration in Victoria, Australia: change is in the wind, and none too soon. *Restoration Ecology* 15(3): 584–591.

Brouwer, R., Bliem, M., Getzner, M., et al. (2016). Valuation and transferability of the non-market benefits of river restoration in the Danube river basin using a choice experiment. *Ecological Engineering* 87: 20–29.

Brouwer, R. and Sheremet, O. (2017). The economic value of river restoration. *Water Resources and Economics* 17: 1–8.

Buijs, A.E. (2009). Public support for river restoration: a mixed-method study into local residents' support for and framing of river management and ecological restoration in the Dutch floodplains. *Journal of Environmental Management* 90(8): 2680–2689.

Cairns Jr, J. (1995). Ecosocietal restoration reestablishing humanity's relationship with natural systems. *Environment: Science and Policy for Sustainable Development* 37(5): 4–33.

Carah, J.K., Blencowe, C.C., Wright, D.W., et al. (2014). Low-cost restoration techniques for rapidly increasing wood cover in coastal coho salmon streams. *North American Journal of Fisheries Management* 34(5): 1003–1013.

Cardinale, B.J., Duffy, J.E., Gonzalez, A., et al. (2012). Biodiversity loss and its impact on humanity. *Nature* 486: 59–67.

Castree, N., Demeritt, D., Liverman, D., et al. (eds) (2016). *A Companion to Environmental Geography*. Chichester: Wiley.

Che, Y., Yang, K., Chen, T., et al. (2012). Assessing a riverfront rehabilitation project using the comprehensive index of public accessibility. *Ecological Engineering* 40: 80–87.

Chen, W.Y. (2017). Environmental externalities of urban river pollution and restoration: a hedonic analysis in Guangzhou (China). *Landscape and Urban Planning* 157: 170–179.

Chen, W.Y. and Cho, F.H.T. (2019). Environmental information disclosure and societal preferences for urban river restoration: latent class modelling of a discrete-choice experiment. *Journal of Cleaner Production* 231: 1294–1306.

Chen, W.Y., Liekens, I., and Broekx, S. (2017). Identifying societal preferences for river restoration in a densely populated urban environment: evidence from a discrete choice experiment in central Brussels. *Environmental Management* 60(2): 263–279.

Chen, X., Wang, D., Tian, F., et al. (2016). From channelization to restoration: sociohydrologic modeling with changing community preferences in the Kissimmee River Basin, Florida. *Water Resources Research* 52(2): 1227–1244.

Cheng, B., Li, H., Yue, S., et al. (2019). Compensation for agricultural economic losses caused by restoration of healthy eco-hydrological sequences of rivers. *Water* 11(6): 1155.

Chin, A., Daniels, M.D., Urban, M.A., et al. (2008). Perceptions of wood in rivers and challenges for stream restoration in the United States. *Environmental Management* 41(6): 893–903.

Choi, Y.D. (2007). Restoration ecology to the future: a call for new paradigm. *Restoration Ecology* 15(2): 351–353.

Chou, R.J. (2016). Achieving successful river restoration in dense urban areas: lessons from Taiwan. *Sustainability* 8(11): 1159.

Clewell, A.F. and Aronson, J. (2006). Motivations for the restoration of ecosystems. *Conservation Biology* 20(2): 420–428.

Clewell, A.F. and Aronson, J. (2013). *Ecological Restoration: Principles, Values, and Structure of an Emerging Profession*. Washington DC: Island Press.

Connelly, N.A., Knuth, B.A., and Kay, D.L. (2002). Public support for ecosystem restoration in the Hudson River Valley, USA. *Environmental Management* 29(4): 467–476.

Costanza, R., d'Arge, R., De Groot, R., et al. (1997). The value of the world's ecosystem services and natural capital. *Nature* 387(6630): 253–260.

Crutzen P.J., Stoermer E.F. (2000). The "Anthropocene." *IGPB Newsletter* 41: 17–18.

Daily, G.C. (1997). *Nature's Services: Volume 3*. Washington DC: Island Press.

Davenport, M.A., Bridges, C.A., Mangun, J.C., et al. (2010). Building local community commitment to wetlands restoration: a case study of the Cache River wetlands in southern Illinois, USA. *Environmental Management* 45(4): 711–722.

Davis, M.A. and Slobodkin, L.B. (2004). The science and values of restoration ecology. *Restoration Ecology* 12(1): 1–3.

Deffner, J. and Haase, P. (2018). The societal relevance of river restoration. *Ecology and Society* 23(4): 35.

Depraz, S. (2005). Le concept d'«Akzeptanz» et son utilité en géographie sociale. *L'Espace Géographique* 34(1): 1–16.

Downs, P.W. and Kondolf, G.M. (2002). Post-project appraisals in adaptive management of river channel restoration. *Environmental Management* 29(4): 477–496.

Downs, P.W. and Thorne, C.R. (2000). Rehabilitation of a lowland river: reconciling flood defence with habitat diversity and geomorphological sustainability. *Journal of Environmental Management* 58(4): 249–268.

Doyle, M.W., Singh, J., Lave, R., et al. (2015). The morphology of streams restored for market and nonmarket purposes: insights from a mixed natural-social science approach. *Water Resources Research* 51(7): 5603–5622.

Drouineau, H., Carter, C., Rambonilaza, M., et al. (2018). River continuity restoration and diadromous fishes: much more than an ecological issue. *Environmental Management* 61(4): 671–686.

Druschke, C.G., Lundberg, E., Drapier, L., et al. (2017). Centring fish agency in coastal dam removal and river restoration. *Water Alternatives* 10(3): 724–743.

Dudgeon, D., Arthington, A.H., Gessner, M.O., et al. (2005). Freshwater biodiversity: importance, threats, status and conservation challenges. *Biological Reviews* 81: 163–182.

Dufour, S. and Piégay, H. (2009). From the myth of a lost paradise to targeted river restoration: forget natural references and focus on human benefits. *River Research and Applications* 25(5): 568–581.

Dunlap, R.E., Buttel, F.H., Dickens, P., et al. (eds) (2002). *Sociological Theory and the Environment: Classical Foundations, Contemporary Insights*. Oxford: Rowman & Littlefield.

Eden, S. and Tunstall, S. (2006). Ecological versus social restoration? How urban river restoration challenges but also fails to challenge the science–policy nexus in the United Kingdom. *Environment and Planning C: Government and Policy* 24(5): 661–680.

Eden, S., Tunstall, S., and Tapsell, S. (2000). Translating nature: river restoration as nature-culture. *Environment and Planning D: Society and Space* 18: 257–273.

Edwards, P.M., Shaloum, G., and Bedell, D. (2018). A unique role for citizen science in ecological restoration: a case study in streams. *Restoration Ecology* 26(1): 29–35.

Elliot, R. (1982). Faking nature. *Inquiry: An Interdisciplinary Journal of Philosophy* 25(1): 81–93.

Emmett, R.S. and Nye, D.E. (2017). *The Environmental Humanities: A Critical Introduction*. Cambridge, MA: MIT Press.

Feng, Q., Miao, Z., Li, Z., et al. (2015). Public perception of an ecological rehabilitation project in inland river basins in northern China: success or failure. *Environmental Research* 139: 20–30.

Fox, C.A., Magilligan, F.J., and Sneddon, C.S. (2016). "You kill the dam, you are killing a part of me": dam removal and the environmental politics of river restoration. *Geoforum* 70: 93–104.

Fox, C.A., Reo, N.J., Turner, D.A., et al. (2017). "The river is us; the river is in our veins": re-defining river restoration in three Indigenous communities. *Sustainability Science* 12(4): 521–533.

Francis, R.A. (2014). Urban rivers: novel ecosystems, new challenges. *Wiley Interdisciplinary Reviews: Water* 1(1): 19–29.

Gallie, W.B. (1956). Essentially contested concepts. *Proceedings of the Aristotelian Society: New Series* 56: 167–198.

Garcia, X., Corominas, L., Pargament, D., et al. (2016). Is river rehabilitation economically viable in water-scarce basins? *Environmental Science & Policy* 61: 154–164.

Gerlak, A.K., Eden, S., Megdal, S., et al. (2009). Restoration and river management in the arid Southwestern USA: exploring project design trends and features. *Water Policy* 11(4): 461–480.

Gerner, N.V., Nafo, I., Winking, C., et al. (2018). Large-scale river restoration pays off: a case study of ecosystem service valuation for the Emscher restoration generation project. *Ecosystem Services* 30: 327–338.

Gideon, L. (2012). *Handbook of Survey Methodology for the Social Sciences*. New York: Springer.

Gifford, R. and Sussman, R. (2012). Environmental attitudes. In: *The Oxford Handbook of Environmental and Conservation Psychology* (ed. S.D. Clayton). Oxford: Oxford University Press.

Gobster, P.H. and Hull, R.B. (eds) (2000). *Restoring Nature: Perspectives from the Social Sciences and Humanities*. Washington DC: Island Press.

Gobster, P.H., Nassauer, J.I., Daniel, T.C., et al. (2007). The shared landscape: what does aesthetics have to do with ecology? *Landscape Ecology* 22(7): 959–972.

Golet, G.H., Roberts, M.D., Larsen, E.W., et al. (2006). Assessing societal impacts when planning restoration of large alluvial rivers: a case study of the Sacramento River project, California. *Environmental Management* 37(6): 862–879.

Gómez, C.M., Pérez-Blanco, C.D., and Batalla, R.J. (2014). Tradeoffs in river restoration: flushing flows vs. hydropower generation in the Lower Ebro River, Spain. *Journal of Hydrology* 518: 130–139.

Gore, J.A. (1985). *Restoration of Rivers and Streams*. Oxford: Butterworth-Heinemann.

Grizzetti, B., Pistocchi, A., Liquete, C., et al. (2017). Human pressures and ecological status of European rivers. *Scientific Reports* 7: 205.

Grossmann, M. (2012). Economic value of the nutrient retention function of restored floodplain wetlands in the Elbe River basin. *Ecological Economics* 83: 108–117.

Guerrin, J. (2015). A floodplain restoration project on the River Rhône (France): analyzing challenges to its implementation. *Regional Environmental Change* 15(3): 559–568.

Haines-Young, R. and Potschin, M. (2010). The links between biodiversity, ecosystem services and human well-being. *Ecosystem Ecology: A New Synthesis* 1: 110–139.

Hannigan, J. (2006). *Environmental Sociology*. New York: Routledge, Taylor & Francis.

Haque, A.K.E., Faisal, I., and Bayes, A. (1997). Welfare costs of environmental pollution from the tanning industry in Dhaka: an EIA study. Paper presented at the Mid-term Review Workshop in Yogyakarta, Indonesia (3–8 September 1997).

Hausman, D.M. (2011). *Preference, Value, Choice, and Welfare*. Cambridge: Cambridge University Press.

Heise, U.K., Christensen, J., and Niemann, M. (eds) (2017). *The Routledge Companion to the Environmental Humanities*. New York: Routledge, Taylor & Francis.

Heldt, S., Budryte, P., Ingensiep, H.W., et al. (2016). Social pitfalls for river restoration: how public participation uncovers problems with public acceptance. *Environmental Earth Sciences* 75(13): 1053.

Hidalgo, M.C. and Hernández, B. (2001). Place attachment: conceptual and empirical questions. *Journal of Environmental Psychology* 21(3): 273–281.

Hill, M.R., McMaster, D.G., Harrison, T., et al. (2011). A reverse auction for wetland restoration in the Assiniboine River Watershed, Saskatchewan. *Canadian Journal of Agricultural Economics* 59(2): 245–258.

Holmes, N.T. and Nielsen, M.B. (1998). Restoration of the rivers Brede, Cole and Skerne: a joint Danish and British EU-LIFE demonstration project, I: setting up and delivery of the project. *Aquatic Conservation: Marine and Freshwater Ecosystems* 8(1): 185–196.

Honey-Rosés, J., Acuña, V., Bardina, M., et al. (2013). Examining the demand for ecosystem services: the value of stream restoration for drinking water treatment managers in the Llobregat river, Spain. *Ecological Economics* 90: 196–205.

Hong, C. and Chun, H. (2018). Barriers, challenges, conflicts, and facilitators in environmental decision-making: a case of An'Yang Stream restoration. *River Research and Applications* 34(5): 472–480.

Hong, C.Y., Chang, H., and Chung, E.S. (2019). Comparing the functional recognition of aesthetics, hydrology, and quality in urban stream restoration through the framework of environmental perception. *River Research and Applications* 35(6): 543–552.

Hull, R.B. and Robertson, D.P. (2000). Which nature? In: *Restoring Nature: Perspectives from the Social Sciences and Humanities* (eds P.H. Gobster and R.B. Hull), 299–307. Washington DC: Island Press.

IPBES (Intergovernmental Science-Policy Platform on Biodiversity and Ecosystem Services) (2020). Ecosystem services: glossary. https://ipbes.net/glossary/ecosystem-services (accessed 8 December 2020).

Jähnig, S.C., Lorenz, A.W., Hering, D., et al. (2011). River restoration success: a question of perception. *Ecological Applications* 21(6): 2007–2015.

Januchowski-Hartley, S.R., Moon, K., Stoeckl, N., et al. (2012). Social factors and private benefits influence landholders' riverine restoration priorities in tropical Australia. *Journal of Environmental Management* 110: 20–26.

Jia, H., Dong, N., and Ma, H. (2010). Evaluation of aquatic rehabilitation technologies for polluted urban rivers and the case study of the Foshan Channel. *Frontiers of Environmental Science & Engineering in China* 4(2): 213–220.

Jordan, W.R., III (2000). Restoration, community, and wilderness. In: *Restoring Nature: Perspectives from the Social Sciences and Humanities* (eds P.H. Gobster and R.B. Hull), 23–36. Washington DC: Island Press.

Junker, B. and Buchecker, M. (2008). Aesthetic preferences versus ecological objectives in river restorations. *Landscape and Urban Planning* 85(3–4): 141–154.

Junker, B., Buchecker, M., and Müller-Böker, U. (2007). Objectives of public participation: which actors should be involved in the decision making for river restorations? *Water Resources Research* 43(10). https://doi.org/10.1029/2006WR005584.

Kail, J., Hering, D., Muhar, S., et al. (2007). The use of large wood in stream restoration: experiences from 50 projects in Germany and Austria. *Journal of Applied Ecology* 44(6): 1145–1155.

Kaiser, F.G., Wölfing, S., and Fuhrer, U. (1999). Environmental attitude and ecological behaviour. *Journal of Environmental Psychology* 19(1): 1–19.

Kaplan, S. (1987). Aesthetics, affect, and cognition: environmental preference from an evolutionary perspective. *Environment and Behavior* 19(1): 3–32.

Kendy, E., Aylward, B., Ziemer, L.S., et al. (2018). Water transactions for streamflow restoration, water supply reliability, and rural economic vitality in the western United States. *Journal of the American Water Resources Association* 54(2): 487–504.

Kenney, M.A., Wilcock, P.R., Hobbs, B.F., et al. (2012). Is urban stream restoration worth it? *Journal of the American Water Resources Association* 48(3): 603–615.

Kern, K. (1992). Rehabilitation of streams in south-west Germany. In: *River Conservation and Management* (eds P.J. Boon and P. Calow, G.E. Petts), 321–335. Chichester: Wiley.

Kim, J.H., Kim, H.J., and Yoo, S.H. (2018). A cost–benefit analysis of restoring the ecological integrity of Jeonju Stream, Korea. *Water and Environment Journal* 32(3): 476–480.

Kim, M., Gim, T.H.T., and Sung, J.S. (2017). Applying the concept of perceived restoration to the case of Cheonggyecheon Stream Park in Seoul, Korea. *Sustainability* 9(8): 1368.

Kondolf, G.M. (2000). Process vs form in restoration of rivers and streams. *Annual Meeting Proceedings of the American Society of Landscape Architects*. Washington DC: American Society of Landscape Architects.

Kristensen, E.A., Baattrup-Pedersen, A., Jensen, P.N., et al. (2012). Selection, implementation and cost of restorations in lowland streams: a basis for identifying restoration priorities. *Environmental Science & Policy* 23: 1–11.

Lamouroux, N., Gore, J.A., Lepori, F., et al. (2015). The ecological restoration of large rivers needs science-based, predictive tools meeting public expectations: an overview of the Rhône project. *Freshwater Biology* 60(6): 1069–1084.

Langhans, S.D., Hermoso, V., Linke, S., et al. (2014). Cost-effective river rehabilitation planning: optimizing for morphological benefits at large spatial scales. *Journal of Environmental Management* 132: 296–303.

Larned, S.T., Suren, A.M., Flanagan, M., et al. (2006). Macrophytes in urban stream rehabilitation: establishment, ecological effects, and public perception. *Restoration Ecology* 14(3): 429–440.

Lave, R. (2016). Stream restoration and the surprisingly social dynamics of science. *Wiley Interdisciplinary Reviews: Water* 3(1): 75–81.

Lave, R., Doyle, M., and Robertson, M. (2010). Privatizing stream restoration in the US. *Social Studies of Science* 40(5): 677–703.

Lave, R., Wilson, M.W., Barron, E., et al. (2014). Critical physical geography. *The Canadian Geographer* 58(1): 1–10.

Lebart, L., Salem, A., and Berry, L. (1998). *Exploring Textual Data*. Dordrecht, The Netherlands: Kluwer Academic Publishers.

Lee, J.S. (2012). Measuring the economic benefits of the Youngsan River Restoration Project in Kwangju, Korea, using contingent valuation. *Water International* 37(7): 859–870.

Lee, M. and Jung, I. (2016). Assessment of an urban stream restoration project by cost–benefit analysis: the case of Cheonggyecheon stream in Seoul, South Korea. *KSCE Journal of Civil Engineering* 20(1): 152–162.

Lee, S. and Choi, G.W. (2012). Governance in a river restoration project in South Korea: the case of Incheon. *Water Resources Management* 26(5): 1165–1182.

Le Floc'h, É. and Aronson, J. (1995). Ecologie de la restauration: définition de quelques concepts de base. *Natures Sciences Sociétés* 3: 29–35.

Lester, J.P. (ed.) (1995). *Environmental Politics and Policy: Theories and Evidence*. Durham: Duke University Press.

Lewicka, M. (2011). Place attachment: how far have we come in the last 40 years? *Journal of Environmental Psychology* 31(3): 207–230.

Lewis, L.Y. and Landry, C.E. (2017). River restoration and hedonic property value analyses: guidance for effective benefit transfer. *Water Resources and Economics* 17: 20–31.

Light, A. and Higgs, E.S. (1996). The politics of ecological restoration. *Environmental Ethics* 18(3): 227–247.

Logar, I., Brouwer, R., and Paillex, A. (2019). Do the societal benefits of river restoration outweigh their costs? A cost–benefit analysis. *Journal of Environmental Management* 232: 1075–1085.

Loomis, J. (1996). Measuring the economic benefits of removing dams and restoring the Elwha River: results of a contingent valuation survey. *Water Resources Research* 32(2): 441–447.

Loomis, J. (2002). Quantifying recreation use values from removing dams and restoring free-flowing rivers: a contingent behavior travel cost demand model for the Lower Snake River. *Water Resources Research* 38(6). https://doi.org/10.1029/2000WR000136.

Loomis, J., Kent, P., Strange, L., et al. (2000). Measuring the total economic value of restoring ecosystem services in an impaired river basin: results from a contingent valuation survey. *Ecological Economics* 33(1): 103–117.

Mansourian, S. (2017). Governance and restoration. In: *Routledge Handbook of Ecological and Environmental Restoration* (eds S. Allison and S. Murphy), 401–413. New York: Routledge, Taylor & Francis.

Martin, D.M. (2017). Ecological restoration should be redefined for the twenty-first century. *Restoration Ecology* 25(5): 668–673.

McCormick, A., Fisher, K., and Brierley, G. (2015). Quantitative assessment of the relationships among ecological, morphological and aesthetic values in a river rehabilitation initiative. *Journal of Environmental Management* 153: 60–67.

McIver, J. and Starr, L. (2001). Restoration of degraded lands in the interior Columbia River basin: passive vs. active approaches. *Forest Ecology and Management* 153(1–3): 15–28.

MEA (Millennium Ecosystem Assessment) (2005). *Ecosystems and Human Well-being: Synthesis*. Washington DC: Island Press.

Mingers, J. and Leydesdorff, L. (2015). A review of theory and practice in scientometrics. *European Journal of Operational Research* 246(1): 1–19.

Mitsch, W.J. (2012). What is ecological engineering? *Ecological Engineering* 45: 5–12.

Morandi, B., Kail, J., Toedter, A., et al. (2017). Diverse approaches to implement and monitor river restoration: a comparative perspective in France and Germany. *Environmental Management* 60(5): 931–946.

Moscovici, S. (2001). *Social Representations: Essays in Social Psychology*. New York: NYU Press.

Nakamura K., Tockner, K., and Amano, K. (2006). River and wetland restoration: lessons from Japan. *Bioscience* 56(5): 419–429.

Nassauer, J.I. (1992). The appearance of ecological systems as a matter of policy. *Landscape Ecology* 6(4): 239–250.

Netusil, N.R., Jarrad, M., and Moeltner, K. (2019). Research note: the effect of stream restoration project attributes on property sale prices. *Landscape and Urban Planning* 185: 158–162.

Normile, D. (2010). Restoration or devastation? *Science* 327(5973): 1568–1570.

NRC (National Research Council) (1992). *Restoration of Aquatic Ecosystems: Science, Technology, and Public Policy*. Washington DC: National Academies Press.

OECD (Organization for Economic Cooperation and Development) (2015). *Frascati Manual: Guidelines for Collecting and Reporting Data on Research and Experimental Development:*

The Measurement of Scientific, Technological and Innovation Activities. Paris: OECD Publishing. http://dx.doi.org/10.1787/9789264239012-en.

Ormerod, S.J. (2004). A golden age of river restoration science? *Aquatic Conservation: Marine and Freshwater Ecosystems* 14(6): 543–549.

Pahl-Wostl, C. (2006). The importance of social learning in restoring the multifunctionality of rivers and floodplains. *Ecology and Society* 11(1): 10.

Palmer, M.A. and Allan, J.D. (2006). Restoring rivers. *Issues in Science and Technology* 22(2): 40–48.

Palmer, M.A. and Bernhardt, E.S. (2006). Hydroecology and river restoration: ripe for research and synthesis. *Water Resources Research* 42(3). https://doi.org/10.1029/2005WR004354.

Palmer, M.A., Bernhardt, E.S., Allan, J.D., et al. (2005). Standards for ecologically successful river restoration. *Journal of Applied Ecology* 42: 208–217.

Palmer, M.A., Bernhardt, E.S., Chornesky, E., et al. (2004). Ecology for a crowded planet. *Science* 304: 1251–1252.

Palmer, M.A., Hondula, K.L., and Koch, B.J. (2014). Ecological restoration of streams and rivers: shifting strategies and shifting goals. *Annual Review of Ecology, Evolution, and Systematics* 45: 247–269.

Palmer, M.A., Menninger, H.L., and Bernhardt, E.S. (2010). River restoration, habitat heterogeneity and biodiversity: a failure of theory or practice? *Freshwater Biology* 55: 205–222.

Perni, Á., Martínez-Paz, J., and Martínez-Carrasco, F. (2012). Social preferences and economic valuation for water quality and river restoration: the Segura River, Spain. *Water and Environment Journal* 26(2): 274–284.

Piégay, H., Gregory, K.J., Bondarev, V., et al. (2005). Public perception as a barrier to introducing wood in rivers for restoration purposes. *Environmental Management* 36(5): 665–674.

Polizzi, C., Simonetto, M., Barausse, A., et al. (2015). Is ecosystem restoration worth the effort? The rehabilitation of a Finnish river affects recreational ecosystem services. *Ecosystem Services* 14: 158–169.

Purcell, A.H., Friedrich, C., and Resh, V.H. (2002). An assessment of a small urban stream restoration project in northern California. *Restoration Ecology* 10(4): 685–694.

Randall, A. (1987). Total economic value as a basis for policy. *Transactions of the American Fisheries Society* 116(3): 325–335.

Redman, C.L., Grove, J.M., and Kuby, L.H. (2004). Integrating social science into the Long-Term Ecological Research (LTER) network: social dimensions of ecological change and ecological dimensions of social change. *Ecosystems* 7(2): 161–171.

Roni, P. and Beechie, T. (2013). Introduction to restoration: key steps for designing effective programs and projects. In: *Stream and Watershed Restoration: A Guide to Restoring Riverine Processes and Habitats* (eds P. Roni and T. Beechie), 1–11. Chichester: Wiley.

Saarinen, T.F. and Sell, J.L. (1980). Environmental perception. *Progress in Human Geography* 4(4): 525–548.

Schläpfer, F. and Witzig, P.J. (2006). Public support for river restoration funding in relation to local river ecomorphology, population density, and mean income. *Water Resources Research* 42(12). https://doi.org/10.1029/2006WR004940.

Seidl, R. and Stauffacher, M. (2013). Evaluation of river restoration by local residents. *Water Resources Research* 49(10): 7077–7087.

Shields, F.D., Cooper Jr, C.M., Knight, S.S., et al. (2003). Stream corridor restoration research: a long and winding road. *Ecological Engineering* 20(5): 441–454.

Smith, B., Clifford, N.J., and Mant, J. (2014). The changing nature of river restoration. *Wiley Interdisciplinary Reviews: Water* 1(3): 249–261.

Sneddon, C.S., Barraud, R., and Germaine, M.A. (2017). Dam removals and river restoration in international perspective. *Water Alternatives* 10(3): 648–654.

Soto-Montes de Oca, G. and Ramirez-Fuentes, A. (2019). Value of river restoration when living near and far: the Atoyac Basin in Puebla, Mexico. *Tecnología y Ciencias del Agua* 10(1). https://doi.org/10.24850/j-tyca-2019-01-07.

Stanford, J.A., Ward, J.V., Liss, W.J., et al. (1996). A general protocol for restoration of regulated rivers. *Regulated Rivers: Research & Management* 12(4–5): 391–413.

Steffen, W., Crutzen, P.J., and McNeill, J.R. (2007). The anthropocene: are humans now overwhelming the great forces of nature? *Ambio* 36(8): 214–221.

Sudding, K.N. (2011). Toward an era of restoration in ecology: successes, failures, and opportunities ahead. *Annual Review of Ecology, Evolution, and Systematics*, 42: 465–487.

Szałkiewicz, E., Jusik, S., and Grygoruk, M. (2018). Status of and perspectives on river restoration in Europe: 310,000 euros per hectare of restored river. *Sustainability* 10(1): 129.

Tanaka, A. (2006). Stakeholder analysis of river restoration activity for eight years in a river channel. In: *Human Exploitation and Biodiversity Conservation* (eds A.T. Bull and D.L. Hawksworth), 447–471. Dordrecht, The Netherlands: Springer.

Thorel, M., Piégay, H., Barthelemy, C., et al. (2018). Socio-environmental implications of process-based restoration strategies in large rivers: should we remove novel ecosystems along the Rhône (France)? *Regional Environmental Change* 18(7): 2019–2031.

Trumble, W.R. and Stevenson, A. (eds) (2002). *Shorter Oxford English Dictionary: Volume 1*. New York: Oxford University Press.

Tunstall, S.M., Penning-Rowsell, E.C., Tapsell, S.M., et al. (2000). River restoration: public attitudes and expectations. *Water and Environment Journal* 14(5): 363–370.

Tunstall, S.M., Tapsell, S.M., and Eden, S. (1999). How stable are public responses to changing local environments? A "before" and "after" case study of river restoration. *Journal of Environmental Planning and Management* 42(4): 527–545.

Turner, R.E. and Boyer, M.E. (1997). Mississippi River diversions, coastal wetland restoration/creation and an economy of scale. *Ecological Engineering* 8(2): 117–128.

Turner, R.K., Pearce, D., and Bateman, I. (1994). *Environmental Economics: An Elementary Introduction*. Hemel Hempstead: Harvester Wheatsheaf.

van den Berg, A.E., Koole, S.L., and van der Wulp, N.Y. (2003). Environmental preference and restoration: (how) are they related? *Journal of Environmental Psychology* 23(2): 135–146.

Vásquez, W.F. and de Rezende, C.E. (2019). Willingness to pay for the restoration of the Paraíba do Sul River: a contingent valuation study from Brazil. *Ecohydrology & Hydrobiology* 19(4): 610–619.

Vaughan, I.P., Diamond, M., Gurnell, A.M., et al. (2009). Integrating ecology with hydromorphology: a priority for river science and management. *Aquatic Conservation: Marine and Freshwater Ecosystems* 19(1): 113–125.

Verbrugge, L. and van den Born, R. (2018). The role of place attachment in public perceptions of a re-landscaping intervention in the river Waal (the Netherlands). *Landscape and Urban Planning* 177: 241–250.

Vermaat, J.E., Wagtendonk, A.J., Brouwer, R., et al. (2016). Assessing the societal benefits of river restoration using the ecosystem services approach. *Hydrobiologia* 769(1): 121–135.

Vining, J., Tyler, E., and Kweon B.S. (2000). Public values, opinions, and emotions in restoration controversies. In: *Restoring Nature: Perspectives from the Social Sciences and Humanities* (eds P.H. Gobster and R.B. Hull), 143–162. Washington DC: Island Press.

Vörösmarty, C.J., McIntyre, P.B., Gessner, M.O., et al. (2010). Global threats to human water security and river biodiversity. *Nature* 467(7315): 555–561.

Weber, C., Åberg, U., Buijse, A.D., et al. (2018). Goals and principles for programmatic river restoration monitoring and evaluation: collaborative learning across multiple projects. *Wiley Interdisciplinary Reviews: Water* 5(1): 1257.

Weber, M.A. and Stewart, S. (2009). Public values for river restoration options on the Middle Rio Grande. *Restoration Ecology* 17(6): 762–771.

Wells, H.B.M. and Dougill, A.J. (2019). The importance of long term social ecological: research for the future of restoration ecology. *Restoration Ecology* 27(5): 929–933.

Westling, E.L., Surridge, B.W., Sharp, L., et al. (2014). Making sense of landscape change: long-term perceptions among local residents following river restoration. *Journal of Hydrology* 519: 2613–2623.

Wilson, E.O. (1992). *Diversity of Life*. Cambridge, MA: Harvard University Press.

Woelfle-Erskine, C. (2017). Collaborative approaches to flow restoration in intermittent salmon-bearing streams: Salmon Creek, CA, USA. *Water* 9(3): 217.

Wohl, E., Lane, S.N., and Wilcox, A.C. (2015). The science and practice of river restoration. *Water Resources Research* 51(8): 5974–5997.

Wohl, E., Angermeier, P.L., Bledsoe, B., et al. (2005). River restoration. *Water Resources Research* 41(10). https://doi.org/10.1029/2005WR003985.

Wortley, L., Hero, J.M., and Howes, M. (2013). Evaluating ecological restoration success: a review of the literature. *Restoration Ecology* 21(5): 537–543.

Yun, S.J. (2014). Experts' social responsibility in the process of large-scale nature-transforming national projects: focusing on the case of the Four Major Rivers Restoration Project in Korea. *Development and Society* 43(1): 109–141.

Zhang, Y., Sheng, L., and Luo, Y. (2019). Evaluation of households' willingness to accept the ecological restoration of rivers flowing in china. *Environmental Progress & Sustainable Energy* 38(4): 13094.

Zhao, J., Liu, Q., Lin, L., et al. (2013). Assessing the comprehensive restoration of an urban river: an integrated application of contingent valuation in Shanghai, China. *Science of the Total Environment* 458: 517–526.

Zingraff-Hamed, A., Noack, M., Greulich, S., et al. (2018). Model-based evaluation of urban river restoration: conflicts between sensitive fish species and recreational users. *Sustainability* 10(6): 1747.

Zube, E.H. (1999). Environmental perception. In: *Encyclopedia of Earth Science* (eds D.E. Alexander and R.W. Fairbridge), 214–216. Dordrecht, The Netherlands: Springer.

Zube, E.H., Sell, J.L., and Taylor, J.G. (1982). Landscape perception: research, application and theory. *Landscape Planning* 9(1): 1–33.

Part II

People–River Relationships: From Ethics to Politics

2

Ethics of River Restoration: The Imitationist Paradigm

Henry Dicks

Faculté de Philosophie, Université Jean Moulin Lyon, Lyon, France

2.1   Introduction

Restoration ecology is sometimes considered a science (Bradshaw 1993; Palmer et al. 1997). As Higgs (1994, p. 134) notes, this promotes a problematic view of restoration ecology as "austere," "disengaged," and thus essentially independent of ethics. Further, even the very briefest examination of the academic literature on the subject shows that ethical considerations are of central importance to restoration ecology. In particular, the literature abounds with radical uncertainties as to the basic goal or goals of restoration. Should we be restoring past ecosystems and, if so, what precise moment in the past should we take as our reference point (Diamond 1987)? Given the essentially dynamic nature of ecosystems and the high likelihood of significant anthropogenic climate change in the near future, would it not make more sense to make restoration "forward-looking" (Rolston 2000; Hobbs and Harris 2001)? Or should restorations be modeled rather on intact ecosystems of the present that share similar traits to the destroyed or degraded pre-disturbance ecosystem we are looking to restore (Rheinhardt et al. 1999)? Further, once we have decided on our model ecosystem, what are the salient features on which restoration should concentrate: functions, species composition, structure, or perhaps all of these (Light and Higgs 1996; Palmer et al. 1997)? And what about our basic reasons for undertaking restoration in the first place? Should we be focusing on the human benefits of restoration, including an increase in our scientific understanding of ecology (Jordan et al. 1987) – in which case one is likely to see significant departure from the initial model – or should our aim be rather that of fidelity to whatever natural model we have selected (Light and Higgs 1996; Egan 2006)?

In the literature on restoration ecology, these and other fundamental questions are often treated in one of two ways. The first, progressive, approach consists in placing them in the category of future research (Palmer et al. 1997), a view which tends to suppose that the science of restoration ecology is still in its "infancy" (Palmer et al. 1997, p. 191), but could, as it matures, provide answers to these questions. The second, Kuhnian, approach sees restoration ecology in terms of successive paradigms, such that, to take a well-known example,

River Restoration: Political, Social, and Economic Perspectives, First Edition. Edited by Bertrand Morandi, Marylise Cottet, and Hervé Piégay.
© 2022 John Wiley & Sons Ltd. Published 2022 by John Wiley & Sons Ltd.

a new forward-looking paradigm may be said to have replaced the old backward-looking one (Choi 2004, 2007). But whether one sees restoration ecology as a young science capable of making cumulative progress or in terms of a succession of different paradigms, the common view of restoration ecology as a science runs the significant risk of implying that fundamental philosophical questions – including those about values and ethics – are extraneous to that science. Following existing arguments against this position put forward by others (Higgs 1994, 1997; Davis and Slobodkin 2004), I argue in this chapter that ethics is not something secondary and external to restoration ecology; to practice ecological restoration is not only to do applied science but also to do applied ethics.

That branch of philosophy that deals with our ethical relationship to nature is called "environmental ethics" (Palmer et al. 2014). Within environmental ethics, ecological restoration has been the subject of intense debate ever since the publication of a famous article on the subject by Elliott (1982) entitled "Faking Nature." Elliott's basic argument was that one cannot ever fully restore an ecosystem, not because it is a technical impossibility (though it may also be a technical impossibility) but rather because an important value associated with the original ecosystem – the value of naturalness itself, where naturalness is understood in terms of having a genesis independent of human action – is necessarily lost. For Elliott, even when the restoration produces an ecosystem that is so successful that a trained eye would struggle to distinguish it from the original, the result is analogous to a fake work of art, whose value – despite the deceptive precision of the reproduction – is necessarily less than the original.

Elliott's critique of ecological restoration was later followed up by Katz, who in a number of articles argued that the assumption – taken as characteristic of restoration ecology in general – that humans can destroy and recreate ecosystems at will is a simple continuation of the traditional anthropocentric goal of dominating and controlling nature (Katz 1996, 1997). Moreover, there are various concrete restoration scenarios that do seem to support Katz's criticism. One of these is the case on which Elliott's argument is based: a mining company seeking to use the claim that no value would be lost in the case of a perfect restoration in order to justify the temporary destruction of the obstructing ecosystem in the present – an argument that Elliott (1982, p. 81) calls the "restoration thesis." Another example is the project of transporting the Zurich Airport wetland by cutting it up into a number of pieces and reassembling them in a different location (Loucks 1994). Similarly, it is not uncommon for streams and rivers to be destroyed and reconstructed elsewhere so as to better fit in with development objectives (Prager and McPhillips 2012). Nature, in all these cases, appears as almost infinitely malleable, as something that can be destroyed to make way for human economic activity, while also being reproduced and even moved around at will.

A wide range of responses have been put forward within environmental ethics to Elliott and Katz's critiques of ecological restoration. Rolston (2000), for example, has put forward the eminently sensible argument that a better analogy for most ecological restorations is not with faking a work of art but rather with restoring a work of art. And Light (2000, p. 98) has argued in favor of distinguishing "malicious restorations," in which the possibility of future restoration is used as an argument to justify an ecosystem's destruction in the present, from "benevolent restorations," which, far from exemplifying the anthropocentric goal of dominating or controlling nature, instead foster a harmonious and cooperative relation between humans and nature. Yet another criticism, advanced by Hourdequin and

Havlick (2013), is that Elliott and Katz's basic assumption about what restoration is – the attempt to recreate a clearly delimited portion of wild nature – does not, in the context of increasingly hybrid (i.e. natural and cultural) landscapes, well describe most restoration activities. This criticism ties in with increased recognition that the identification of nature with wilderness characteristic of much North American environmental philosophy is ill suited to a European context characterized by significant long-term interactions between humans and nature (Eden 2006). Drawing above all on this third response to the arguments of Light and Katz, this contribution will seek to answer two key research questions about the ethics of river restoration.

1) What are the ethically significant challenges of river restoration?
2) Is it possible to put forward an ethics of river restoration capable of responding appropriately to these challenges?

In responding to these questions, I argue that the challenges of river restoration are such that it cannot be carried out in isolation but must instead be integrated as one component within a broader approach, the central concept of which is "imitating nature," or biomimicry. In short, I argue that river restoration is one form of biomimicry among many and must be articulated with other instances of biomimicry if it is to be genuinely successful. This in turn implies a need to go beyond both the preservationist and restorationist paradigms in environmental philosophy and to integrate ecological restoration within a new "imitationist" paradigm, which, I further argue, requires the development of a new approach to environmental ethics.

2.2 Three challenges of river restoration

As noted in Section 2.1, the case discussed by Elliott (1982) at the outset of "Faking Nature" is that of a mining company using the argument that a perfect or near-perfect reconstruction of an ecosystem is possible to justify the almost total destruction of the ecosystem in the present. In this case, it seems reasonable to say that the restored ecosystem will in many respects be a human artefact, for it will have been designed and put together by human beings on a now barren landscape, albeit on the basis of a model provided by nature. But not all ecological restorations fit this pattern. In the case where restoration consists only of reintroducing a lost species, for example, it is far less clear that the resulting ecosystem counts as an artefact. Likewise, there are forms of restoration in which human intervention is conceived and practiced as merely complementing or accelerating natural processes of self-regeneration, in which case it is again far from clear that the resulting ecosystem is an artefact (Palmer et al. 1997). In response to objections such as these, Katz (2012) argues that ecological restorations always sit somewhere on a continuum. In some cases, where human intervention is maximal, as in the case of restoration after strip-mining or the displacement and reconstruction of a river section, the resulting ecosystem is largely artificial. In other cases, where human intervention is minimal, as is the case when volunteers remove garbage from a local stream, the resulting ecosystem remains largely natural. In every case, however, Katz insists that the act of restoration introduces an element of artificiality and a corresponding loss of naturalness, and therewith also of value.

But even if one accepts the idea of a continuum of more or less natural restorations, Elliott and Katz's basic assumption about what restoration is – the attempt to reconstruct a clearly delimited portion of wild nature – does not fit well with typical cases of river restoration. Indeed, there are three key attributes of river restoration that set it apart from the paradigmatic cases discussed by Elliott and Katz and which together give rise to what I consider the three most ethically significant challenges of river restoration.

2.2.1 Challenge 1: the connectedness of rivers

The first challenge pertains to the fact that rivers are not easy to isolate from the ecosystems that surround them. Of course, one can easily identify the channel of the river, as well as the banks, but it is in the nature of rivers to flow from their source to the sea, while at the same time receiving water from other channels. The river, in other words, cannot be easily separated either from its watershed or from the receiving body of water into which it flows. This in turn has huge significance for restoration projects. If a river is polluted or subject to problematic flooding as a result of agricultural, industrial, or urban runoff, then focusing only on restoring the main channel may prove ineffective (Bernhardt and Palmer 2007). In this respect, the situation would appear to be quite different from the restoration of a portion of wild forest following an isolated mining operation. Moreover, it is interesting in this context to consider a thought experiment suggested by Elliott (1997), in which we are asked to imagine two apparently identical islands, the first of which is natural, and the second of which has been constructed by environmental engineers. Elliott's claim is that the value of the former exceeds that of the latter, for the latter is a fake. Whatever one may think of this argument, it is clear that a basic presupposition of the thought experiment is that the thing being restored is radically cut off from the rest of nature, like a painting whose frame demarcates it from the rest of the world. But the static island – cut off and framed by the surrounding ocean – is in this respect almost the exact opposite of the fluid river, intricately connected to its surrounding ecosystems.[1]

2.2.2 Challenge 2: human habitation

A second challenge of river restoration concerns human habitation. In the examples discussed by Elliott and Katz, it is supposed that what is to be restored is an isolated fragment of wilderness. In the case of mining operations, this will often be the case, for valuable mineral deposits are often located far from human civilization in areas of more or less untouched wilderness. Similarly, the two islands of Elliott's thought experiment are clearly desert islands. In most cases, however, river restorations will not be of wild rivers that have been temporarily destroyed or degraded but could in theory be fully restored to their wild state, for permanent human settlement – in the form of agriculture, industry, cities, and so on – will typically have established itself along the banks and in the surrounding catchment area, a situation that significantly affects restoration projects on such rivers as the Cole, the Brent, and the Alt in the United Kingdom (Eden et al. 2000; Eden 2006). When combined with the first key attribute of rivers – their connectedness to the surrounding watershed – this will typically make the idea of restoring them to a wild state completely out of the question (Eden et al. 2000; Eden 2006).

2.2.3 Challenge 3: multiple stakeholders

This brings us to a third important challenge: rivers often constitute focal points for human activity, such that their use and degradation – and potentially also their restoration – involves a multiplicity of different actors. So, whereas in other cases – especially the paradigmatic instances of "malicious restorations" criticized by Light (2000, p. 98) – the destruction or degradation of the original ecosystem is the result of one party, acting for one purpose, over one specific period of time, in the case of rivers there will often be a multiplicity of parties using and degrading the river in a large number of ways, for a large number of reasons, and over extended and ill-defined periods of time, as is perhaps most obviously the case regarding such large and historically important rivers as the Ganges in India or the Yangtze in China. With this in mind, it is also instructive to consider the defense put forward by Light and Higgs (1996, p. 236) of restoration as having an "inherent democratic potential," for it allows humans collectively to participate in nature. While this may not be true in all cases – it is surely not unreasonable, for example, to expect a company whose strip mine has destroyed a wild ecosystem to undertake the restoration efforts itself rather than expecting local volunteers to do it – in the case of river restorations it is clear that there will usually be significant potential for something like the sort of participative democratic restoration advocated by Light and Higgs. Of course, this is not necessarily to say that the restoration will always – or even usually – be able to embody the sort of democratic equality between participants one finds when members of a local community all muck in to clean up a local stream (Light 2001). Particularly when we factor in the first two challenges discussed above, it is clear that river restoration will often require the involvement of a large number of stakeholders, contributing in a wide number of different ways, including potentially complex engineering work necessarily carried out by specialists (dam removal, re-meandering, etc.), but it is nevertheless true that river restoration has the potential to act as a focal point for a wide number of stakeholders to act together toward a common goal.

2.3 Restoration ecology as a type of biomimicry

Given the three challenges outlined in Section 2.2, what sort of ethic is appropriate to river restoration? With a view to answering this question, let us briefly consider the long-running debate in environmental ethics between preservationism and restorationism (Hettinger 2012). To most people, preservation and restoration are simply different types of environmentally benign activity. It is also possible, however, to elevate each of these activities to the status of what Jordan (1994, p. 17) calls an "environmental paradigm," understood as a specific way of relating to the environment that involves strong philosophical and ethical suppositions, hence the presence of the suffix "-ism." According to preservationists, like Elliott and Katz, the basic objective of environmental ethics should be to preserve nature, for restorations, as the deliberate result of human agency, simply add more artefacts to an increasingly unnatural and therefore less valuable planet. Restorationists, like Jordan (1994) and Turner (1994), see a variety of problems with this position, of which four are particularly noteworthy:

- The first problem is the assumption that humans always have a negative impact on nature (for even when restoring nature they diminish the value of naturalness).

- The second problem is the radical separation between humans and nature that underlies this claim, and entails that deliberate human actions can only ever have nonnatural results.
- The third problem is that preservation only allows for a very limited number of ways of interacting with nature; to preserve nature we typically do little more than designate it as out of bounds for development and then restrict human use to, at most, recreational and scientific activities.
- The fourth problem is that relatively little wilderness now remains, in which case there is relatively little genuine preservation work left to be done.

According to its advocates, the paradigm of restorationism, which holds that restoration is the "central challenge" of ecology (Jordan et al. 1987, p. 15), makes it possible to overcome these four problems. In restoration, human impact on nature is positive; it overcomes the separateness from nature presupposed by preservationists; it allows us to interact with nature in complex and meaningful ways, thereby also deepening our ecological understanding (Jordan et al. 1987); and it provides much more scope than preservation for future environmental activity.

There are, however, also major limitations to restorationism. For a start, the positive human impact on nature is limited to restoration work. Other activities, such as how we produce the goods and services that sustain our own existence, many of which depend directly on rivers or take place within their watersheds, are overlooked. The result is a continued separation between humans and nature, with preserved or restored nature on one side of the divide and human culture – agriculture, industry, cities, etc. – on the other. This in turn limits the possible scope for meaningful interactions with nature. How we produce the food, goods, buildings, and infrastructure we require would appear to have little or nothing to do with nature. Lastly, while there is no doubt much scope for restoring degraded ecosystems no longer exploited by humans, increasing swathes of the planet are in direct human use, in which case there is clearly much more environmental work to be done transforming this use than there is in restoring portions of degraded nature to something like their wild state. This is not, of course, to deny the possibility of restoring a site that is and will remain in use. The restoration of grazing meadowlands in the United Kingdom is an obvious example. But it is also true that, if such restoration projects are to be sustainable, they must also be accompanied by a transformation in the way the site is used, for otherwise there will be nothing to stop degradation reoccurring.

These limitations of restoration as an activity, and restorationism as an environmental paradigm, are particularly apparent in the case of river restoration. As noted in Section 2.2.1 and 2.2.2, rivers are not isolatable from their surrounding ecosystems and their catchment area is often a site of extensive and permanent human settlement and use. In many cases, this means that it will be insufficient simply to restore the river and its banks, let alone just a mere stretch of the river, for human settlement and use of the catchment area will, at least as long as it persists in its current form, continue to have a strong negative impact on the river (Palmer et al. 2005). To some extent river restorationists have acknowledged these problems and have thus tried to expand the scope of river restoration to the level of the watershed (Bohn and Kershner 2002). But, given extensive and permanent human settlement and land use, it will in many cases not be possible to restore the surrounding watershed to anything like its predevelopment state (Eden et al. 2000; Palmer et al. 2005). When a river is negatively impacted by a dense city of several million people straddling its banks and surrounded by thousands of acres of farmland and industry, the restoration of its watershed – where restoration is seen as a return to the predevelopment state – will at best be limited to restoring the "gaps" in the landscape not directly exploited by humans, typically in the form of "buffer zones" and "corridors linking established fragments" of undeveloped land (Hobbs and Norton 1996, p. 104). Moreover, even when one extends the conceptual framework of restoration to the landscape level (Naveh 1994) – an approach which typically extends the scope of restoration ecology to productive land (Hobbs and Norton 1996) – the resulting restoration will not necessarily put an end to the negative impact that future exploitation of the landscape will have on the river. To restore the soils of degraded farmland, for example, will not in itself stop renewed agricultural activity impacting negatively on the river and it could even increase that negative impact (e.g. by making possible continued applications of pesticides and fertilizers).

Given all these limitations, it is instructive to consider a powerful criticism of the "traditional project of environmentalism" put forward by Mathews (2011, p. 364). According to Mathews, environmentalism has traditionally focused on the objective of protecting nature from negative human impacts. In opposition to this, she argues that the emerging concept of biomimicry, which she sees as a "turning point in Western thinking" (Mathews 2011, p. 368), represents a radically different form of environmentalism. So what is biomimicry? Biomimicry is often defined in terms of the transfer of function from biology to engineering, a famous example of which is Velcro, which was modeled on the fastening mechanism used by the *Arctium lappa* thistle to propagate its seeds. According to Marshall and Loveza (2009, p. 2), however, biomimicry is better understood not as a novel method for design but rather as an "organizing concept" capable of bringing together a variety of different fields of design and innovation, including obviously related fields like bionics, biomimetics, or bio-inspiration, but extending also to such areas as permaculture, analogue forestry, ecological engineering, ecological design, industrial ecology, and thus as including also the imitation of ecosystems (ecomimicry or ecosystem biomimicry) (Dicks 2017a).

There is, I believe, another important sense in which biomimicry may be seen as an organizing concept. This second sense concerns biomimicry's own internal organization, which is structured according to four distinct principles, each of which corresponds to a different branch of philosophical inquiry (Dicks 2016). This framework may be represented as in Figure 2.1.

The first principle, nature as model, consists in taking nature as model for human designs. The second principle, nature as measure, holds that there are ecological standards against which the "rightness" of these designs should be evaluated. And the third principle, nature as mentor, holds that nature is not so much something *about* which we should learn as something *from* which we should learn. Lastly, the principle of nature as *physis* consists in the interpretation of nature as self-producing, that is to say as applying to beings that bring themselves into and maintain themselves in existence, including, at least in some cases, by repairing or healing themselves (Dicks 2016). Biomimicry, it follows, is also an organizing concept in the sense that these four principles may be applied to any given

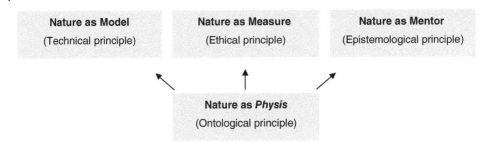

Figure 2.1 The philosophical framework of biomimicry. *Source:* Based on Dicks, H. (2016). The philosophy of biomimicry. *Philosophy & Technology* 29(3): 223–243.

applied field of biomimicry – industrial production, industrial systems, agriculture, architecture, urbanism, and so on – thus structuring the basic theoretical framework of that field.

It is not hard to see that ecological restoration may be understood as a type of biomimicry (Merchant 1986). Restoration ecology has almost always assumed that some sort of model ecosystem is required and that the restored ecosystem will ultimately be an imitation of this model (Jordan 1985; Turner 1987, 1994; Jordan et al. 1988). Moreover, just as in biomimicry, ecological standards have been put forward as criteria against which biomimetic designs and innovations are to be evaluated, so the same is very often true in ecological restoration (Ewel 1987; Brinson and Rheinhardt 1996), river restoration included (Giller 2005; Palmer et al. 2005). Likewise, just as biomimicry sees nature not as an object of human knowledge, something about which we may learn things, but as a source of knowledge, from which we may learn important lessons about how to do things, so the same is often also the case in restoration ecology. In observing the process of ecological succession in nature, for example, we may learn not just certain facts about how ecosystems regenerate themselves but also how we might go about restoring them (Dobson et al. 1997). River restorationists, for example, may learn how to emulate processes of natural succession in their choice of the vegetation planted on riverbanks, with fast-growing pioneer species used initially to stabilize the soil and slower-growing but more ecologically enriching species used later on (Wilke 1994).[2] Lastly, the question of what nature is, including the question of what ecosystems are and what rivers are, is also of fundamental importance to restoration ecology. If, for example, we see nature in terms of self-production (*physis*) then this will likely underpin a view of restoration efforts in terms of helping degraded ecosystems repair and maintain themselves, in which case the role of the restorationist will be analogous not to that of a craftsman or builder but rather to that of a doctor, enabling a patient to heal themselves and thus regain their former autonomy.

Seeing ecological restoration as just one field encompassed by the broader organizing concept of biomimicry provides a coherent response to many of the problems posed by restorationism in general, and river restoration in particular. We noted earlier that restorationism claims to go beyond preservationism in that it considers the human impact on nature to be potentially positive, in that it overcomes the traditional separation between humans and nature, in that it makes for a wider range of possible human interactions with nature, and in that it provides greater scope for future environmental activity. All of this is true a fortiori of biomimicry. In biomimicry, it is not just restoration efforts that may have a positive impact on nature but other fields of human activity as well, from agriculture and industry to architecture and urbanism. As Mathews points out, the aim of biomimicry "is not so much to *reduce* our impact as to make that impact *generative* for nature" (Mathews 2011, pp. 366–367). This in turn may give rise to a world that is much less segregated even than the one implied by restorationism, which, in its advocacy of restoring degraded nature, has little to say about how it is that we should go about conducting the basic activities by and through which human civilization sustains itself. Further, in taking nature as model, measure, and mentor not just for restoration activities but also for farming, industry, architecture, urbanism, and so on, the scope for meaningful interactions with nature increases significantly. And finally, precisely because so much of the earth is now taken up with these activities, their transformation in accordance with the biomimetic principles of nature as model, measure, and mentor provides much greater scope for environmental activity than does restoration alone.

Seeing restoration as but one type of biomimicry also allows one to address the three challenges of river restoration outlined in Section 2.2. In particular, by seeing such human activities as farming, industrial production, and urbanization as modeled on nature, many of the problems faced by river restorationists could also be addressed, for the generalization of natural models would completely transform – in a manner highly complementary to the work of dedicated river restorationists – the workings of those parts of the catchment area that are settled and used by humans. Taking nature as model for a city, for example, could lead to the generalization of permeable soils which allow water to infiltrate directly, as opposed to being channeled through a sewer system (Chocat 2013). And taking nature as model for agriculture, as is the case in the pioneering experiments of Jackson (2011), could involve much higher levels of nutrient cycling and soil retention, as well as the replacement of synthetic pesticides by bio-inspired alternatives (natural predators, genetic diversity, etc.), thus allowing much higher-quality water to flow from the farm to the river. In all these cases, however, human activity goes far beyond the limited scope of restoration work, at least as usually conceived. Taking a forest as model for a city, for example, could involve designing or retrofitting buildings such that they play similar ecological and hydrological roles to trees, including facilitating stormwater infiltration, evapotranspiration, and soil stabilization. The result would not, however, be a restored forest, but rather a "city like a forest" (Braungart and McDonough 2009), which, precisely because of its similarity to a forest, could potentially play a beneficial role in the overall hydrological regime of the river basin to which it belongs.[3] Further, the availability of a new environmental paradigm – imitationism – applicable not just to restoration work but also to other fields of human activity, including agriculture, industry, and urbanism, could potentially help produce the shared beliefs and values necessary for multiple stakeholders to work together in a coherent and complementary manner. If civil engineers, designers, planners, and so on, also take imitating nature as a starting point then their values and objectives would likely dovetail much more easily with those of river restorationists than would otherwise be the case.

Adopting imitationism as a new environmental paradigm would in turn involve the adoption of a new ethical principle: nature as measure. As I argue elsewhere (Dicks 2017b), the key feature of this principle is the idea that, unlike in traditional environmental ethics, nature is not seen as an object of ethics, something toward which human subjects have duties and obligations, as is the case when ecological restoration is seen as a form of "restitution" (Taylor 1986; Basl 2010), righting past wrongs committed toward nature, but rather as a source of ethics, something whose inner workings may be translated into a set of normative, ethical constraints on how we humans should carry out our various activities. From this perspective, the attempt to locate standards in nature for judging the success of restoration projects is already to adopt a specific approach within environmental ethics, and, in our case, a specific restoration ethic, for it is to make the assumption that the restoration efforts should be evaluated against nature's ecological standards. This approach may in turn be integrated into broader biomimicry objectives, as is the case regarding the proposal of Pedersen Zari (2017) for the transformation of the city of Wellington in New Zealand according to ecological standards derived from the predevelopment ecosystem, and which, as she points out, would involve significant restoration work.

Another important feature of the ethical principle of biomimicry is that it avoids an important problem in approaches to ecological restoration grounded in Latour's actor network theory (ANT). According to Eden et al. (2000), a major advantage of ANT is that is does not suppose a radical ontological separation between humans and nature. The problem with ANT, however, is that seeing humans and nonhumans as imbricated in a hybrid network provides no evaluative or ethical criteria that allow us to distinguish bad imbrications from good ones (Eden et al. 2000). Like ANT, biomimicry accepts the inevitable imbrication of humans and nature, and, in keeping with this, it does not seek to preserve or restore a pristine nature free from all human interference. Where it differs from ANT, however, is in the provision of evaluative criteria to judge successful imbrications. This does, interestingly, involve a certain work of purification, for it requires us to consider how nature does things independently of us. Applied to a river basin, this involves asking how nature would do things were we not there, a question which may be answered in various different ways: by considering the general configuration of the river basin before we arrived, by inspecting parcels of more or less wild nature within the basin and extrapolating to other parts, or by developing computer models of the basin based on various input data (geomorphological, climatic, ecological, etc.) and potentially capable of predicting future changes and developments. But this purification does not serve the aim of keeping nature pure or of returning it to a state of purity but rather of enabling humans to learn from the models and measures nature is able to provide, and thus of integrating into our products and our systems those traits of nature that are eminently desirable, especially its effectiveness, appropriateness, and sustainability (Benyus 1997). This does not, of course, imply that all of our ethics may be derived from nature in this way; the ecological standards nature provides give rise to an ecological ethics, an ethic governing our relation to nature, but not an ethic governing interhuman relations. In this respect, biomimicry leaves a significant space open for humans to introduce other ethical criteria and aspirational objectives both into dedicated restoration work and into other complementary biomimetic activities.

Finally, it is also significant that the imitationist paradigm I am proposing lends itself to articulation with certain non-Western philosophical frameworks, especially Daoism (Mathews 2016). A concrete example of this in relation to rivers is the Dujiangyang irrigation scheme, developed along the Min River in 256 BCE, which, instead of damming the river, as is very often the case in contemporary China, took natural flood patterns as a model for a network of irrigation channels that "harmlessly and productively dispersed the flood waters across the flood plain" and whose sustainability is manifest in the fact that it is still in use today, over 2000 years later (Mathews 2019, p. 35). Here, then, we have an example of a river that, precisely because of the ecologically benign but economically and socially productive way its watershed was developed in the first place, is not even in need of restoration.

2.4 Conclusions

If we are to see river restoration not simply as a technical endeavor theorized by the applied science of restoration ecology but rather as itself an activity to which questions of ethics are central then we must engage with environmental ethics. But discussions of restoration in environmental ethics often presuppose a vision of ecological restoration – the restoration of a clearly delimited portion of wilderness – that is of little relevance to typical cases of river restoration. Taking various significant attributes of river restoration into account, it further becomes clear that successful river restoration will ultimately depend not just on restoration work itself but also on the transformation of human activity present in the surrounding watershed. For the most part, however, theorists of river restoration have concentrated only on activities that belong within the clear rubric of restoration and so have not adequately considered the necessity for a broader environmental ethic that extends to such fundamental human activities as farming, industrial production, and the planning and design of cities, all of which significantly affect the character of rivers. This broader ethic, the basic principle of which is to take nature as measure, may be found in biomimicry, understood as an organizing concept capable of bringing together and providing a basic theoretical framework for a wide range of human activities, ecological restoration included, thus resolving the fundamental problems present in preservationism, restorationism, and ANT, while also lending itself to articulation with certain non-Western philosophical traditions, notably Daoism.

As far as policy is concerned, the principal recommendation that follows from the position presented in this chapter is that river restoration should not be seen as an isolated and self-contained activity but rather as a part of a broader systemic change, one that embraces the "Biomimicry Revolution" proposed by Benyus (1997, p. 2), and, in so doing, adopts the new environmental paradigm of imitationism. This approach would both require and facilitate collaborations among a wider range of stakeholders than is typically the case when river restoration is seen as an isolated activity cut off from what takes place in other areas of the river basin.

A second important policy recommendation concerns the need to articulate the ecological standards that derive from nature with other standards deriving from both interhuman ethical constraints (justice, participation, etc.) and contingent human demands and aspirations. Applied to river restoration, this means that, while there should be broad acceptance of the need to derive ecological standards from nature, in this instance the river (and its surrounding watershed) in something like its natural state, these standards must be articulated with other standards – social, economic, etc. – that derive from its present human inhabitants. If done intelligently, this will not necessarily involve a clash of standards and values, as is the case when preservationists lament human activity in general for its negative consequences on natural value, for respecting nature's standards would not only constrain but also enable the attainment of other specifically human standards, as exemplified by the ancient Dujiangyang irrigation scheme in China, which, in imitating natural flood patterns, allowed that floodplain to become "the richest agricultural area in China" (Mathews 2019, p. 35).

Notes

1 This is not to say, of course, that islands actually possess the radical ontological separateness that Katz's argument implies. A strong case could no doubt be made for saying that islands are ultimately like rivers with respect to their connectedness to and dependence on their surrounding environment.

2 In such instances, river restoration comes into close proximity to bioengineering (Evette et al. 2009) and ecological engineering (Mitsch 2012).

3 Barrett (1994) puts forward a similar argument to the one developed here, albeit without attempting an explicit articulation of the theoretical frameworks of biomimicry and ecological restoration.

References

Barrett, G.W. (1994). Restoration ecology: lessons yet to be learned. In: *Beyond Preservation: Restoring and Inventing Landscapes* (eds D.A. Baldwin, J. de Luce, and C. Pletsch), 113–126. Minneapolis: University of Minnesota Press.

Basl, J. (2010). Restitutive restoration: new motivations for ecological restoration. *Environmental Ethics* 32(2): 135–147.

Benyus, J. (1997). *Biomimicry: Innovation Inspired by Nature*. New York: Harper Perennial.

Bernhardt, E. and Palmer, M. (2007). Restoring streams in an urbanizing world. *Freshwater Biology* 52: 738–751.

Bohn, B.A. and Kershner, J.L. (2002). Establishing aquatic restoration priorities using a watershed approach. *Journal of Environmental Management* 64(4): 355–363.

Bradshaw, A.D. (1993). Restoration ecology as a science. *Restoration Ecology* 1(2): 71–73.

Braungart, M. and McDonough, W. (2009). *Cradle to Cradle: Re-Making the Way We Make Things*. London: Vintage.

Brinson M.M. and Rheinhardt, R. (1996). The role of reference wetlands in functional assessment and mitigation. *Ecological Applications* 6(1): 69–76.

Chocat, B. (2013). Un nouveau paradigme pour les eaux pluviales urbaines. *Techniques, Sciences, Méthodes* 6: 14–15.

Choi, Y.D. (2004). Theories for ecological restoration in changing environment: toward "futuristic" restoration. *Ecological Research* 19(1): 75–81.

Choi, Y.D. (2007). Restoration ecology to the future: a call for new paradigm. *Restoration Ecology* 15(2): 351–353.

Davis, M.A. and Slobodkin, L.B. (2004). The science and values of restoration ecology. *Restoration Ecology* 12(1): 1–3.

Diamond, J. (1987). Reflections on goals and on the relationship between theory and practice. In: *Restoration Ecology: A Synthetic Approach to Ecological Research* (eds W.R. Jordan III, M.E. Gilpin, and J.D. Aber), 329–336. Cambridge: Cambridge University Press.

Dicks, H. (2016). The philosophy of biomimicry. *Philosophy & Technology* 29(3): 223–243.

Dicks, H. (2017a). A new way of valuing nature: articulating biomimicry and ecosystem services. *Environmental Ethics* 39(3): 281–299.

Dicks, H. (2017b). Environmental ethics and biomimetic ethics: nature as object of ethics and nature as source of ethics. *Journal of Agricultural and Environmental Ethics* 30(2): 255–274.

Dobson, A.P., Bradshaw, A.D., and Baker, A.J.M. (1997). Hopes for the future: restoration ecology and conservation biology. *Science* 277: 515–522.

Eden, S. (2006). Ecological versus social restoration? How urban river restoration challenges but also fails to challenge the science–policy nexus in the United Kingdom. *Environment and Planning C: Government and Policy* 24: 661–680.

Eden, S., Tunstall, S.M., and Tapsell, S.M. (2000). Translating nature: river restoration as nature – culture. *Environment and Planning D: Society and Space* 18: 257–273.

Egan, D. (2006). Authentic ecological restoration. *Ecological Restoration* 24(4): 223–224.

Elliott, R. (1982). Faking nature. *Inquiry* 25(1): 85–93.

Elliott, R. (1997). *Faking Nature: The Ethics of Environmental Restoration*. London: Routledge.

Evette, A., Labonne, S., Rey, F., et al. (2009). History of bioengineering techniques for erosion control in rivers in Western Europe. *Environmental Management* 43(6): 972–984.

Ewel, J.J. (1987). Restoration is the ultimate test of ecological theory. In: *Restoration Ecology: A Synthetic Approach to Ecological Research* (eds W.R. Jordan III, M.E. Gilpin, and J.D. Aber), 31–33. Cambridge: Cambridge University Press.

Giller, P.S. (2005). River restoration: seeking ecological standards. *Journal of Applied Ecology* 42: 201–207.

Hettinger, N. (2012). Nature restoration as a paradigm for the human relationship with nature. In: *Ethical Adaptation to Climate Change* (eds A. Thompson and J. Bendik-Keymer), 27–46. Cambridge, MA: MIT Press.

Higgs, E. (1994). Expanding the scope of ecological restoration. *Restoration Ecology* 2(3): 137–145.

Higgs, E. (1997). What is good ecological restoration? *Conservation Biology* 11(2): 338–348.

Hobbs, R.J. and Harris, J.A. (2001). Restoration ecology: repairing the earth's ecosystems in the new millennium. *Restoration Ecology* 9(2): 239–246.

Hobbs, R.J. and Norton, D.A. (1996). Towards a conceptual framework for restoration ecology. *Restoration Ecology* 4: 93–110.

Hourdequin, M. and Havlick, D.G. (2013). Restoration and authenticity revisited. *Environmental Ethics* 35: 75–93.

Jackson, W. (2011). *Nature as Measure: The Selected Essays of Wes Jackson*. Berkeley, CA: Counterpoint.

Jordan, W.R., III (1985). On the imitation of nature. *Restoration and Management Notes* 3: 2–3.

Jordan, W.R., III (1994). Sunflower forest: ecological restoration as the basis for a new environmental paradigm. In: *Beyond Preservation: Restoring and Inventing Landscapes* (eds D.A. Baldwin, J. de Luce, and C. Pletsch), 17–34. Minneapolis: University of Minnesota Press.

Jordan, W.R., III, Gilpin, M.E., and Aber, J.D. (1987). Restoration ecology: ecological restoration as a technique for basic research. In: *Restoration Ecology: A Synthetic Approach to Ecological Research* (eds W.R. Jordan III, M.E. Gilpin, and J.D. Aber), 3–22. Cambridge: Cambridge University Press.

Jordan, W.R., III, Peters, R.L., and Allen, E.B. (1988). Ecological restoration as a strategy for conserving biological diversity. *Environmental Management* 12(1): 55–72.

Katz, E. (1996). The problem of ecological restoration. *Environmental Ethics* 18(2): 222–224.

Katz, E. (1997). The big lie: human restoration of nature. In: *Nature as Subject: Human Obligation and Natural Community* (ed. E. Katz), 93–107. London: Rowman and Littlefield.

Katz, E. (2012). Further adventures in the case against restoration. *Environmental Ethics* 34(1): 67–97.

Light, A. (2000). Restoration or domination? A reply to Katz. In: *Environmental Restoration: Ethics, Theory, and Practice* (ed. W. Throop), 95–111. New York: Humanity Books.

Light, A. (2001). The urban blind spot in environmental ethics. *Environmental Politics* 10(1): 7–35.

Light, A. and Higgs, E. (1996). The politics of ecological restoration. *Environmental Ethics* 18(3): 227–247.

Loucks, O.L. (1994). Art and insight in remnant native ecosystems. In: *Beyond Preservation: Restoring and Inventing Landscapes* (eds D.A. Baldwin, J. de Luce, and C. Pletsch), 127–135. Minneapolis: University of Minnesota Press.

Marshall, A. and Loveza, S. (2009). Questioning the theory and practice of biomimetics. *International Journal of Design & Nature and Ecodynamics* 1(4): 1–10.

Mathews, F. (2011). Towards a deeper philosophy of biomimicry. *Organization & Environment* 24(4): 364–387.

Mathews, F. (2016). Do the deepest roots of a future ecological civilization lie in Chinese soil? In: *Learning from the Other: Australian and Chinese Perspectives on Philosophy* (ed. J. Makeham), 15–27. Canberra: Australian Academy of the Humanities.

Mathews, F. (2019). Biomimicry and the problem of praxis. *Environmental Values* 28(5): 573–599.

Merchant, C. (1986). Restoration and reunion with nature. *Restoration and Management Notes* 4(2): 68–70.

Mitsch, W.J. (2012). What is ecological engineering? *Ecological Engineering* 45: 5–12.

Naveh, Z., (1994). From biodiversity to ecodiversity: a landscape ecology approach to conservation and restoration. *Restoration Ecology* 2(3): 180–189.

Palmer, M.A., Ambrose, R.F., and Poff, N.L. (1997). Ecological theory and community restoration ecology. *Restoration Ecology* 5(4): 291–300.

Palmer, M.A., Bernhardt, E.S., Allan, J.D., et al. (2005). Standards for ecologically successful river restoration. *Journal of Applied Ecology* 42: 208–217.

Palmer, C., McShane K., and Sandler, R. (2014). Environmental ethics. *Annual Review of Environmental Resources* 39: 419–42.

Pedersen Zari, M. (2017). Biomimetic urban design: ecosystem service provision of water and energy. *Buildings* 7(1): https://doi.org/10.3390/buildings7010021.

Prager, R. and McPhillips, M. (2012). Ethics in river and stream restoration: biomimicry or charade? In: *World Environmental and Water Resources Congress* (ed. R. Graham), 1–7. Reston, VA: American Society of Civil Engineers. https://doi.org/10.1061/40856(200)342.

Rheinhardt, R.D., Rheinhardt, M.C., Brinson, M.M., et al. (1999). Application of reference data for assessing and restoring headwater ecosystems. *Restoration Ecology* 7(3): 241–251.

Rolston, H., III (2000). Restoration. In: *Environmental Restoration: Ethics, Theory, and Practice* (ed. W. Throop), 127–132. New York: Humanity Books.

Taylor, P.W. (1986). *Respect for Nature: A Theory of Environmental Ethics*. Princeton, NJ: Princeton University Press.

Turner, F. (1987). The self-effacing art: restoration as imitation of nature. In: *Restoration Ecology: A Synthetic Approach to Ecological Research* (eds W.R. Jordan III, M.E. Gilpin, and J.D. Aber), 47–50. Cambridge: Cambridge University Press.

Turner, F. (1994). The invented landscape. In: *Beyond Preservation: Restoring and Inventing Landscapes* (eds D.A. Baldwin, J. de Luce, and C. Pletsch), 35–66. Minneapolis: University of Minnesota Press.

Wilke, G.E. (1994). Landscape restoration: more than ritual and gardening. In: *Beyond Preservation: Restoring and Inventing Landscapes* (eds D.A. Baldwin, J. de Luce, and C. Pletsch), 90–96. Minneapolis: University of Minnesota Press.

3

Restoring Sociocultural Relationships with Rivers: Experiments in Fluvial Pluralism

Dan Hikuroa[1], Gary Brierley[2], Marc Tadaki[3], Brendon Blue[2],
and Anne Salmond[1]

[1] Te Wānanga o Waipapa, University of Auckland, Auckland, New Zealand
[2] School of Environment, University of Auckland, Auckland, New Zealand
[3] Cawthron Institute, Nelson, New Zealand

3.1   Introduction

Selecting which rivers to restore, how and why involves complex and deeply value-laden decisions. Often understood as the responsibility of scientific and technical expertise, choices regarding which forms, processes, and ecologies merit restoration, and to what condition, are the subjective products of cultural values and assumptions (Bliss and Fischer 2011; Dufour et al. 2017; Egan et al. 2011; Gobster and Hull 2000; Sneddon et al. 2017). Aspirations for rivers and freshwater are not preexisting problems waiting for a solution but contested ideals in need of negotiation (Lave 2012). Confronting this fact requires river restorationists to ask two interrelated questions:

1) To what ideal standards should rivers be restored?
2) Who should make these decisions?

While scientists and science have an important role to play in responding to both of these questions, their roles need to be understood alongside other ways of knowing and valuing river systems.

Restoration ecology has at times been criticized for attempting to create an imaginary nature (Hilderbrand et al. 2005). For example, many river restoration practices seek to construct notions of fluvial forms that mimic predetermined visions of how rivers should look (Dufour and Piégay 2009; Palmer et al. 2010). Many rivers, in a wide range of geopolitical and geomorphologic settings, have been designed to reflect European pastoral ideals of stable, single-thread meandering channels, even though such rivers are not necessarily representative of that particular region (Kondolf 2006). Researchers have pointed to the need for interventions that reflect site-specific attributes, framed in terms of objectives that are realistically achievable given the spatial context and historical evolution of a river (Brierley et al. 2013; Simon et al. 2007). In turn, however, these scientific

River Restoration: Political, Social, and Economic Perspectives, First Edition. Edited by Bertrand Morandi, Marylise Cottet, and Hervé Piégay.

considerations themselves entail presuppositions regarding what rivers could and should be (Ashmore 2015; Blue and Brierley 2016; Koppes and King 2020; Lave 2016).

Arguments for participatory approaches to restoration have arisen alongside and in response to these concerns for physical and ecological appropriateness (Eden and Tunstall 2006; Eden et al. 2000; Egan et al. 2011; Hillman 2006). The case for community engagement in river restoration has been made on the basis of a range of potential benefits for both people (e.g. fostering connections between people and the landscape, creating employment opportunities, facilitating knowledge transfer and environmental justice), and for the projects they work on (e.g. cost savings through volunteer labor, improving community buy-in, and facilitating ongoing monitoring) (Fore et al. 2001; Overdevest et al. 2004). Yet, while participatory and inclusive approaches to river restoration are increasingly acknowledged as best practice (Smith et al. 2014), in reality, community participation is often rendered instrumental and tokenistic, failing to provide meaningful challenges to prevailing assumptions or to reconnect people with their rivers (Parsons et al. 2017; Spink et al. 2009, 2010).

A key outcome of the participatory turn has been the recognition that river restoration should not be considered a purely technical task framed in relation to ecological goals. The problem confronting river restorationists is not only a degraded ecosystem awaiting ecological improvement but also the fraught and complex relationships between people and these ecosystems through time. When a river is degraded to the point of ecological catastrophe, this often reflects a wider disconnect between the people who live with and value the river and the forces driving ecological decline. As such, restoration is not simply about ecological improvement, but is also about relationships both between people, and between people and the environment.

Just as there are multiple ways of relating to rivers, there are many ways to undertake their restoration. All too often, scientifically and technologically framed approaches to river restoration impose an anthropocentric perspective upon rivers through application of engineering principles within a command-and-control ethos (Brierley and Fryirs 2005). Certain visions of nature – and people's places within it – have been historically prioritized over others (Eden 2017; Eden et al. 2000). Attempts to restore the environment without restoring people's relationships to it are self-limiting in the long term. They also raise important questions of equity, influence, and justice: whose ideals should guide restoration (Bliss and Fischer 2011; Castleden et al. 2017; Egan et al. 2011)? Taking sociocultural restoration seriously requires confronting questions of whose rivers have been degraded, whose should be restored, and who gets to decide what, exactly, "restoration" means.

In this chapter, we explore novel indigenous-led governance experiments in *Aotearoa* New Zealand to consider different ways of practicing a more socioculturally responsible river restoration. Initiatives driven by *Māori*, the collective name of the indigenous kin groups of *Aotearoa* New Zealand, have created spaces for thinking about rivers pluralistically, valuing rivers as holistic, historical, and cultural agents with lives and rights of their own (Harmsworth et al. 2016; Parsons et al. 2019). Seeking to repair both human–human and human–environment relations at the same time, these initiatives foreground cultural assumptions, values, and aspirations instead of ignoring them. Building upon relational understandings of rivers as ancestral beings that are more ancient and powerful than people, they treat water and rivers as the lifeblood of society and the land (Fox et al. 2017;

Smith 2017; Thomas 2015; Wilcock et al. 2013; Wilkinson et al. 2020; Yates et al. 2017). Such reconceptualizations show how river restoration can seek relations of care and justice among humans as well as between humans and landscapes. As *Māori* perspectives realign the guiding ideals of river restoration and management, they offer a more ethically grounded basis for environmental governance (e.g. Fisher and Parsons 2020; Parsons et al. 2021).

After reviewing the meaning of river restoration – and our use of the term – we briefly outline how sociocultural contexts influence restoration prospects. We then explore how three experiments in river governance encompass a different relational framing with which to approach restoration practice in *Aotearoa* New Zealand. After briefly reviewing some integrative concepts found in *Māori* cosmologies, we sketch key features of *Aotearoa* New Zealand's governance that have evolved to recognize and incorporate *Māori* values, knowledge, and thinking in approaches to river and environmental management. We focus on recent developments in national policy that have embedded *Māori* concepts and platforms for participation in freshwater management, and new governance institutions that have emerged to make decisions about the Waikato River and the Whanganui River. For each of these developments, we reflect on how – and how well – they enable truly plural and integrative understandings of river systems. Finally, we discuss the implications of this work, presenting a critique of emerging approaches to co-management, associated policy developments, and assertions of river rights.

3.2 What is river restoration?

By definition, restoration entails returning something to a former condition. While this might usefully describe the process of returning an artwork or a heritage building to its former glory, its application to rivers, landscapes, and ecosystems is questionable (Brierley 2020). As landscapes and ecosystems change over time, irrespective of human intervention, reinstating a particular past condition is often neither possible nor necessarily desirable (Brierley 2020).

Attempts to restore nature can represent a particular kind of human arrogance: the assumption that we can simply "fix" parts of the world to absolve ourselves of responsibility for the destruction wrought upon it (Eden 2017). Philosophers have argued that ecological restoration as a universal objective is at best misguided; at worst it is inauthentic, fraudulent, and malicious (Katz 1996; Light 2000). Elliot (1997) describes this as "faking nature" through a replacement thesis, wherein the interests of development take precedence over environmental protection, trading off concerns for one place with another. Regardless of such contestations, river restoration is a rapidly growing industry in various parts of the world, with consulting companies, government agencies, and nonprofit groups competing for restoration funds as profit-motivated, commercial ventures create and sell environmental commodities (Kondolf and Yang 2008; Lave 2018).

Improvements to water quality in the latter decades of the twentieth century have enhanced societal reconnections to rivers in many places, re-establishing them as potentially safe and desirable environments (Ziegler 2014). Once this reconnection has been established, it is unlikely that people will want to go back to previous, more degraded conditions (Pauly 1995). Although industrial development and modern capitalism have alienated people from landscapes and waterscapes, people across many nations want to foster restorative – if not regenerative – relationships with ecosystems.

Despite wide-ranging criticisms of river restoration as a management goal, the term resonates powerfully with many people (Kelly et al. 2018; Kondolf and Yang 2008). Increasingly, however, river management practices and aspirations have a future focus, rather than looking backwards: they are less concerned with trying to restore a river to what it was than maximizing socioecological functionality into the future (Brierley and Fryirs 2008, Dufour and Piégay 2009; Everard and Moggridge 2012). This recognizes that, while insights from the past can guide management endeavors, the future will be different in ways we do not and cannot necessarily know (Brierley and Fryirs 2016). Viewed in this way, the concept of restoration can support the articulation, incorporation, and application of sociocultural relationships with rivers (Johnston et al. 2012; Strang 2004). Rather than taking for granted particular ideals of what a river should be, careful negotiation is required to determine what is considered desirable for a given river (Blue 2018; Egan et al. 2011; Mould et al. 2018).

3.3 Placing river restoration in its biophysical and sociocultural contexts

The variable condition of rivers worldwide reflects a complex interplay between physical settings and processes, ecological dynamics, and human activities. Human activities imprint specific values, ideologies, and structures upon the landscape, such that landscapes become enduring reflections of cultural values, past and present, as well as their evolutionary histories (Ashmore 2015; Nassauer 1995; Wilcock et al. 2013). While increasing talk of the Anthropocene suggests the emergence of a human-dominated world (Wohl 2013), these impacts are not universal: variations in human imprints influence prospects for river restoration along a spectrum that ranges from extensively modified conditions through to places that have been subjected to relatively minor human impacts (Kondolf and Yang 2008). Across this spectrum, prospects for river restoration and the processes through which restoration might be practiced are very different.

For less-impacted rivers, conservation rather than restoration may be the priority. In highly domesticated landscapes, however, the boundary conditions within which rivers operate have been so modified that path dependencies constrain future management options (Beattie and Morgan 2017). Despite these limitations, there are many high-profile examples of effective restoration, whether in an urban context (Lee et al. 2014) or for major industrial rivers (Verweij 2017). Although only minor physical adjustments to these rivers are possible, improvements in water quality, ecological functionality, and the accessibility and capacity for public use of these riverscapes has engendered significant societal renewal, regeneration, and reconnection (Everard and Moggridge 2012; Petts 2007; Smith et al. 2014). As enhanced societal relationships to riverscapes regenerate the river, improved conditions enhance physical and psychological connections to, and interactions with, the river.

Rivers that fall somewhere in between the most- and least-impacted situations present the greatest opportunities for significant biophysical transformation through restoration efforts. The wide spectrum of approaches to restoration practice within this category ranges from hands-on societal engagement through community-led riparian revegetation and associated weed management programs to freedom space or space to move initiatives that support ecological enhancement of rivers that have been modified by engineering practices (e.g. reconnection of former channels or manipulation of instream habitat while allowing the channel to adjust within an erodible corridor) (Biron et al. 2014; Buffin-Bélanger et al. 2015; Buijse et al. 2002; Piégay et al. 2005). Such practices reconceptualize societal relations to rivers, as demonstrated by Pahl-Wostl (2006, p. 1) in her overview of the social learning programs accompanying a shift in management practices from "fighting against water" to "living with floods and give room to water" in the Netherlands.

These efforts involve articulating and pursuing desirable socio-natural relationships within place-based contexts that are historically situated and biophysically constrained. In most instances, however, restoration practices reproduce the artificial separation of humans and nature, envisaging restoration as a technical task that leaves the design and implementation to experts. Many of these endeavors continue to assert human authority over the river in efforts to meet human needs, or simply replace one biophysical objective with another, failing to reconstitute the way that human communities and practices incorporate and connect with the river (or not). All too often, such approaches not only fail to engage productively with the regenerative potential of inclusive participatory practices but also sometimes further alienate some groups from their rivers (Eden and Tunstall 2006; Parsons and Fisher 2020; Spink et al. 2010), thereby overlooking opportunities to rethink and address concerns for fundamental relations between nature and society. Quite different sociocultural relations emerge through collective efforts to live with the river in harmonious ways, relative to external imposition of particular perspectives upon the river.

Indigenous worldviews potentially offer a more generative way of viewing relationships between nature and society, framing humans as part of nature. While caution must be taken to avoid oversimplification and overgeneralization, many indigenous knowledge systems emphasize concepts of reciprocity in which interactions with the land impact (in) directly on human society and people, land and rivers are existentially entangled (Salmond 2014; Kimmerer 2011; Wehi and Lord 2017). In this sense, restoration of land and of people are inseparable: sustaining the land sustains human communities (Fox et al. 2017; Thomas 2015). Reciprocal restoration recognizes that it is not just the land that is broken but our relationships with it, and with each other. Many indigenous viewpoints extend notions of care to the more-than-human realm, recognizing that "we live in a moral landscape governed by relationships of mutual responsibility, which are simultaneously material and spiritual" (Kimmerer 2011, p. 268).

While indigenous knowledges might challenge dominant understandings of landscapes, they do not necessarily contradict scientific knowledge. For example, Wilcock et al. (2013) demonstrate alignment in scientific and indigenous knowledges of river landscapes, highlighting shared concerns for spatial relations, evolutionary trajectories, and notions of emergence. Building upon such shared yet multiple knowledges, collaborations between scientists and indigenous peoples can conceive, design, implement, and monitor river restoration and management activities that encompass new ways of seeing and valuing rivers (Brierley et al. 2019; Fisher and Parsons 2020; Harmsworth et al. 2016; Hudson et al. 2016; Memon and Kirk 2012; Mould et al. 2018, Parsons et al. 2019; Wilkinson et al. 2020). In *Aotearoa* New Zealand and other settler colonial contexts, the histories of colonialism and the life worlds of indigenous peoples provide an intriguing context to appraise questions of how and for whom rivers ought to be restored and managed. In Section 3.4 we consider some recent developments in what we term "fluvial pluralism" (Lyver et al. 2016). This term encompasses explicit recognition that there are different ways of knowing and managing rivers. We contend that recent developments in *Aotearoa* New Zealand present opportunities to productively reconceptualize and implement approaches to sociocultural and socio-natural restoration within the same frame of reference.

3.4 Emerging approaches to knowing and valuing rivers differently in Aotearoa New Zealand

Aotearoa New Zealand is in a freshwater crisis, with research and data showing an alarming and overwhelming trend toward degraded water quality, lost wetlands, exhausted or polluted aquifers, erosion and sedimentation problems, and habitat loss induced by intensive land modification (Elston et al. 2015; Gluckman 2017; Joy 2015; Joy et al. 2019; Kelly et al. 2014, 2017; Larned et al. 2020; Ministry for the Environment & Stats NZ 2020; Quinn et al. 2017). The OECD (2017) indicated that *Aotearoa* New Zealand's economic growth is approaching its environmental limits, and that pollution of freshwater is spreading. The contamination of Havelock North's drinking water supply in August 2016 led to approximately 5500 residents becoming infected with campylobacteriosis, likely causing four deaths (DIA 2017). Furthermore, the DIA has found substantial risks to drinking water, estimating that between 18 000 and 100 000 cases of waterborne illness occur each year in *Aotearoa* New Zealand (DIA 2017). Additionally, 70% of rivers do not meet Ministry of Health recreational contact guidelines (McBride and Soller 2017). Reliance on legislation such as the Resource Management Act 1991, the purpose of which is to promote the sustainable management of natural and physical resources by avoiding, remedying, or mitigating adverse effects of activities, has categorically failed to protect waterways (Joy and Canning 2020).

Like most parts of the world, river restoration in *Aotearoa* New Zealand primarily aims to repair changes resulting from excessive water extraction and nutrient leaching, damaging land uses, and the legacy of past river management practices. Other than occasional high-profile daylighting projects (Neale and Moffett 2016), fish ladder and regulation structures (Franklin and Bartels 2012), forestry management practices (Quinn and Wright-Stow 2008), and site-specific restoration initiatives (Caruso 2006a, 2006b; Ginders et al. 2016), contemporary approaches to river restoration in *Aotearoa* New Zealand are overwhelmingly dominated by community-led riparian vegetation and weed management initiatives (Collins et al. 2013; Daigneault et al. 2017; Death and Collier 2010; Greenwood et al. 2012; McKergow et al. 2016; Parkyn et al. 2003; Peters et al. 2015; Quinn et al. 2017; Storey and Wright-Stow 2017; Storey et al. 2016; Sullivan and Molles 2016).

Critically, many of these practices are part of co-governance and co-management arrangements with *Māori*, striving to link cultural and scientific values (Fisher and Parsons 2020; Harmsworth et al. 2011, 2016; Parsons et al. 2019; Tipa et al. 2017). These interventions build directly upon formal treaty arrangements that date back to the nineteenth century. The Treaty of Waitangi (the English draft) and *Te Tiriti o Waitangi* (in *Māori*), an agreement signed in 1840 between the British Crown (now represented by the

New Zealand government) and the leaders of *Māori* kin groups, established an alliance between Queen Victoria and the *rangatira* (kin group leaders) and recognition of indigenous rights and governance, providing the basis for collaborative partnership and engagement between *Māori* and the Crown (Harmsworth et al. 2016). Increasingly, attempts to enact restorative justice related to the Treaty of Waitangi have begun to reshape approaches to land and water management in *Aotearoa* New Zealand (Harmsworth et al. 2016; Hudson et al. 2016; Memon and Kirk 2012; Parsons et al. 2021; Parsons and Fisher 2020; Tipa et al. 2017). These transformational processes gained international prominence in 2017 when, for the first time in the world and following the example set by the passing of *Te Urewera* Act in 2014, the Whanganui River became a legal entity (*Te Awa Tupua* – the river which is an ancestral being) with associated rights (Boyd 2017; Brierley et al. 2019; Charpleix 2018; Ruru 2018).

3.4.1 *Te Ao Māori*, the *Māori* world

In *Te Ao Māori*, the *Māori* world, humans exist in a kin-based relationship with *Te Taiao* – the earth, universe, and everything in it – tracing shared descent from the primal parents *Ranginui* (Sky Father) and *Papatūānuku* (Earth Mother) (Hikuroa 2017). This relational understanding, known as *whakapapa*, positions people as just one element in a cosmic kin network, linked with everything through their shared descent from the primal parents (Salmond 2014). As direct descendants of *Papatūānuku* and *Ranginui*, *Māori* see themselves not merely "of the land" but "as the land" (Te Aho 2010, p. 285). In this relational schema, ancestors are literally planted in the earth. For this reason, *Māori* identify as *tāngata whenua*: people of the land. Kin-based networks of people, plants, and animals live together in intricate patterns of seasonal movement and exchange, governed by various kinds of *tapu* or ancestral restrictions (Salmond 2018). Kinship is tied to place through reciprocal relationships in which an individual is considered a part of that place. In the ontology of *Māori* language, people, land, and ancestors existentially overlap (Salmond 2014). In this way of being, rivers may be conceptualized as plaited ropes that entwine genealogical lines, tying land, people, and ancestors together.

Through *Māori* perspectives, rivers are inextricably linked to tribal identities, yet at the same time they have their own life force, their own energy, and their own powerful identities, authority, prestige, and sacredness (Te Aho 2010). Rivers such as the Whanganui and the Waikato, for instance, are understood as ancestral beings, with their own life and power. When first meeting, *Māori* might ask: "*Na wai koe?*" ("Of whose waters are you?") (Morgan 2006a). *Māori* have always valued water as a *taonga* – something to be treasured, owing to its life-giving properties and importance in sustaining aquatic environments from which food was sourced through *mahinga kai* (food gathering practices). Carefully prescribed codes of practice based on detailed observations made over the generations maintained water quality and sustainable levels of *tuna* (eels), *piharau/kanakana* (lamprey), *ika* (fish), *koura* (freshwater crayfish), and other freshwater species (Harmsworth and Awatere 2013). Designated places were put aside for activities such as washing and religious ceremonies. Restrictions were often placed on the use of water, whether temporary (*rāhui*) or more permanent by placement of *owheo* (Williams 2006).

Viewed through a *Māori* lens, the economy is a wholly owned subsidiary of the environment, expressed in the *whakataukī* (proverb) "*Toi tū te whenua, whatungarongaro te tangata*" ("the land endures, while people come and go"). Custodial linkages are expressed through *kaitiakitanga* (guardianship), with deep respect for ancestral connections that inherently frame people as part of their landscapes and ecosystems. This is expressed in common sayings such as: "Harm the river and you harm my ancestors," "Take care of the land, and the land will take care of you," "If the river is dying, so am I." A reciprocal relationship relates *manaaki whenua* (caring for the land) and *manaaki tangata* (caring for people) (Harmsworth et al. 2016).

Te Ao Māori positions the well-being of the river – its *mauri* (life force) – as intertwined with the well-being of its people. Ideally, this maintains a state of *ora*: peace, prosperity, and well-being for people, plants, and animals, as well as the river. If human activities deplete the *mauri* of a river, those who depend on it will ultimately suffer through a state known as *mate* (ill health, dysfunction). Shifts toward a state of *mate* and associated declines in the state of *mauri* have been induced through loss of habitat, water quality impacts, altered flow regimes, changes to flora and fauna populations, and the poor condition of ecosystems and resources. This has had a negative impact on *mahinga kai* (food gathering practices) and *taonga* species (biota of special significance and importance to *Māori*).

3.4.2 Colonization, transformation, and the case for restorative justice

In the initial years following the signing of the Treaty of Waitangi (1840), *Māori* still controlled much of the land that was notionally sold to colonists. However, a significant influx of European immigrants, especially after the 1860s, brought a progressive decline in *Māori* control of land and resources, as land was notionally purchased or confiscated if tribes refused to sell. Even before the first parties of settlers arrived in New Zealand, surveyors began to map out and subdivide the land, using grids based on latitude and longitude to define sections and settlements in what *Māori* bitterly referred to as "cutting up the land" (Salmond 2014). As *Papatūānuku*, the earth mother, was divided into bounded blocks to be owned and sold by people, the genealogical networks of relationship with land and sea, mountains and rivers were also severed. This shift was profound, alienating *Māori* kin groups from much of their ancestral land and water resources, and also from each other (Harmsworth and Awatere 2013). *Iwi* and *hapū* not only lost their living connection with their *awa* (rivers) but were also set at odds with each other and excluded from decision-making processes that would determine the management of rivers: in effect, they lost their voice and *mana* (authority, ancestral power). This had a fundamental impact on the *Māori* economy, way of life, and identity, initiating a state of *mate* that affected both *Māori* kin groups and their ancestral rivers (Parsons et al. 2021).

Colonial settlement not only brought about the wholesale clearance of bush, the drainage of wetlands, and the creation of large grassland areas for pastoral farming on steep, erodible hills as well as foothills and plains, it also often reconceptualized river systems as drains or sewers: conduits for the disposal of waste with an assumed limitless capacity for self-cleansing and self-renewal (Knight 2016). Cumulative responses to bush or forest clearance, sawmilling plants, drainage of wetlands, flax-milling, local mining activities, pastoral farming, the operation of tanneries, dairy factories and meat works, and incursions of exotic species transformed rivers across most of the country. These impacts were compounded further by various command and control engineering practices: construction of dams for hydroelectricity and irrigation schemes, wastewater facilities, emplacement of stop-banks (artificial levees) for flood control, and piping or channelizing streams (especially in urban areas). Institutional framings asserted and embedded the values of a 'rivered Eden' (Beattie and Morgan 2017). Civil engineers were tasked with harnessing the powers of nature for human benefit, and keeping rivers away from people (Knight 2016). Agriculture has been the mainstay of the economy – initially sheep, but increasingly dairy production. Industrial systems of forestry, fishing, and farming have caused significant biodiversity losses, as waterways and harbors have become choked with sediment, pollutants, and contaminants, and aquifers and streams have been depleted beyond sustainable limits.

In 1975, the Waitangi Tribunal was established to hear grievances related to historic Crown actions that breached Treaty promises. Many claims relate to the loss or degradation of ancestral rivers, lakes, springs, wetlands, estuaries, and other waterways (Morgan 2006b). This process has directly driven the creation of Treaty settlement legislation – often passed with cross-parliament support – that includes a formal apology to affected *Māori* on behalf of the Crown, as well as economic and cultural redress (Ruru 2009). Treaty settlements often emphasize the ancestral and cultural significance of freshwater to *Māori*, recognizing that the Treaty of Waitangi guaranteed *Māori tino rangatiratanga* (absolute authority) over their lands, resources, and *taonga* (Waitangi Tribunal 2011). The ability of *hapū* and *iwi* to exercise *rangatiratanga* (chiefly authority) in water management is pivotal to enacting the Treaty's guarantee *of tino rangatiratanga* (Ruru 2012). Incrementally, this is being supported through national freshwater policy and Treaty settlement legislation (Harmsworth et al. 2016).

Approaches to river restoration practice in *Aotearoa* New Zealand are in a state of flux, prospectively in the midst of a transformative reframing. To highlight the potential that is offered and the challenges that are faced, we focus upon three examples of fluvial pluralism: emerging approaches to co-management and co-governance arrangements for the Waikato River, national-scale policy developments, and assertions of river rights for the Whanganui River.

3.4.3 Co-management plans for restoration of the Waikato River that build on *kaitiakitanga* (guardianship)

The Waikato River is New Zealand's largest river in terms of flow volume and length (425 km) (Figure 3.1). The catchment has been subject to myriad contested uses including land clearance and drainage of wetlands for agricultural use, construction of eight dams for hydropower generation, disposal of various human, industrial and agricultural wastes, and provision of domestic water for Auckland. Many of these developments were undertaken with little or no regard for *Māori* concerns and interests, in what Te Aho (2010) refers to as a paradigm of exclusion (see also Knight 2016; Muru-Lanning 2016; Young 2013). For *Waikato-Tainui iwi*, the Waikato River is conceptualized as a living ancestor and is recognized as having its own *mauri* and spiritual integrity. The river's very origins are said to contain the life-giving water sent by one ancestral mountain, *Tongariro*, to heal another,

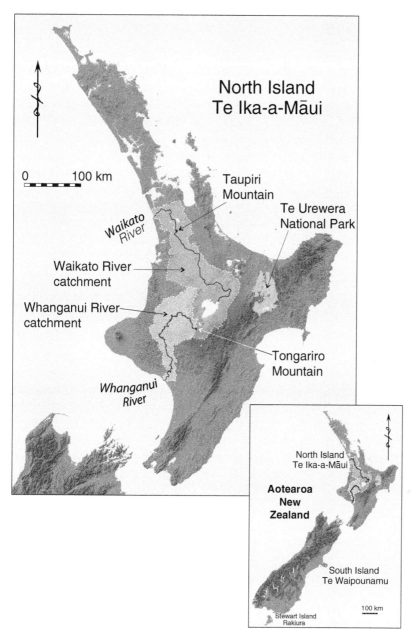

Figure 3.1 Map of Waikato River and Whanganui River catchments, *Aotearoa* New Zealand. *Source:* Louise Cotterall.

the maiden mountain *Taupiri* (Te Aho 2010) (located on Figure 3.1). To *Waikato-Tainui* the Waikato River is a *tupuna* (ancestor) which has its own *mana* and in turn represents the *mana* and *mauri* of the tribe. The river is a single indivisible being. Disregard for *Māori* cultural heritage and sacred sites was directly acknowledged by the Waitangi Tribunal

(1993): "the failure to acknowledge the significance of *wāhi tapu* (sacred place) in *Māori* terms has contributed to a sense of powerlessness and grievance among *Māori* people" (quoted in Knight 2016, p. 141).

In 2008, *Waikato-Tainui iwi* and the Crown established a new co-governance structure in which *Waikato-Tainui* members became the guardians of the Waikato River (Harmsworth et al. 2016). The *Waikato-Tainui Raupatu* Claims (Waikato River) Settlement Act 2010 also ushered in a new era of co-management arrangements that seek to restore and protect the health and wellbeing of the Waikato River, addressing the social, cultural, and economic injustices of alienating *Māori* from the Waikato River and its surrounding lands (Muru-Lanning 2016, p. 142). In the Deed of Settlement the Crown accepts that it failed to respect, provide for, and protect the special relationship *Waikato-Tainui* have with the river as their *tupuna*, and accepts responsibility for the degradation of the river that has occurred while the Crown has had authority over the river. The Waikato River Act 2010 recognizes two important principles:

1) *Te mana o te awa* reflects that, to *Waikato-Tainui*, the Waikato River is an ancestor that possesses *mana* and *mauri*,
2) *Mana whakahaere* (mandate or authority) embodies the authority that *Waikato-Tainui* and other river *iwi* have established over many generations to exercise control, access to, and management of the river and its resources in accordance with *tikanga* (values, ethics, and norms of conduct).

Governance is the responsibility of the newly formed Waikato River Authority, made up of equal numbers of Crown and *iwi*-appointed members, including some of the other *iwi* with interests along the river, with the Crown retaining the overall decision-making ability in the case of disagreement. The authority administers a clean-up fund for restoring and protecting the health and well-being of the river that has allocated $38 million to over 250 projects since it was established. One funded program is the *Waipā* Catchment Plan, the implementation of which spans 35 properties and includes the retirement of 477 hectares of land from grazing, fencing-off stock from 57 km of streams, and riparian revegetation using native species (88 700 plants). A concerted emphasis has been placed upon newly formed relationships between and amongst landowners, *iwi* members, local government staff, and the general community – united in their passion and effort to clean up the river (Fisher and Parsons 2020; Parsons et al. 2021). Similarly, *Ngā Kaitiaki o te awa o Pūniu* (Guardians of Puniu River) is a five-year project initiated in 2018 which aims to improve water quality, the *mauri* of the *awa*, as well as helping to restore *tuna* (eel) habitat and terrestrial biodiversity. Activities for this project include fencing 32 km of bank length to exclude stock, erosion management works, planting 160 000 native trees, and the creation of a *marae*[3]-based restoration guide to enable growth in the development of other new organizations and groups to undertake similar activities. Stream-fencing and riparian revegetation programs support efforts at socioeconomic and cultural renewal based upon relations to the river (Waikato River Plan 2018). *Iwi*-appointed commissioners are involved in any decisions relating to applications for resource consents for activities affecting the river, including taking, using, damming or diverting water, and discharges to the river.

However, these deliberations have been divisive among *Māori*, as the government repositioned and relabeled *Māori* interests as stakeholders, rather than rights-holders, reclassifying and relabeling particular *Māori* with newly defined roles and rights (Muru-Lanning 2016). The *iwi* representation in the Waikato River Authority is disproportionate and deeply contested, with the largest *iwi* having the same representation and decision-making power as the smallest, thereby failing to provide for equitable proportional representation based on population and extent of tribal *rohe* (territory) (Muru-Lanning 2016; Te Aho 2010). This reshaping of identities has transformed the cultural landscapes of the *Waikato*, masking the "fluidity and political importance of other forms of *Māori* descent and the socio-political importance of other *Māori* descent and socio-political groupings" (Muru-Lanning 2016, p. 119).

3.4.4 Re-framing relations to the river through emerging policy frameworks

Recent developments in national freshwater policy have incorporated *Māori* concepts and included provisions for *Māori* participation in water management in *Aotearoa* New Zealand (New Zealand Government 2014, 2017). In 2014 the National Policy Statement on Freshwater Management[4] was launched. It included a statement detailing *Te Mana o te Wai* – the integrated and holistic well-being of a freshwater body – and described it as a key purpose of the policy. The *Te Mana o te Wai* statement was informed by engagement between the Crown and the Iwi Leaders Group, a collective of tribal representatives. Upholding *Te Mana o te Wai* requires that the use of water also provides for *Te Hauora o te Wai* (health and *mauri* of socioecological systems), *Te Hauora o te Taiao* (health and *mauri* of the environment), and *Te Hauora o te Tangata* (the health and *mauri* of the people). The revised 2017 National Policy Statement on Freshwater Management comments that: "The health and well-being of our freshwater bodies is vital for the health and well-being of our land, our resources (including fisheries, flora, and fauna) and our communities" (New Zealand Government 2017, p. 7). Specifically, *Te Mana o te Wai* incorporates principles of *mātauranga Māori* in efforts to maintain or enhance a healthy ecosystem that is appropriate to each freshwater body type (river, lake, wetland, or aquifer) (Salmond et al. 2017). This emphasizes concerns for the integrity of the catchment and biota, the health and *mauri* of the natural form and character of the environment (including iconic values and aesthetic features), *mahinga kai* (gathering of food that is safe to eat), and the protection of *wai tapu* (sacred waters) and the health and *mauri* of the people.

These *Māori* concepts and values have been embedded within the National Policy Statement as a part of the government recognizing its responsibility to honor the guarantees of *Te Tiriti o Waitangi* (the Treaty of Waitangi). While the language of *Te Mana o te Wai* is aspirational and integrative, however, it does not provide detail and is not given specific priority over other objectives. Objective A1 requires regional authorities "To consider and recognise *Te Mana o te Wai* in the management of fresh water" (p. 11). In contrast, Objective A4 – which was added in 2017 – requires regional authorities "To enable communities to provide for their economic well-being, including productive economic opportunities" (p. 12). Although *Te Mana o te Wai* adds integrative language to the National Policy Statement, it also sits in a contradictory relation to other elements of the policy, and indeed other policy priorities of the government. In this way, the extent to which *Te Mana o te Wai* provides a substantive mechanism for *Māori* to exercise authority over waterways remains an open question, to be worked out through interpretative battles in the legal and planning systems. Despite widespread support for the intent of *Te Mana o te Wai*, regional and local governments have thus far largely failed to deliver on upholding and implementing associated practices. As a consequence, the transformative potential therein is yet to be realized. Nevertheless, the evolution of such policies is promising and has considerable restorative potential.

3.4.5 The rights of the Whanganui River: the agency of the river

Recent Treaty settlements (e.g. *Rukutia te Mana Ngāti Rangi* Deed of Settlement 2018 and Pare Hauraki Settlement Redress 2018) potentially mark ground-breaking developments in emerging sociocultural, political, and governance arrangements in the management and restoration of rivers in *Aotearoa* New Zealand. In contrast to the Waikato River settlement, which centered on the Waikato River Authority as a mechanism for *iwi* to exercise authority over the river, the recent Whanganui settlement legislation is built around granting the Whanganui River the rights of legal personhood (Charpleix 2018; see Figure 3.1). In March 2017, the *Te Awa Tupua* (Whanganui River Claims Settlement) Act was passed by the New Zealand Parliament. This is one of only a few instances in the world where a natural entity has been recognized as having its own legal personality and rights (Boyd 2017). In the Act, the Whanganui River is described as "an indivisible and living whole, comprising the Whanganui River from the mountains to the sea, incorporating all its physical and metaphysical elements." Designating the river as a "living being" and recognizing the river as a legal entity with an independent voice placed the Whanganui River in a new set of relationships with human beings, more closely aligning with ancestral ways of knowing and being. The river's own needs and rights are given legal protection. Within the Act, *Tupua te Kawa*, the values that define the essence of the river, describe the Whanganui River as the source of *ora* (life, health, and well-being). This is expressed in a *whakatauāki* (proverb) often used by Whanganui people, "*E rere kau mai te awa nui mai i Te Kāhui Maunga ki Tangaroa*; *ko au te awa, ko te awa ko au*" – "The great river flows from the Mountains to the Sea; I am the river, the river is me" (Rangiwaiata Rangitihi Tahuparae in Wilson 2010) and is often shortened to "*ko au te awa, ko te awa ko au*." Such framings reflect a reciprocal merging of rights and obligations. Brierley et al. (2019) explore potential implications of river rights in reconceptualizing the incorporation of scientific principles to guide river management practices in *Aotearoa* New Zealand (c.f. Wilkinson et al. 2020). They argue that river scientists can act as allies for indigenous peoples and as advocates for local communities, endorsing others' aspirations and support progressive visions of living with/in nature, concluding that partnerships that incorporate such framings acknowledge the fundamental importance of cultural identity, building upon understanding and respect for multiple knowledge systems, values, and perspectives (Brierley et al. 2019).

   The origins of the *Te Awa Tupua* Act can be traced back to US law academic Christopher Stone, and his book *Should Trees Have Standing?* (Stone 1974), in which he argues that forests, oceans, and the environment should have the ability to be heard and have someone speak on their behalf. Linking such principles to *Te Tiriti o Waitangi*, Morris and Ruru (2010, p. 50) propose that "to afford legal personality to New Zealand's rivers would create an exciting link between the *Māori* legal system and the state legal system," noting further that "the legal personality concept aligns with the *Māori* legal concept of a personified natural world" (p. 50). The legal argument provided the opportunity for the Government to give consideration to a solution for the claim by *Te-Atihaunui-a-Paparangi* (Whanganui River people), who first petitioned the Government for the restoration of an active relationship with their ancestral river in 1887, and had been persistently seeking resolution thereafter.

Although *Te Awa Tupua* Act declares the Whanganui river to be a legal person, this only roughly approximates ancestral realities. A *tupua* is not a person but a powerful being from the dark, ancestral realm; an *awa* is not an individual, but a living community of fish, plants, people, ancestors, and water, linked by *whakapapa* (Salmond 2017). In the deed of settlement for the Whanganui River Treaty claim, two people, mutually chosen by the Crown and the Whanganui tribes, are established as *Te Pou Tupua*, "the human (living) face" of the river, acting in its name (representing its voice) and in its interests (protecting its rights) and administering *Te Korotete* (literally, a storage basket for food from the river), a fund of $30 million to support the health and well-being of the river, the criteria for which are being developed. In addition, a payment of $80 million is made to *Whanganui iwi* as redress for breaches of their rights in relation to the river under the Treaty of Waitangi. It is yet to be seen whether making the *Whanganui* a legal person with its own rights is merely a modernist device that ultimately undermines conventional *Māori* values, or whether it underpins a regime in which both people and the river have rights and responsibilities, aimed at returning the river to a state of *ora* (Ruru 2018; Salmond 2017).

3.4.6 Reconceptualizing river restoration in *Aotearoa* New Zealand

These case studies point to several key issues that emerge from the reconceptualization of rivers as a focal point for their restoration in *Aotearoa* New Zealand. A river-centric perspective moves beyond imposition of an anthropocentric lens, embracing concerns for more-than-human relations to the river (Brierley 2020). This emphasizes the prevention and minimization of harm through implementation of a duty of care. Equivalent approaches to working with the river are outlined in a UN report on nature-based solutions for water management problems (UN Water 2018). Typically, this entails a whole of system approach, taking conscious decisions to live with the river as a living entity (Brierley et al. 2019). Such framings reconceptualize restoration practice, extending beyond the realms of scientific and technical experts to embrace more inclusive approaches to participatory practices. However, many issues are yet to be addressed in efforts to maximize the potential of such experiments in fluvial pluralism (cf. Parsons et al. 2019, 2021).

3.5 Conclusions

Administrative and governance arrangements in *Aotearoa* New Zealand are being reconceptualized and reconfigured in ways that encapsulate and encompass *Māori* values, knowledge, and perspectives. This offers prospects to forge new, albeit based upon old, relations to rivers, wherein human activities including restoration practices are understood as part of an emergent, evolutionary nature. Through these negotiations, we are beginning to re-know rivers in different ways, reframing socio-natures in ways that build upon indigenous (*Māori*) knowledges (Parsons et al. 2017; Wilcock et al. 2013; Wilkinson et al. 2020). A wide range of new approaches to managing and restoring rivers builds upon these emerging relationships

(Awatere et al. 2017; Fisher and Parsons 2020; Harmsworth et al. 2016; Hikuroa et al. 2018; Hudson et al. 2016; Tipa et al. 2017). Such prospects are considered here as a form of fluvial pluralism – a way of embedding different relations between humans and nature through a platform for restorative justice (Charpleix 2018; Lyver et al. 2016).

These developments are starting to dismantle visions of river restoration which are rooted in the past, viewing people as integral parts of contemporary living riverscapes (e.g. *Te Awa Tupua* [Whanganui River Claims Settlement] Act in 2017). Emerging approaches to restorative justice offer promise for better outcomes through re-imagined and remembered relationships between people and freshwater. *Māori* understandings of rivers can guide the constructive renegotiation of river restoration efforts, recognizing the independent life and power of rivers and the interdependence – rather than the separation – of people and the more-than-human (Brierley 2020). Prospectively, restoration provides a platform to promote human–environment relations that are inclusive of differing cultural identities as well as different species (Ziegler 2014), seeking to address issues of power and privilege (Egan et al. 2011; Fox et al. 2017; Magallanes 2018).

In such deliberations it is inappropriate to simply present indigenous peoples as natural protectors of the environment, or to merely conflate their interests with "the local" (Coombes et al. 2012, 2013). Rather, this chapter has argued that incorporation of *Māori* ideas into contemporary management regimes reconceptualizes river restoration as a simultaneously biophysical and sociocultural process (Mould et al. 2018; Wilcock et al. 2013). We have explored three mechanisms through which *Māori* and the Crown have tried to enact a fluvial pluralism, and identified practical, scientific, and political tensions with their implementation. The freshwater policies discussed herein map a progression of increased influence of *mātauranga Māori*, each step celebrated for its novelty and innovation. All three policies involve creating space to reappraise what constitutes successful river management (cf. Le Heron et al. 2021). The real challenge lies in the implementation of these innovations.

While our examples are indelibly New Zealand-specific, our examination can offer insight for river restorationists in other settings. At a basic level, these three mechanisms can be conceptualized as different attempts to incorporate holistic concepts of rivers into environmental management and environmental policy. To abstract from our case, others may be interested in the form and function of these innovative governance arrangements. Institutions matter, and it is likely that new decision-making structures based on strong values with specific mandates (as exemplified by the Waikato and Whanganui co-governance arrangements) may be able to realize better outcomes than vague, symbolic language of national policy. At the same time, however, every place and people must find their own path, recognizing their histories and ancestors, giving effect to their values, and providing a set of shared aspirations for the future.

In the context of *Aotearoa* New Zealand, recent developments guided by *mātauranga Māori* offer prospects for reciprocal and regenerative restoration that re-engages people with land and waterways, renews human–place connections, and enables people to reclaim responsibility for sustaining the land and waterways that sustain them. Such fundamental reconceptualizations do not envisage restoration of waterways as a trade-off between the economy and the environment; rather, they envisage prospects that everyone acts as an agent of change, wherein we all have a shared responsibility for these *taonga* (treasures) for future generations. To date, far too many restoration efforts have been little more than symbolic in their aspirations and applications, encompassing middle- to upper-class values in both their orientation and benefit accrual. Engagements with fluvial pluralism offer prospects that restoration can achieve much more than this.

Acknowledgments

Initial conceptualizations of this work were developed as part of the *Te Awaroa* project coordinated by Anne Salmond, with support from the Tindall and NEXT Foundations and a SRIF Grant from the University of Auckland (UoA). Ongoing work by Dan Hikuroa on this topic is supported by *Ngā Pae o te Māramatanga* at UoA. Gary Brierley gratefully acknowledges study leave support from UoA and support from ENS in Lyon to support his efforts to conduct this work. Brendon Blue gratefully acknowledges funding from the George Mason Centre for the Natural Environment, UoA. The authors thank Louise Cotterall for drafting Figure 3.1. and Nga Remu Tahuparae for assistance with attributing *whakatauāki*.

Notes

1 *Aotearoa* is a *Māori* name for New Zealand's North Island, and the combination of *Aotearoa* New Zealand is often used as a name for the entire country to reflect its bicultural history.
2 Although *Māori* is the collective name of the indigenous kin groups of *Aotearoa* New Zealand, *iwi* (tribal nations) and *hapū* (subtribal groups) are the key groupings that underpin *Māori* identity. *Māori* people share common ancestry and a language and culture that are closely related to those of peoples inhabiting the Polynesian triangle, delineated by the Hawai'ian Islands in the north, Rapanui (Easter Island) in the east, and *Aotearoa* New Zealand in the south and all those in between.
3 A *mārae* is a *Māori* community institution, usually comprising an ancestral house and a food preparation and dining house. They are a bastion of *Māori* culture and serve communal, social, and religious functions.
4 The National Policy Statement on Freshwater Management is a key instrument of the Resource Management Act 1991, that provides water quality and quantity objectives and policies that must be given effect in regional policy statements, and regional and district plans.

References

Ashmore, P. (2015). Towards a sociogeomorphology of rivers. *Geomorphology* 251: 149–156.
Awatere, S., Robb, M., Taura, Y., et al. (2017). *Wai Ora Wai Māori*: a *kaupapa Māori* assessment tool. http://www.envirolink.govt.nz/assets/Envirolink/PB19-Wai-Ora-Wai-Māori-June-2017.pdf (accessed 21 April 2018).
Beattie, J. and Morgan, R. (2017). Engineering Edens on this "Rivered Earth"? A review article on water management and hydro-resilience in the British Empire, 1860–1940s. *Environment and History* 23(1): 39–63.
Biron, P.M., Buffin-Bélanger, T., Larocque, M., et al. (2014). Freedom space for rivers: a sustainable management approach to enhance river resilience. *Environmental Management* 54(5): 1056–1073.

Bliss, J.C. and Fischer, A.P. (2011). Toward a political ecology of ecosystem restoration. In: *Human Dimensions of Ecological Restoration* (eds D. Egan, E.E. Hjerpe, and J. Abrams), 135–148. Washington DC: Island Press.

Blue, B. (2018). What's wrong with healthy rivers? Promise and practice in the search for a guiding ideal of freshwater management. *Progress in Physical Geography* 42(4): 462–477.

Blue, B. and Brierley, G. (2016). "But what do you measure?" Prospects for a constructive critical physical geography. *Area* 48(2): 190–197.

Boyd, D.R. (2017). *The Rights of Nature: A Legal Revolution that Could Save the World*. Toronto, Canada: ECW Press.

Brierley, G.J. (2020). *Finding the Voice of the River: Beyond Restoration and Management*. Cham, Switzerland: Palgrave Pivot.

Brierley, G.J. and Fryirs, K.A. (2005). *Geomorphology and River Management: Applications of the River Styles Framework*. Chichester: Wiley.

Brierley, G.J. and Fryirs, K.A. (2008). *River Futures*. Washington DC: Island Press.

Brierley, G.J. and Fryirs, K.A. (2016). The use of evolutionary trajectories to guide "moving targets" in the management of river futures. *River Research and Applications* 32(5): 823–835.

Brierley, G.J., Fryirs, K.A., Cullum, C., et al. (2013). Reading the landscape: integrating the theory and practice of geomorphology to develop place-based understandings of river systems. *Progress in Physical Geography* 37(5): 601–621.

Brierley, G.J., Tadaki, M., Hikuroa, D., et al. (2019). A geomorphic perspective on the rights of the river in Aotearoa New Zealand. *River Research and Applications* 35(10): 1640–1651.

Buffin-Bélanger, T., Biron, P.M., Larocque, M., et al. (2015). Freedom space for rivers: an economically viable river management concept in a changing climate. *Geomorphology* 251: 137–148.

Buijse, A.D., Coops, H., Staras, M., et al. (2002). Restoration strategies for river floodplains along large lowland rivers in Europe. *Freshwater Biology* 47(4): 889–907.

Caruso, B.S. (2006a). Project river recovery: restoration of braided gravel-bed river habitat in New Zealand's high country. *Environmental Management* 37(6): 840–861.

Caruso, B.S. (2006b). Effectiveness of braided, gravel-bed river restoration in the Upper Waitaki Basin, New Zealand. *River Research and Applications* 22(8): 905–922.

Castleden, H.E., Hart, C., Harper, S., et al. (2017). Implementing Indigenous and western knowledge systems in water research and management (part 1): a systematic realist review to inform water policy and governance in Canada. *The International Indigenous Policy Journal* 8(4): 7.

Charpleix, L. (2018). The Whanganui River as *Te Awa Tupua*: place-based law in a legally pluralistic society. *Geographical Journal* 184(1): 19–30.

Collins, K.E., Doscher, C., Rennie, H.G., and Ross, J.G. (2013). The effectiveness of riparian "restoration" on water quality: a case study of lowland streams in Canterbury, New Zealand. *Restoration Ecology* 21(1): 40–48.

Coombes, B., Johnson, J.T., and Howitt, R. (2012). Indigenous geographies I: mere resource conflicts? The complexities in Indigenous land and environmental claims. *Progress in Human Geography* 36(6): 810–821.

Coombes, B., Johnson, J.T., and Howitt, R. (2013). Indigenous geographies II: the aspirational spaces in postcolonial politics: reconciliation, belonging and social provision. *Progress in Human Geography* 37(5): 691–700.

Daigneault, A., Eppink, F.V., and Lee, W.G. (2017). A national riparian restoration programme in New Zealand: is it value for money? *Journal of Environmental Management* 187: 166–177.

Death, R.G. and Collier, K.J. (2010). Measuring stream macroinvertebrate responses to gradients of vegetation cover: when is enough enough? *Freshwater Biology* 55(7): 1447–1464.

DIA (Department of Internal Affairs) (2017). Report of the Havelock North drinking water inquiry: Stage 2. http://www.dia.govt.nz/Government-Inquiry-Into-Havelock-North-Drinking-Water (accessed 9 December 2020).

Dufour, S. and Piégay, H. (2009). From the myth of a lost paradise to targeted river restoration: forget natural references and focus on human benefits. *River Research and Applications* 25(5): 568–581.

Dufour, S., Rollet, A.J., Chapuis, M., et al. (2017). On the political roles of freshwater science in studying dam and weir removal policies: a critical physical geography approach. *Water Alternatives* 10(3): 853–869.

Eden, S. (2017). Environmental restoration. *The International Encyclopedia of Geography.* Chichester: Wiley. https://doi.org/10.1002/9781118786352.wbieg0622.

Eden, S. and Tunstall, S.M (2006). Ecological versus social restoration? How urban river restoration challenges but also fails to challenge the science–policy nexus in the United Kingdom. *Environment and Planning C: Government and Policy* 24(5): 661–680.

Eden, S., Tunstall, S.M., and Tapsell, S.M. (2000). Translating nature: river restoration as nature-culture. *Environment and Planning D: Society and Space* 18(2): 258–273.

Egan, D., Abrams, J., and Hjerpe, E.E. (eds) (2011). *Human Dimensions of Ecological Restoration.* Washington DC: Island Press.

Elliot, R. (1997). *Faking Nature: The Ethics of Environmental Restoration.* London: Routledge.

Elston, E., Anderson-Lederer, R., Death, R.G., et al. (2015). The Plight of New Zealand's Freshwater Biodiversity. Conservation Science Statement No. 1. Society for Conservation Biology (Oceania), Sydney.

Everard, M. and Moggridge, H.L. (2012). Rediscovering the value of urban rivers. *Urban Ecosystems* 15(2): 293–314.

Fisher, K. and Parsons, M. (2020). River co-governance and co-management in *Aotearoa* New Zealand: enabling Indigenous ways of knowing and being. *Transnational Environmental Law* 9(3): 455–480.

Fore, L.S., Paulsen, K., and O'Laughlin, K. (2001). Assessing the performance of volunteers in monitoring streams. *Freshwater Biology* 46(1): 109–123.

Fox, C.A., Reo, N.J., Turner, D.A., et al. (2017). "The river is us; the river is in our veins": re-defining river restoration in three indigenous communities. *Sustainability Science* 12(4): 521–533.

Franklin, P.A. and Bartels, B. (2012). Restoring connectivity for migratory native fish in a New Zealand stream: effectiveness of retrofitting a pipe culvert. *Aquatic Conservation: Marine and Freshwater Ecosystems* 22(4): 489–497.

Ginders, M.A., Collier, K.J., Duggan, I.C., et al. (2016). Influence of hydrological connectivity on plankton communities in natural and reconstructed side-arms of a large New Zealand river. *River Research and Applications* 32(8): 1675–1686.

Gluckman, P. (2017). *New Zealand's Fresh Waters: Values, State, Trends and Human Impacts.* Wellington, New Zealand: Office of the Prime Minister's Chief Science Advisor.

Gobster, P.H. and Hull, R.B. (eds) (2000). *Restoring Nature: Perspectives from the Social Sciences and Humanities.* Washington DC: Island Press.

Greenwood, M.J., Harding, J.S., Niyogi, D.K., et al. (2012). Improving the effectiveness of riparian management for aquatic invertebrates in a degraded agricultural landscape: stream size and land-use legacies. *Journal of Applied Ecology* 49(1): 213–222.

Harmsworth, G.R. and Awatere, S. (2013). Indigenous Māori knowledge and perspectives of ecosystems. In: *Ecosystem Services in New Zealand: Conditions and Trends* (ed. J.R. Dymond), 274–286. Lincoln, New Zealand: Manaaki Whenua Press.

Harmsworth, G., Awatere, S., and Robb, M. (2016). Indigenous Māori values and perspectives to inform freshwater management in Aotearoa-New Zealand. *Ecology and Society* 21(4): 9.

Harmsworth, G., Young, R., Walker, D., et al. (2011). Linkages between cultural and scientific indicators of river and stream health. *New Zealand Journal of Marine and Freshwater Research* 45(3): 423–436.

Hikuroa, D. (2017). *Mātauranga Māori*: the *ūkaipō* of knowledge in New Zealand. *Journal of the Royal Society of New Zealand* 47(1): 5–10.

Hikuroa, D., Clark, J., Olsen, A., et al. (2018). Severed at the head: towards revitalizing the *mauri* of *Te Awa o te Atua*. *New Zealand Journal of Marine and Freshwater Research* 52(4): 643–656.

Hilderbrand, R.H., Watts, A.C., and Randle, A.M. (2005). The myths of restoration ecology. *Ecology and Society* 10(1): 19.

Hillman, M. (2006). Situated justice in environmental decision-making: lessons from river management in Southeastern Australia. *Geoforum* 37(5): 695–707.

Hudson, M., Collier, K., Awatere, S., et al. (2016). Integrating Indigenous knowledge and freshwater management: an Aotearoa/New Zealand case study. *International Journal of Science in Society* 8(1): 1–14.

Johnston, R.J., Schultz, E.T., Segerson, K., et al. (2012). Enhancing the content validity of stated preference valuation: the structure and function of ecological indicators. *Land Economics* 88(1): 102–120.

Joy, M. (2015). *Polluted Inheritance: New Zealand's Freshwater Crisis* (Vol. 36). Wellington, New Zealand: Bridget Williams Books.

Joy, M.K. and Canning, A.D. (2020). Shifting baselines and political expediency in New Zealand's freshwater management. *Marine and Freshwater Research* 72(4): 456–461.

Joy, M.K., Foote, K.J., McNie, P., and Piria, M. (2019). Decline in New Zealand's freshwater fish fauna: effect of land use. *Marine and Freshwater Research* 70(1): 114–124.

Katz, E. (1996). The problem of ecological restoration. *Environmental Ethics* 18(2): 222–224.

Kelly, J.M., Scarpino, P., Berry, H., et al. (2018). *Rivers of the Anthropocene*. Oakland, CA: University of California Press.

Kelly, S., Sim-Smith, C., Bartley, J., et al. (2017). State of our Gulf: *Hauraki Gulf – Tīkapa Moana – Te Moana-nui-a-Toi*. State of the Environment Report 2017. Hauraki Gulf Forum.

Kelly, S., Sim-Smith, C., Faire, S., et al. (2014). State of our Gulf: *Hauraki Gulf – Tikapa Moana/Te Moananui a Toi*. State of the Environment Report 2011. Hauraki Gulf Forum.

Kimmerer, R. (2011). Restoration and reciprocity: the contributions of traditional ecological knowledge. In: *Human Dimensions of Ecological Restoration* (eds D. Egan, E.E. Hjerpe, and J. Abrams), 257–276. Washington DC: Island Press.

Knight, C. (2016). *New Zealand's Rivers: An Environmental History*. Christchurch, New Zealand: Canterbury University Press.

Kondolf, G.M. (2006). River restoration and meanders. *Ecology and Society* 11(2): 42.

Kondolf, G.M. and Yang, C.N. (2008). Planning river restoration projects: social and cultural dimensions. In: *River Restoration: Managing the Uncertainty in Restoring Physical Habitat* (eds S. Darby and D. Sear), 43–60. Chichester: Wiley.

Koppes, M. and King, L. (2020). Beyond x, y, z (t): navigating new landscapes of science in the science of landscapes. *Journal of Geophysical Research: Earth Surface* 125(9): e2020JF005588.

Lave, R. (2012). *Fields and Streams: Stream Restoration, Neoliberalism, and the Future of Environmental Science*. Athens, GA: University of Georgia Press.

Lave, R. (2016). Stream restoration and the surprisingly social dynamics of science. *Wiley Interdisciplinary Reviews: Water* 3(1): 75–81.

Lave, R. (2018). Stream mitigation banking. *Wiley Interdisciplinary Reviews: Water* 5(3): 1279.

Lee, YK., Lee, C-K., Choi, J., et al. (2014). Tourism's role in urban regeneration: examining the impact of environmental cues on emotion, satisfaction, loyalty, and support for Seoul's revitalized Cheonggyecheon stream district. *Journal of Sustainable Tourism* 22(5): 726–749.

Le Heron, E., Allen, W., Le Heron, R., et al. (2021). What does success look like? An indicative rubric to assess and guide the performance of marine participatory processes. *Ecology and Society* 26(1): 29.

Light, A. (2000). Ecological restoration and the culture of nature: a pragmatic perspective. In: *Restoring Nature: Perspectives from the Social Sciences and Humanities* (eds P.H. Gobster and R.B. Hull), 49–70. Washington DC: Island Press.

Lyver, P.O.B., Akins, A., Phipps, H., et al. (2016). Key biocultural values to guide restoration action and planning in New Zealand. *Restoration Ecology* 24(3): 314–323.

Magallanes, C.J.I. (2018). Improving the global environmental rule of law by upholding Indigenous rights: examples from Aotearoa New Zealand. *Global Journal of Comparative Law* 7(1), 61–90.

McBride, G. and Soller, J. (2017). Technical Background for 2017 MfE "Clean Water" Swimmability Proposals for Rivers. NIWA Report: FWWQ1722.

McKergow, L.A., Matheson, F.E., and Quinn, J.M. (2016). Riparian management: a restoration tool for New Zealand streams. *Ecological Management & Restoration* 17(3): 218–227.

Memon, P.A. and Kirk, N. (2012). Role of indigenous Māori people in collaborative water governance in Aotearoa/New Zealand. *Journal of Environmental Planning and Management* 55(7): 941–959.

Ministry for the Environment & Stats NZ (2020). New Zealand's Environmental Reporting Series: our freshwater 2020. https://environment.govt.nz/assets/Publications/Files/our-freshwater-report-2020.pdf.

Morgan, T.K.K.B. (2006a). Decision-support tools and the indigenous paradigm. *Proceedings of the Institution of Civil Engineers-Engineering Sustainability* 159(4): 169–177.

Morgan, T.K.K.B. (2006b). An indigenous perspective on water recycling. *Desalination* 187: 127–136.

Morris, J.D. and Ruru, J. (2010). Giving voice to rivers: legal personality as a vehicle for recognizing Indigenous peoples' relationships to water? *Australian Indigenous Law Review* 14(2): 49–62.

Mould, S.A., Fryirs, K., and Howitt, R. (2018). Practicing sociogeomorphology: relationships and dialog in river research and management. *Society & Natural Resources* 31(1): 106–120.

Muru-Lanning, M. (2016). *Tupuna Awa: People and Politics of the Waikato River*. Auckland: Auckland University Press.

Nassauer, J.I. (1995). Culture and changing landscape structure. *Landscape Ecology* 10(4): 229–237.

Neale, M.W. and Moffett, E.R. (2016). Re-engineering buried urban streams: daylighting results in rapid changes in stream invertebrate communities. *Ecological Engineering* 87: 175–184.

New Zealand Government (2014). National Policy Statement for Freshwater Management. Sourced from: www.mfe.govt.nz (accessed 30 April 2018).

New Zealand Government (2017). National Policy Statement for Freshwater Management. Sourced from: www.mfe.govt.nz. (accessed 30 April 2018).

OECD (Organization for Economic Cooperation and Development) (2017). Environmental Performance Reviews: New Zealand 2017. https://www.oecd-ilibrary.org/environment/oecd-environmental-performance-reviews-new-zealand-2017_9789264268203-en (accessed 30 April 2018).

Overdevest, C., Orr, C.H., and Stepenuck, K. (2004). Volunteer stream monitoring and local participation in natural resource issues. *Human Ecology Review* 11(2): 177–185.

Pahl-Wostl, C. (2006). The importance of social learning in restoring the multifunctionality of rivers and floodplains. *Ecology and Society* 11(1): 10.

Palmer, M.A., Menninger, H.L., and Bernhardt, E. (2010). River restoration, habitat heterogeneity and biodiversity: a failure of theory or practice? *Freshwater Biology* 55 (s1): 205–222.

Parkyn, S.M., Davies-Colley, R.J., Halliday, N.J., et al. (2003). Planted riparian buffer zones in New Zealand: do they live up to expectations? *Restoration Ecology* 11(4): 436–447.

Parsons, M. and Fisher, K. (2020). Decolonising settler hazardscapes of the Waipā: Māori and Pākehā remembering of flooding in the Waikato 1900–1950. In: *Disasters in Australia and New Zealand* (eds S. McKinnon and M. Cook), 159–177. Singapore: Palgrave Macmillan.

Parsons, M., Fisher, K., and Crease, R.P. (2021). *Decolonising Blue Spaces in the Anthropocene: Freshwater Management in Aotearoa New Zealand*. London: Palgrave Macmillan.

Parsons, M., Nalau, J., and Fisher, K. (2017). Alternative perspectives on sustainability: Indigenous knowledge and methodologies. *Challenges in Sustainability* 5(1). https://doi.org/10.12924/cis2017.05010007.

Parsons, M., Nalau, J., Fisher, K., and Brown, C. (2019). Disrupting path dependency: making room for Indigenous knowledge in river management. *Global Environmental Change* 56: 95–113.

Pauly, D. (1995). Anecdotes and the shifting baseline syndrome of fisheries. *Trends in Ecology & Evolution* 10(10): 430.

Peters, M.A., Hamilton, D., and Eames, C. (2015). Action on the ground: a review of community environmental groups' restoration objectives, activities and partnerships in New Zealand. *New Zealand Journal of Ecology* 39(2): 179–189.

Petts, J. (2007). Learning about learning: lessons from public engagement and deliberation on urban river restoration. *The Geographical Journal* 173(4): 300–311.

Piégay, H., Darby, S.E., Mosselman, E., et al. (2005). A review of techniques available for delimiting the erodible river corridor: a sustainable approach to managing bank erosion. *River Research and Applications* 21(7): 773–789.

Quinn, J.M. and Wright-Stow, A.E. (2008). Stream size influences stream temperature impacts and recovery rates after clearfell logging. *Forest Ecology and Management* 256(12): 2101–2109.

Quinn, J.M., Green, M.O., Schallenberg, M., et al. (2017). Management and rehabilitation of aquatic ecosystems: introduction and synthesis. *New Zealand Journal of Marine and Freshwater Research* 51: 1–16.

Ruru, J. (2009). Undefined and unresolved: exploring Indigenous rights in Aotearoa New Zealand's freshwater legal regime. *Journal of Water Law* 20(5–6): 236–242.

Ruru, J. (2012). The right to water as the right to identity: legal struggles of indigenous peoples of Aotearoa New Zealand. In: *The Right to Water: Politics, Governance and Social Struggles* (eds F. Sultana and A. Loftus), 110–122. Abingdon: Earthscan.

Ruru, J. (2018). Listening to *Papatūānuku*: a call to reform water law. *Journal of the Royal Society of New Zealand* 48(2–3): 215–224.

Salmond, A. (2014). Tears of Rangi: people, water and power in New Zealand. *Hau: Journal of Ethnographic Theory* 4(3): 285–309.

Salmond, A. (2017). *Tears of Rangi: Experiments across Worlds*. Auckland: Auckland University Press.

Salmond, A. (2018). Lifeblood of the land: rights, responsibilities and the governance of waterways in New Zealand. In: *ResponsAbility, Law and Governance for Living Well with the Earth* (eds L. Te Aho, M. Humphries, and B. Martin), 183–192. London: Routledge.

Simon, A., Doyle, M., Kondolf, M., et al. (2007). Critical evaluation of how the Rosgen classification and associated "natural channel design" methods fail to integrate and quantify fluvial processes and channel response. *JAWRA Journal of the American Water Resources Association* 43(5): 1117–1131.

Smith, B., Clifford, N.J., and Mant, J. (2014). The changing nature of river restoration: changing nature of river restoration. *Wiley Interdisciplinary Reviews: Water* 1(3): 249–261.

Smith, J.L. (2017). I, river? New materialism, riparian non-human agency and the scale of democratic reform. *Asia Pacific Viewpoint* 58(1): 99–111.

Sneddon, C., Barraud, R., and Germaine, M.A. (2017). Dam removals and river restoration in international perspective. *Water Alternatives* 10(3): 648.

Spink, A., Fryirs, K., and Brierley, G. (2009). The relationship between geomorphic river adjustment and management actions over the last 50 years in the upper Hunter catchment, NSW, Australia. *River Research and Applications* 25(7): 904–928.

Spink, A., Hillman, M., Fryirs, K., et al. (2010). Has river rehabilitation begun? Social perspectives from the Upper Hunter catchment, New South Wales, Australia. *Geoforum* 41(3): 399–409.

Stone, C.D. (1974). *Should Trees Have Standing? Toward Legal Rights for Natural Objects*. Los Altos, CA: Kaufmann.

Storey, R.G. and Wright-Stow, A. (2017). Community-based monitoring of New Zealand stream macroinvertebrates: agreement between volunteer and professional assessments and performance of volunteer indices. *New Zealand Journal of Marine and Freshwater Research* 51(1): 60–77.

Storey, R.G., Wright-Stow, A., Kin, E., et al. (2016). Volunteer stream monitoring: do the data quality and monitoring experience support increased community involvement in freshwater decision making? *Ecology and Society* 21(4). https://doi.org/10.5751/ES-08934-210432.

Strang, V. (2004). *The Meaning of Water*. Oxford: Berg.

Sullivan, J.J. and Molles, L.E. (2016). Biodiversity monitoring by community-based restoration groups in New Zealand. *Ecological Management & Restoration* 17(3): 210–217.

Te Aho, L. (2010). Indigenous challenges to enhance freshwater governance and management in Aotearoa New Zealand: the Waikato River settlement. *Journal of Water Law* 20(5): 285–292.

Thomas, A.C. (2015). Indigenous more-than-humanisms: relational ethics with the Hurunui River in Aotearoa New Zealand. *Social & Cultural Geography* 16(8): 974–990.

Tipa, G.T., Williams, E.K., van Schravendijk-Goodman, C., et al. (2017). Using environmental report cards to monitor implementation of iwi plans and strategies, including restoration plans. *New Zealand Journal of Marine and Freshwater Research* 51: 21–43.

UN Water (2018). Nature-based Solutions for Water: 2018 UN World Water Development Report. UNESCO.

Verweij, M. (2017). The remarkable restoration of the Rhine: plural rationalities in regional water politics. *Water International* 42(2): 207–221.

Waikato River Plan (2018). Sourced from: www.hamilton.govt.nz (accessed 13 June 2018).

Waitangi Tribunal (1993). Interim Report and Recommendation in Respect of the Whanganui River Claim. Sourced from: www.waitangitribunal.govt.nz (accessed 13 June 2018).

Waitangi Tribunal (2011). *Ko Aotearoa tēnei*: a report into claims concerning New Zealand law and policy affecting *Māori* culture and identity. Wellington, New Zealand: Waitangi Tribunal.

Wehi, P.M., Lord, J.M. (2017). Importance of including cultural practices in ecological restoration. *Conservation Biology* 31(5): 1109–1118.

Wilcock, D., Brierley, G., Howitt, R. (2013). Ethnogeomorphology. *Progress in Physical Geography* 37(5): 573–600.

Wilkinson, C., Hikuroa, D.C., Macfarlane, A.H., and Hughes, M.W. (2020). *Mātauranga Māori* in geomorphology: existing frameworks, case studies, and recommendations for incorporating Indigenous knowledge in Earth science. *Earth Surface Dynamics* 8(3): 595–618.

Williams, J. (2006). Resource management and Māori attitudes to water in southern New Zealand. *New Zealand Geographer* 62: 72–80

Wilson, C. (2010). *Ngā hau o tua, ngā ia o uta, ngā rere o tai: Ngā erenga kōrero, kīanga, kupu rehe, whakataukī, whakatauāki, pepeha hoki o Whanganui: A Whanganui reo phrase book: Sayings, phrases & proverbs*. Whanganui, New Zealand: Te Puna Mātauranga o Whanganui.

Wohl, E. (2013). Wilderness is dead: whither critical zone studies and geomorphology in the Anthropocene? *Anthropocene* 2: 4–15.

Yates, J.S., Harris, L.M., and Wilson, N.J. (2017). Multiple ontologies of water: politics, conflict and implications for governance. *Environment and Planning D: Society and Space* 35(5): 797–815.

Young, D. 2013. *Rivers: New Zealand's Shared Legacy*. Auckland: Random House.

Ziegler, R. (2014). Reconciliation with the river: analysis of a concept emerging from practice. *Environmental Values* 23(4): 399–421.

4

Political Ecology and River Restoration

Jamie Linton

Geolab UMR 6042 CNRS, Université de Limoges, Limoges, France

4.1   Introduction

Wohl et al. (2015, p. 5974) identify river restoration as "one of the most prominent areas of applied water-resources science, supporting a multibillion dollar industry across many countries and helping to drive fundamental river research." Given this prominence, it is not surprising, as many researchers – including the contributors to this volume – have pointed out, that river restoration science and its application give rise to numerous questions, problems, controversies, and conflicts as is evidently the case with every environmental issue. This chapter considers how researchers applying a political ecology perspective have approached some of these questions.

People have always modified rivers and streams in accordance with what they perceive as their interests, whether it be by diverting, damming, or dredging the stream or streambed, or by altering the banks. This history of river modification reflects social power; it is one aspect of the well-studied theme tracing the links between water and power, between the control of water and social relations (Wittfogel 1957; Worster 1985; Swyngedouw 2004, 2015). Even though river restoration would seem to be a departure from the historic practice of modifying rivers, consistent with a political ecology approach, this chapter considers policies and practices of river restoration as further examples of how people impact rivers, and how this gives rise to an uneven distribution of benefits and costs (Katz 1992).

River restoration involves a variety of interventions, including bank stabilization, channel reconfiguration, fish passage, in-stream habitat restoration or construction, floodplain and abandoned channel reconnection, dam and weir removal. . . A great diversity of projects that comes under the rubric of river restoration is undertaken at widely different scales and degrees of complexity, and in different environments. The ultimate goals for these projects also have a wide range, including ecosystem restoration, habitat restoration for biodiversity, flood management, sediment management, water quality improvement, ecosystem service provision, and provision of aesthetic and recreational amenities (Emery et al. 2013). If we can generalize from research on European cases, since approximately the

1990s, ecological restoration goals, such as restoring ecosystem functions, have predominated over hydraulic (such as flood management) and landscape priorities (Morandi 2014). Such interventions would appear to be distinct from previous forms of river modification in that they are not ostensibly done for economic gain, for industrial development, or to make rivers part of the infrastructure of the state. Rather, the proponents and agents of river restoration are considered to be acting in the interests of the environment, or the river itself, so as to allow for more natural river processes and functions. River restoration might thus be considered a postindustrial form of river modification, reflecting the ecological knowledge, values, and ethics of typically wealthy, Western, postindustrial societies. However, despite its scientific, ecological basis and rationale, this form of intervention in rivers is no exception to the rule that modifying rivers benefits some people more than, if not to the exclusion of, others. River restoration, like every other aspect of river and water management generally, is political (Linton and Budds 2014). This makes it a suitable subject for political ecology, even if relatively little political ecological research to date has been focused on the question of river restoration. Robertson (2000, p. 464) declared that "Ecological restoration is finding its way timidly into the literature on political ecology and the cultural construction of nature," and according to a recent survey, "Only a minority of political ecology of water works (about fifteen to our knowledge) chose to study a river or its watershed" (Lafaye de Micheaux 2019, p. 99).

The chapter begins (Section 4.2) with a description of political ecology, a growing field of social science that examines the political dimensions of various concepts of nature as well as environmental issues, problems, and controversies. In Section 4.3 we consider some examples of how researchers are using a political ecology approach to examine issues related to river restoration with a focus on cases from the United States. Section 4.4 introduces the ecological continuity of rivers (ECR), which has become an important, if increasingly contested, component of river restoration in Europe as well as in North America. Here we consider how researchers working with political ecology in France are developing approaches to understanding controversies involved in the restoration of ecological continuity.

Political ecology, as discussed in Section 4.2, is fundamentally a practice of critical scholarship: studying the genealogy of environmental concepts, explaining the controversies involved, and analyzing the power relations cultural differences that animate differing positions with respect to environmental issues and conflicts. And yet political ecology also manifests an engaged approach to research, with its practitioners often seeking to improve environmental practices by advancing democratic and socially just solutions to the problems they investigate (Bryant 2015). In light of this, the chapter concludes with an assessment of what political ecology has to offer by way of clarifying what is at stake with differing goals and interpretations of river restoration, and how political ecology might contribute to improved practices of restoring rivers.

4.2 Political ecology: A critical approach to environmental issues

Political ecology can be described as an academic field dedicated to understanding how environmental problems and issues are formulated, represented, studied, and managed. It is practiced by researchers in a variety of disciplines, mainly but not exclusively in the social sciences.[1] Developed for the most part within Anglo-American academia, political ecology is critical in the sense that it stresses how knowledges and practices involving the environment and nature – including scientific knowledges and practices – are not politically and socially neutral; they reflect not only different visions of the world but also the intellectual presuppositions, cultural dispositions, and often the material interests of their proponents (Forsyth 2003). With specific reference to restoration, including river restoration, it has been observed, "Most fundamentally, political ecology is interested in the economic, cultural and political forces, factors and power structures that underlie ecological restoration" (Robertson 2000, p. 464).

Political ecology has been described recently (albeit by some of its own practitioners) as "experiencing a meteoric rise" (Perrault et al. 2015, p. 3). This growth is particularly manifest in Anglophone geography in North America (including Canada), but it is also evident in Australia and Europe, where in the latter case it is developing complementary streams of inquiry, for example in Francophone and Spanish academic traditions. Along with its rise in popularity, political ecology has diversified from an initial focus on criticizing environmental management and environmental discourse (ways of knowing, thinking, and representing) that produces uneven access to resources, and is now applied to a wide diversity of issues falling into the very large category of questions involving the relationship between human society and the nonsocial environment. As the editors of a handbook on political ecology point out, "Even a cursory look at journal titles and conference presentations shows that the label 'political ecology' is applied to research topics as seemingly disparate as water access in India, land grabs in the Amazon, Sahelian pastoralism, lawn care in the United States, fisheries management, wetland markets, indoor air quality, AIDS, and obesity" (Perrault et al. 2015, p. 3).

Robbins (2012a), who has authored what is perhaps the best-known textbook introduction to political ecology, marks an important distinction between what he describes as "a *political* and an *apolitical* ecology. This is the difference," he writes, "between viewing ecological systems as power-laden rather than politically inert; and between taking an explicitly normative approach rather than one that claims the objectivity of disinterest" (Robbins 2012a, p. 13). With theoretical roots in political economy and critical social theory, political ecology can be traced to some founding works in the 1980s and became a recognized field of scholarship in the early 1990s. It has expanded to reflect a wide range of theoretical approaches and subjects of research, characterized by what has been described as "theoretical, conceptual, empirical and methodological eclecticism" (Bryant 2015, p. 20). The geographical scope of its terrain has also shifted, having originally been applied to less developed regions, but now extending to environmental issues in industrialized, Western countries as well.

Methods used by political ecologists are varied and include qualitative and quantitative analysis of data from interviews, surveys, questionnaires, and ethnographic participant observation as well as published work in the social and natural sciences. A basic characteristic of research in political ecology is the foregrounding of the particular social (including economic, political, cultural) dimensions of an environmental issue. With specific reference to environmental restoration, Eden et al. (1999, p. 151) note, "it is difficult to generalize about restoration in principle because its consequences and value are highly contingent in practice. Instead, we need to evaluate each restoration scheme on its own technical and socio-political merits, because each may play a diversity of roles for environmental management and policy."

As noted, while Anglo-American researchers have dominated the field, scholars working in other academic traditions have advanced complementary trends in political ecology. Most relevant for the purposes of this chapter, political ecologists working in France have begun to examine a variety of environmental issues, among which water and river issues have featured prominently.[2] The application of political ecology approaches to water issues in France stems partly from work by Trottier and her students (e.g. Trottier 2008; Trottier and Fernandez 2010; Bouleau and Fernandez 2012; Bouleau 2014; Fernandez 2014;). Furthermore, Blanchon, his students, and colleagues at Paris 10 Nanterre have been influential in bringing a political ecology approach to bear on water issues in France and beyond (e.g. Blanchon and Graefe 2012: Réseau d'Etudes et d'Echanges en Sciences Sociales sur l'Eau 2019). The work of Molle (2008, 2009, 2012) has also been influential in this regard.

Much of this research exemplifies an approach prevalent in political ecology, described by Bouleau (2017, p. 214) in terms of "transcending the opposition between naturalism and constructivism." Another way of putting this is to say that these political ecologists are making an effort to overcome the divisions that arise owing to the constructivist approaches used by critical social scientists (focusing on the environment as discourses, or social representations) and the more naturalist approaches (recognizing phenomena that exist irrespective of how they are represented linguistically or otherwise) employed by natural scientists (Bouleau 2017). In this respect, the project is similar to the initiative, closely associated with political ecology, known as "critical physical geography" (Lave et al. 2014).

Much of the French research discussed in this chapter also pays particular attention to understanding the causes of environmental conflicts and controversies (Gautier and Hautdidier 2015). On this theoretical and empirical basis, and of particular relevance to the subject of this chapter, French political ecology appears to be making a contribution in proposing alternative approaches to environmental problems and controversies. As described by De Sartre et al. (2014, p. 24), "by recognizing both the reality of these problems and the sense in which they are constructed, political ecology opens the door to alternative approaches to these problems – and thus to alternative policies to manage them." This inclination toward offering policy alternatives to help resolve environmental controversies can be identified in work on river restoration done by French political ecologists who argue for broadening the definition of public participation in restoration projects, as discussed in the conclusion of this chapter.

4.3 Political ecology and river restoration

4.3.1 The relevance of political ecology to river restoration

"River restoration is used to describe a variety of modifications of river channels and adjacent riparian zones and floodplains, and of the water, sediment, and solute inputs to rivers. These modifications share the goal of *improving* hydrologic, geomorphic, and/or ecological processes within a *degraded* watershed and replacing *lost, damaged, or compromised* elements of the *natural system*" (Wohl et al. 2015, p. 5974, emphasis added).

Reading this description of river restoration, along with the assertion in Section 4.1 that it supports "a multibillion dollar industry across many countries" (Wohl et al. 2015,

p. 5974), suggests why it is pertinent to examine it from a political ecology perspective. While the definition given might initially seem straightforward and objective, in fact each of the terms in italics is politically charged in the sense that it offers the possibility of being contested by people with different interests or perspectives. The authors themselves declare, "Determining what constitutes improved river conditions is highly subjective. Improvements may focus on protection of property, or esthetic or recreational enhancements that do not necessarily improve ecological functions. Improvements may also focus on creating conditions that are not particularly natural or historically, geomorphically, or ecologically appropriate" (Wohl et al. 2015, p. 5975). By the same token a "degraded watershed" might suggest one thing to a farmer for whom economic conditions determine the reversion of pastureland to scrub, and quite another thing to a hydrologist who studies erosion at the scale of the watershed. What constitutes loss, damage, and compromise is equally contestable, as is – perhaps most critically – the notion of a "natural system."

As already noted, political ecology is a fundamentally critical approach. Its value is in subjecting hegemonic (or dominant) environmental ideas and concepts to critical scrutiny in order to analyze their cultural presuppositions, their social construction, and their social impacts, including their political implications. As noted by Blot and Besteiro (2017, quoting Robbins 2012b), political ecologists "must produce texts which disturb, which render dominant understandings fragile and which 'insist on contradictions and paradoxes'." As an increasingly important field of study, and practice, it is not surprising that various aspects of river restoration have come into the focus of political ecologists. As argued by Fox et al. (2016, pp. 101–102), "Because political ecology focuses attention on complex cultural dynamics and divergent interpretations of nature – issues that have traditionally fallen outside the concerns of ecological restoration – it is a crucial lens for understanding the environmental politics that characterize interventions into highly altered ecosystems."

For the Anglophone literature, Fox et al. (2016) identify a dozen or more recent publications that focus on questions such as the social factors involved in setting river restoration objectives, perceptions of what constitutes successful river restoration, the place of science in river restoration, the role of indigenous communities and indigenous knowledge in river restoration projects, environmental discourse and river restoration, and "the nature-culture interface in river restoration projects" (Fox et al. 2016, p. 95).

4.3.2 Whose interpretation of the river counts?

Reflecting this list of questions, a persistent theme in American and other Anglophone political ecological research is the tendency for river restoration objectives or goals to be set based on abstract standards, norms and methods, often packaged as "toolboxes" and "best practices for planning and implementing river restoration projects" (Perry 2009, p. 13). These often have been found irresponsive local communities' understandings of the health and state of rivers. For example, writing of river projects in New Zealand, Tadaki et al. (2014) argue that river monitoring methods, which provide a framework for restoration projects, are conceived and applied to provide abstract national and international comparisons that are of little value in assigning local social priorities for restoration.

Political ecologists studying river restoration in urban settings have identified similar issues: considering the restoration of the Lower Neponset River in metropolitan Boston,

Massachusetts, Perry (2009, p. 6) notes: "Restoration plans were developed primarily by scientists, engineers, and professional planners. Local citizens were not surveyed to find out what 'restoration' meant to them." She recommends that river managers and decision makers should recognize local citizens as "not only 'stakeholders' representing a 'special interest', but as equal partners with specialized knowledge and interpretations about the local environment" (Perry 2009, p. 14). Thus, instead of merely being consulted or providing input on technical proposals made by restoration experts, Perry and other political ecologists suggest that the very category of expert be expanded to include nonscientists and nonmanagers. Similarly, Weng (2015) investigates the expert–lay hierarchy that characterizes ecological restoration projects in the United States involving citizen science/volunteers, showing how "contrasting visions of science between professional practitioners and volunteers led to conflicts and presented challenges for the institutions to genuinely engage the public in contributing local knowledge and framing management priorities" (Weng 2015, p. 134).

A key theme emerging from this and other research in the political ecology tradition is that different people may have very different "interpretations" of what a river is and what it means, and this implies "different, and sometimes conflicting visions for restoring" rivers (Perry 2009, p. 101). Within this tradition, perhaps the most pertinent question raised is: whose interpretation of the river and its problems counts? Almost invariably, as research in political ecology shows, the interpretation that counts is aligned with dominant social and economic interests. A non-Western example is given by Lee (2015), who has shown how the South Korean "Four Major Rivers Restoration Project," completed in 2012, constitutes what political ecologists describe as an "environmental fix,"[3] by which environmental protection and restoration projects are used as an accumulation strategy (for economic growth). Despite the title, Lee (2015, pp. 353–354) writes, "this Project does not restore the river ecosystems in question but handsomely benefits the politically well-connected construction corporations involved," which is described as "a powerful regional class coalition keen to profit from either the large-scale construction activities involved in river 'restoration' or post-redevelopment river basin business opportunities prevailed."

4.3.3 Identifying underlying economic (structural) determinants of river restoration

Even when river restoration projects are aimed at restoring ecosystem health, political ecology analysis directs attention to the structural (economic) circumstances that make a given interpretation of river restoration salient and compelling. Thus, Fox et al. (2016, p. 96) call for further research showing "how purportedly scientifically designed restoration approaches are always already expressions of social power."

To cite a well-known example of this type of research, Lave et al (2014, p. 4) have explored various aspects of "the interrelated neoliberalization of environmental science and management in stream restoration" in the United States (see also Lave et al. 2010; Lave 2012, 2014, 2015). Considering the new "science regime"[4] as it pertains to river restoration in the United States, Lave and others have shown how river restoration is increasingly dominated by commercial interests and corporate science. "The neoliberal shift towards valorization of privately-produced and commercially applicable knowledge claims" is identified as the main underlying factor in this domination (Lave 2014, p. 242). In a series of papers, Lave has examined the river restoration technique known as "natural channel design" (NCD) developed by Dave Rosgen, who is described as "the most influential person in the American stream restoration field" (Lave 2014, p. 242), revealing its commercial roots. While answering managers' needs to meet regulatory requirements for environmental quality, NCD often produces unsatisfactory results, yet has successfully resisted efforts by academic hydrologists and fluvial geomorphologists to debunk its tenets. "Academics," she argues, "have been notably ineffective at delegitimizing these knowledge claims, in the end changing their own practices rather than Rosgen's . . . This suggests a future of increasing marginalization for academics, sidelined by the claims of commercially interested, private-sector knowledge producers" (Lave 2014, p. 249).

The application of the NCD method contributes to what Lave et al. (2010) identify as "the privatization of stream restoration" in the United States. They note how stream restoration has "catalyzed the development of a substantial market in consulting services to design, implement, and manage restoration projects" (Lave et al. 2010, p. 688). But more fundamentally, they demonstrate "how the particular state and market logics of neoliberalism are shifting both the practice of restoration scientists and the relations between public and private sector science." They argue that neoliberal environmental management regimes have given rise to applied science that can be delivered as a standardized package, be used by agencies to justify decisions, and form the basis for markets in ecosystems (Lave et al. 2010, p. 678).

4.3.4 Whose knowledge counts?

A related theme developed by researchers in political ecology deals with the question of whose knowledge informs projects of river restoration. While ecological restoration as a popular form of citizen participation is often praised for incorporating lay and local knowledges, political ecologists have long examined the politics of participation, asking questions such as how participation is defined, what constitutes the community, or the stakeholders, who decides project goals and implementation strategies, and how procedures of citizen participation are led. From a political ecology perspective, "most studies of the participatory aspect of restoration naively portray ecological restoration as a communal endeavor without addressing the contrasting visions and power dynamics among participants" (Weng 2015, p. 136).

In general, political ecology critically problematizes citizen participation in environmental management. To cite an example, Eden and Tunstall (2006) draw from cases of urban river restoration in England to conclude that, despite much lip service paid to citizen participation in the conception and application of river restoration, in the final analysis, these projects mobilize scientific and technocratic knowledge to the exclusion of local knowledge, treating the public as if they had a knowledge deficit which is in need of fulfillment by experts. "[A]lthough restorationists may be seen as radical in scientific and policy terms because of their challenge to the tradition of the 'hard engineering' of rivers, they are not radical in social science terms because they fail to challenge the tradition of technocratic environmental management of the public and its deficit model" (Eden and Tunstall 2006, p. 661). To provide another example, Weng (2015) investigated the role natural science played in expert–volunteer dynamics at two university arboreta in the American Midwest from 2006 to 2008. Focusing on the role science plays in expert–volunteer dynamics, she found that "the contrasting visions of science between professional practitioners and volunteers" stymied genuine engagement of the public in contributing local knowledge and framing management priorities (Weng 2015, p. 134). Weng's conclusion provides an example of political ecology's frequent call for critical reflection in order to reconstruct more democratic practices of environmental management: "Only by critically examining how different knowledges, values, and practices are shared and exchanged can we begin to theorize ecological restoration as a democratic practice, in which all stakeholders are engaged equally and throughout the restoration processes" (Weng 2015, p. 143).

4.3.5 Whose "nature" counts?

As for their own work, Fox et al. (2016) examine conflicts over dam removal projects in northeastern United States, asking what these conflicts reveal about the politics of river restoration. Their research contributes to that of others cited in this chapter in arguing that more thought should be given to "the cultural and political-economic processes that are often far-removed from restoration advocates' ecological goals, but nevertheless drive restoration activities" (Fox et al. 2016, p. 95). Whereas, for Lave et al. (2010), these processes involve the structural imperatives of neoliberalism, Fox et al. (2016) show how they involve people's attachments to historical landscapes altered by dams, and how these attachments "are critically important to understanding anti-dam sentiments in New England" (Fox et al. 2016, p. 99). Their research highlights the importance of recognizing different understandings of the natural and how these different understandings contribute to environmental conflict: "Often the ecosystems that people become attached to could be described as 'artificial ecosystems' and include non-native species . . . In all of these cases, who gets to interpret what is a 'better' environment is an open question with potentially high socio-ecological stakes and which reinforces the divergence of views on nature in the early 21st century" (Fox et al. 2016, p. 100).

Political ecology highlights this divergence, explaining how it can give rise to conflict between people over what functions or characteristics of a river ought to be restored. Most of the examples given so far in this chapter have come from the United States, where political ecology has been practiced for decades. Section 4.4 gives an example drawn from Europe, focusing on France, where the restoration of ecological continuity has been examined by researchers working in political ecology.

4.4 Restoring the ecological continuity of rivers: Controversies involving different meanings of the river

River restoration occupies an important and growing place in Europe, including in France (Morandi et al. 2016), where perhaps the most relevant policy is known as the ecological continuity of rivers (ECR). Consistent with the European Union (EU) Water Framework Directive (WFD), ECR in France has been described as the "concept (and project) aimed at promoting the circulation of certain aquatic organisms and the transport of sediments within . . . rivers" (Perrin 2018, p. 9). According to some observers, the promotion of eco-logical continuity has become "the flagship tool for achieving good ecological status of rivers" in France, as called for by the EU WFD (Germaine and Barraud 2017, p. 18), and even "the panacea of the policy of river restoration in France" (Bravard 2017, p. 10).

In the name of ECR, France has developed an ambitious program to demolish thousands of small dams and weirs, usually but not always associated with the historic legacy of watermills (Germaine and Barraud 2017). However, this initiative has been met with fierce and largely unexpected opposition from owners and defenders of watermills, people living or owning property along rivers, producers or would-be producers of small-scale hydroe-lectricity, local fishing associations, local and national politicians including many members of parliament and hundreds of mayors, and senior and well-respected water scientists. For example, researchers working in political ecology have detailed how opponents have employed a variety of arguments and strategies to impede implementation of ECR (Germaine and Barraud 2013a; Barraud 2017, Barraud and Le Calvez 2017; Le Calvez 2017; Perrin 2018). "[F]rom a 'political ecology' perspective," writes Barraud (2017, pp. 796–797), "such conflict reveals a yawning gap between social representations and the value systems of experts, local managers and the local population. Most studies highlight local explana-tory drivers of the opposition to dam dismantling (attachment to places, history). They also show the asymmetry of power that may exist between users and experts – the latter often perceived as external to the local scene."[5]

Much of this research reflects the qualities identified in Section 2.2 that characterize emerging themes in political ecology, particularly in France. First, there is emphasis on gaining an understanding of the controversies surrounding this issue and explaining why ECR has become so divisive. Second, this understanding is gained through an approach that transcends constructivism and realism, through an appreciation of rivers as hybrid (both social and natural) objects. And third, there is an effort by French political ecologists to move from the understanding gained through critical analysis toward suggestions for policy alternatives that might help resolve the controversies.

The emphasis on transcending constructivism and realism is perhaps the consequence of studying a terrain that has been radically altered by people through impounding, diverting, and exploiting rivers and streams, most notably for the kinetic energy they afford but also for flood protection, navigation, and irrigation. Especially to those with an appreciation of their history, rivers in France and other parts of Europe (Nones 2016) are patently hybrid objects, internalizing natural and social processes that have been underway for hundreds if not thousands of years (Lespez 2012). This makes any appeal to an ahistorical natural sta-tus as a basis or reference for restoring rivers extremely complicated (Lespez et al. 2015; Dufour and Piégay 2009; Bouleau and Pont 2014, 2015). In fact, it is argued that at least in northwestern France, from the early Middle Ages to the early modern period, the influence of people operating mills and dams produced "socio-environments where an equilibrium was maintained by societies for more than a millennium" (Lespez et al. 2017, p. 38).

As the analysis is brought forward in time, an approach rooted in political ecology allows for diachronic study of how changing meanings and values of rivers, mills, and weirs artic-ulate with state policy (Barraud 2017). Indeed, while almost all of the mills themselves have ceased commercial operation, the associated hydraulic works have proven resilient, and have gone through several cycles of meaning over the last few centuries, belying an ahistorical appraisal of such structures as merely obstacles to ecological continuity: dams and weirs were regarded as an intrinsic part of what has been described as the "productive river" (Lespez et al. 2015) in the late eighteenth and the nineteenth centuries. Thus, as dams – along with overfishing and industrial pollution – threatened migratory fish, laws were passed beginning in the mid-nineteenth century, requiring fish passes on some new dams. However, the dams and weirs themselves were never the issue. Rather, within the paradigm of the productive river, fish culture was the policy of choice, with rivers regarded, much like agricultural fields, as being capable of increased production by means of modern technique (Barthélémy 2013). The development of hydroelectricity in the late nineteenth and early twentieth centuries gave new meaning to the productive river, and imbued dams and weirs with a new kind of value. Nevertheless, as electricity itself contributed to the decline of watermills, weirs and millponds were left untended, forming what many regarded as a hazard to rational hydraulic control. Demolishing and cleaning up unused weirs thus became aspect of modernizing the river (Morandi et al. 2016).

A process of requalification began in the 1970s, whereby old watermills began to be renovated as residences and commercial properties, or for the production of micro-hydro production. Since the 1990s, the conversion-to-heritage of fluvial infrastructure, especially old mills, has been marked, as evidenced by the proliferation of touristic "mill routes" (Germaine and Barraud 2013a, p. 378). Increasingly, local elected officials have reinterpreted mills and millponds as a territorial resource (tourism, heritage, living environment). The controversy over ECR is sustained by their latest, current, cycle of meaning(s) – whereby these works are considered a form of cultural heritage as well as a threat to the ecological integrity of the river (Germaine and Barraud 2013b). Nevertheless, as emphasized in political ecology and historical research on this theme, this is only the latest chapter of what is fundamentally the same story of conflict and controversy that has accompanied every stage of the long history of rivers featuring mills and dams (Barraud 2017). "We have shown that this controversy did not simply emerge ten years ago with the introduction of the concept of ecological continuity . . . The controversy developed from a combination of the legacies of land use and ideas" (Barraud 2017, p. 814).

As noted above, political ecologists are interested in studying how certain policies are underlain by particular constructions of nature or of what is natural. ECR is based on a way of defining the nature of rivers that favors free-flowing watercourses, which in turn favor certain fish species, notably migratory fish including trout and salmon (these species known in France as the "noble fish"). The nonmigratory species (known less nobly as "white fish") thrive in the stiller and deeper waters, such as are formed by dams. But while these fish, and the ecosystems in which they thrive, are just as natural as the so-called noble fish, they appear to have less cultural value. As Bouleau (2017, p. 221) notes: "The reality of rivers does not easily fit into the classifications established by public policies. In French rivers, there is a hybrid reality shaped in part by fishermen, maintained by public policies favorable to migratory fish (in which fishermen participate), restored thanks to actions taken to comply with the EU WFD and in partly unexpected because natural evolution is not entirely predictable." The irony in this, as pointed out by Bouleau and other researchers, is that one of the arguments contributing to the critique of ECR comes from fishing associations, particularly those associated with the practice of still-water fishing. These groups have successfully argued that well managed and maintained mill dams and weirs support healthy ecosystems that are not only beneficial to fish but are conducive to the maintenance of riverbanks and the oxygenation of the water (Barraud and Le Calvez 2017, p. 138).

4.5 Conclusions: What does political ecology have to offer river restoration?

Many authors have observed that it is hard to generalize about river restoration, because its precise definition, its practices, and its consequences are highly variable and contingent (Eden et al. 1999; Wheaton et al. 2008; Emery et al. 2013; Morandi 2014). Indeed, "many recent river management interventions have been presented under the rubric strategy, or philosophy of 'restoration'" (Emery et al. 2013, p. 168). Identifying the various interests at play and analyzing the cultural factors and power relations involved in these diverse interventions is the task of political ecology. Indeed, political ecology is ideally suited to analyzing "the ambiguous definition of restoration and its ideological usage to relate any number of environmental interventions to virtuosity" (Emery et al. 2013, p. 174). So far, this chapter has looked at how political ecologists have critically investigated a variety of cases involving such interventions.

Beyond exposing and analyzing social (i.e. economic, political, and cultural) factors that come into play in any case of river restoration, we want to conclude by considering what political ecology might have to offer by way of improving or contributing to better forms and practices of river restoration. Fox et al. (2016, p. 94) argue a political ecology approach "facilitates a more critical examination of the political and social dimensions of ecological restoration by underscoring hitherto less visible aspects of [restoration]." Arguably, making visible these often-occluded "political and social dimensions" can contribute ultimately to more successful projects. At the very least, political ecology highlights how the technological and environmental science aspects of river restoration projects have political and cultural dimensions that need to be considered for any such project to be considered successful.

As a means of helping ensure such success, most political ecologists would add that the process of river restoration should be fully democratic in the sense of involving a broader range of knowledges, interests, and perspectives than are usually implicated in these projects. The aim is to ensure the involvement of those who are not scientists or administrators or representatives of associations or syndicates but who nevertheless have a stake in a given restoration project by merely living in or nearby the affected river basin. For some obvious reasons, this can help ensure the realization of river restoration projects as the need "to make river restoration a more democratically accountable process . . . may also be needed because of the powers that exist within communities to halt restoration projects" (Wohl et al. 2015, p. 5984). Developing these processes, they argue, "will require a deeper involvement of the social sciences and humanities in river restoration research" (p. 5985).

Furthermore, as we have argued in this chapter, while a subject such as river restoration is presumed by many to be essentially a matter of scientific or technical expertise, political ecologists regard it as essentially a matter of politics, culture, and social relations. And as with any issue involving water management, many political ecologists look for opportunities to improve social relations in what otherwise appears to be an environmental, technical issue (Linton 2015). As Light (2000, pp. 163–164) contends: "The practice of eco-logical restoration contains an inherent democratic potential. By this claim, I meant that at its best the activity of ecological restoration preserves the democratic ideal that public par-ticipation in a public activity increases the value of that activity. This value in restoration is brought out most effectively by those projects that unite local human and natural commu-nities, and that increase the level of local participation in those restoration projects."

Nevertheless, despite the need and the potential, a frequently recognized shortcoming of river restoration projects, at least in the Anglophone literature, has been the failure to effec-tively incorporate the nonscientific and nonadministrative community in planning and implementation (Eden and Tunstall 2006; Eden and Bear 2011; Weng 2015). Even where citizen participation has been a recognized component of ecological restoration, it has been found that where projects are guided by professionals the involvement of nonexperts and volunteers "does not necessarily ensure democratic knowledge exchange and production" (Weng 2015, p. 134).

Similar shortcomings are recognized by political ecologists working in non-Anglophone traditions. As noted by Germaine and Barraud (2013b), despite the EU WFD's encourage-ment to do so, projects of river restoration in France hardly involve the active participation of the public: "Local actors are in a paradoxical situation. The state encourages local authorities to take charge of water management . . . But, at the same time injunctions com-ing from the French state (such as prioritizing lists of hydraulic works for demolition) increasingly constrain [democratic] initiatives taken at the local level" (Germaine and Barraud 2013b, p. 379). Their recommendation is to stop thinking of rivers and valley bot-toms as natural but rather "as geographical objects humanized, constructed, represented and inhabited" and thus moving "from the question of ecological continuity to the develop-ment of real territorial projects" (Germaine and Barraud 2013b, p. 382).

In France, territorial projects by definition involve the participation of local stakehold-ers in their planning. Further elaboration of what a territorial project involving river res-toration might look like is found in two recent theses on the topic from researchers in France who identify with the political ecology tradition. In her thesis whose title trans-lates as "River users confronted with the restoration of ecological continuity in the Brittany region," Le Calvez (2017) shows how ECR is based on a particular view of nature that is radically different from that of users and indeed of many citizens in general. Rivers are often apprehended as lived and familiar spaces – with which affective and practical attach-ments are formed – as opposed to the objective, abstract phenomena rendered and acted upon by scientists and administrators implementing the policy of ECR. While tools for public participation (mediation, translation, etc.) are suggested as a means to resolve the contradiction, Le Calvez (2017) admits that this, in itself, is inadequate in a situation involving such radically different ways of perceiving rivers. What is required is to requalify rivers so as to accommodate both types of perception at the outset of any project of resto-ration. She suggests this can be accomplished through "debate" involving "a broad inclu-sion of actors and the broader population" (Le Calvez 2017, p. 337). Making a similar argument, Perrin (2018) observes that, in most projects to restore ecological continuity, local residents express a desire that nonscientific appraisals of the river play a role in defining a territorial project rather than restoring something that might never have actu-ally existed. "This," he notes, "raises the question of the organization of procedures, the public to be invited to meetings and, more generally, the perimeter of the territory" (Perrin 2018, p. 263). Following draws of Kondolf and Pinto (2017), Perrin's recommendation is to rethink the territory of river governance in terms of the "social connectivity" that the river facilitates. Thus an ad hoc "water territory" might be defined by "mapping the social connectivity" manifested by issues involving the river and "studying the areas from which the participants come for public meetings, public inquiries or petitions or protests . . . Thinking first about social connectivity and not ecological continuity would in fact make it possible to expand the public to other basins, perimeters and politico-administrative structures" (Perrin 2018, pp. 266–267).

The precise means by which such a project of reterritorialization might be put into practice would need to be further defined and tailored to the relevant socio-natural circumstances. But, in conclusion, the point we want to stress is that, through the contribution of such ideas, arising from critical social science assessment, political ecology can contribute to more successful river restoration projects as well as merely bringing to light the social and political dimensions of what are often simply understood as scientific and technical environmental issues.

Notes

1 The physical geographers who associate themselves with the multidisciplinary initiative known as "critical physical geography" would be one exception (Lave et al. 2014).

2 It should be pointed out that a different project – known as "*écologie politique*" – has long been underway in France, predating the adoption of "political ecology" by the French researchers discussed in this chapter (see Chartier and Rodary 2015). This *écologie politique* comprises both intellectual (especially philosophical and sociological) and militant aspects, and is associated with what Anglophones describe as the "environmental movement." As described by Chartier and Rodary, "the notion of *écologie politique* makes reference first and foremost to streams of thought and politics that aim to incorporate ecological issues into political, economic and social action. Until recently, it was backed mainly by the Green Party. . ." (2015, p. 550). It was in the 2000s that researchers in France began to "elaborate a French political ecology; mainly under the English title 'political ecology' that takes explicit cognizance of Anglo-American scholarship" (Gautier and Hautdidier 2015, p. 64; see also Castro-Larranaga 2009; Gautier and Benjaminsen 2012; Blanchon and Graefe 2012; De Sartre et al. 2014).

3 "Environmental fix" is a term used to describe ways that powerful actors in a capitalist system profit from environmental issues and problems. This often involves the creation or production of the problem itself. Environmental fix has been described as "the strategies of capital that are designed to expand or reorganize the subsumption of nature in order to overcome accumulation crises" (Lee 2015, pp. 346–347).

4 The concept of "science regime," taken from D. Pestre's concept of "regimes of knowledge," describes the regime of external political, institutional, and economic forces that impinge on scientific practice in a particular historical instance. Far from an independent sphere of activity, these forces have always shaped scientific practice. Pestre shows how "for at least the last five centuries, what we now call scientific knowledge – be it characterized as pure or

applied . . . has been of crucial interest to the political and economic powers since knowledge led to the material and social techniques of control" (quoted in Lave 2015, p. 245).

5 A special issue of *Water Alternatives* on the theme "Dam removals and river restoration in international perspective" is the product of a workshop held in France in 2016 (Sneddon et al. 2017). The issue reports a series of case studies, all of which "either explicitly or implicitly intersect with recent work in political ecology" (Sneddon et al. 2017, p. 651) to reveal the conflicting science and competing interests that often accompany dam-removal projects. See also Barraud and Germaine (2017).

References

Barraud, R. (2017). Removing mill weirs in France: the structure and dynamics of an environmental controversy. *Water Alternatives* 10: 796–818.

Barraud, R. and Germaine, M.-A. (eds) (2017). *Démanteler les barrages pour restaurer les cours d'eau: controverses et représentations*. Versailles, France: Éditions Quae.

Barraud, R. and Le Calvez, C. (2017). S'opposer aux projets de démantèlement d'ouvrage: rhétorique, valeurs et vision de l'espace. In: *Démanteler les barrages pour restaurer les cours d'eau: controverses et représentations* (eds R. Barraud and M.-A. Germaine), 129–142. Versailles, France: Éditions Quae.

Barthélémy, C. (2013). *La pêche amateur au fil du Rhône et de l'histoire: usages, savoirs et gestions de la nature*. Paris: L'Harmattan.

Blanchon, D. and Graefe, O. (2012). La radical political ecology de l'eau à Khartoum: une approche théorique au-delà de l'étude de cas. *L'Espace Géographique* 1: 35–50.

Blot, F. and Besteiro, A.G. (2017). Francophone geography's contribution to political ecology: two studies of the relationship between societies and underground water in semiarid Spain. *L'Espace Géographique* 3(46): 193–213.

Bouleau, G. (2014). The co-production of science and waterscapes: the case of the Seine and the Rhône rivers. *Geoforum* 57: 248–257.

Bouleau, G. (2017). La catégorisation politique des eaux sous l'angle de la political ecology: le patrimoine piscicole et la pollution en France. *L'Espace Géographique* 46: 214–230.

Bouleau, G. and Fernandez, S. (2012). La Seine, le Rhône et la Garonne: trois grands fleuves et trois représentations scientifiques. In: *Environnement, discours et pouvoir: L'approche Political Ecology* (eds D. Gautier and T.A. Benjaminsen), 201–217. Versailles, France: Éditions Quae.

Bouleau, G. and Pont, D. (2014). Les conditions de référence de la directive cadre européenne sur l'eau face à la dynamique des hydrosystèmes et des usages. *Natures, Sciences, Sociétés* 22: 3–14.

Bouleau, G. and Pont, D. (2015). Did you say reference conditions? Ecological and socio-economic perspectives on the European Water Framework Directive. *Environmental Science and Policy* 47: 32–41.

Bravard, J.-P. (2017). Preface. In: *Démanteler les barrages pour restaurer les cours d'eau: controverses et représentations* (eds R. Barraud and M.-A. Germaine), 9–12. Versailles, France: Éditions Quae.

Bryant, R.L. (2015). Reflecting on political ecology. In: *The International Handbook of Political Ecology* (ed. R.L. Bryant), 14–24. Cheltenham: Edward Elgar.

Castro-Larranaga, M. (2009). Nouvelles questions, nouveaux défis: réponses de la "political ecology." *Natures Sciences Sociétés* 17(1): 12–17.

Chartier, D. and Rodary, E. (2015). Globalizing French écologie politique: a political necessity. In: *The International Handbook of Political Ecology* (ed. R.L. Bryant), 547–560. Cheltenham: Edward Elgar.

De Sartre, X.A., Castro, M., Dufour, S., et al. (2014). *Political Ecology des services écosystémiques*. Brussels: Peter Lang.

Dufour, S. and Piégay, H. (2009). From the myth of a lost paradise to targeted river restoration: forget natural references and focus on human benefits. *River Research and Applications* 25(5): 568–581.

Eden, S. and Bear, C. (2011). Reading the river through "watercraft": environmental engagement through knowledge and practice in freshwater angling. *Cultural Geographies* 18(3): 298–314.

Eden, S. and Tunstall, S.M. (2006). Ecological versus social restoration? How urban river restoration challenges but also fails to challenge the science–policy nexus in the United Kingdom. *Environment and Planning C: Government and Policy* 24: 661–680.

Eden, S., Tunstall, S.M., and Tapsell, S.M. (1999). Environmental restoration: environmental management or environmental threat? *Area* 31(2): 151–159.

Emery, S.B., Perks, M.T., and Bracken, L.J. (2013). Negotiating river restoration: the role of divergent reframing in environmental decision-making. *Geoforum* 47: 167–177.

Fernandez, S. (2014). Much ado about minimum flows. . . unpacking indicators to reveal water politics. *Geoforum* 57: 258–271.

Forsyth, T. (2003). *Critical Political Ecology: The Politics of Environmental Science*. London and New York: Routledge.

Fox, C.A., Magilligan, F.J., and Sneddon, C.S. (2016). "You kill the dam, you are killing a part of me": dam removal and the environmental politics of river restoration. *Geoforum* 70: 93–104.

Gautier, D. and Benjaminsen T.A. (2012). Introduction à la political ecology. In: *Environnement, Discours et Pouvoir: L'approche Political Ecology* (eds D. Gautier and T.A. Benjaminsen), 5–19. Versailles, France: Éditions Quae.

Gautier, D. and Hautdidier, D. (2015). Connecting political ecology and French geography: on tropicality and radical thought. In: *The International Handbook of Political Ecology* (ed. R.L. Bryant), 57–86. Cheltenham: Edward Elgar.

Germaine, M.-A. and Barraud, R. (2013a). Restauration écologique et processus de patrimonialisation des rivières dans l'ouest de la France. *VertigO, Revue électronique en sciences de l'environnement* 16. https://doi.org/10.4000/vertigo.13583.

Germaine, M.-A. and Barraud, R. (2013b). Les rivières de l'ouest de la France sont-elles seulement des infrastructures naturelles? Les modèles de gestion à l'épreuve de la directive-cadre sur l'eau. *Natures Sciences Sociétés* 21: 373–384.

Germaine, M.-A. and Barraud, R. (2017). Introduction. In: *Démanteler les barrages pour restaurer les cours d'eau: controverses et représentations* (eds R. Barraud and M.A. Germaine), 13–23. Versailles, France: Éditions Quae.

Katz, E. (1992). The ethical significance of human intervention in nature. *Restoration and Management Notes* 9: 90–96.

Kondolf, G. M. and Pinto P.J. (2017). The social connectivity of urban rivers. *Geomorphology* 277: 182–196.

Lafaye de Micheaux, F. (2019). Political Ecology of a Sacred River: Hydrosocial Cycle and Governance of the Ganges. Doctoral thesis, Faculté des géosciences et de l'environnement, Institut de géographie et durabilité de l'Université de Lausanne.

Lave, R. (2012). Bridging political ecology and STS: a field analysis of the Rosgen Wars. *Annals of the Association of American Geographer* 102(2): 366–382.

Lave, R. (2014). Freedom and constraint: generative expectations in the US stream restoration field. *Geoforum* 52: 236–244.

Lave, R. (2015). The future of environmental expertise. *Annals of the Association of American Geographers* 105(2): 244–252.

Lave, R., Doyle M., and Robertson, M. (2010). Privatizing stream restoration in the US. *Social Studies of Science* 40(5): 677–703.

Lave, R., Wilson, M.W., Barron, E.S. et al. (2014). Intervention: critical physical geography. *The Canadian Geographer* 58: 1–10.

Le Calvez, C. (2017). *Les usagers confrontés à la restauration de la continuité écologique des cours d'eau: Approche en région Bretagne*. Doctoral thesis. Université Rennes, Haute-Bretagne.

Lee, S. (2015). Assessing South Korea's green growth strategy. In: *The International Handbook of Political Ecology* (ed. R.L. Bryant), 345–358. Cheltenham: Edward Elgar.

Lespez, L. (ed.) (2012). *Paysages et gestion de l'eau: sept millénaires d'histoire des vallées en Normandie*. Caen, France: Presses Universitaires de Caen.

Lespez, L., Beauchamp, A., Germaine, M.-A., et al. (2017). De l'aménagement au déménagement: les temps de l'environnement des systèmes fluviaux ordinaires de l'ouest de la France. In: *Démanteler les barrages pour restaurer les cours d'eau: controverses et représentations* (eds R. Barraud and M.A. Germaine), 27–42. Versailles, France: Éditions Quae.

Lespez, L., Viel, V., Rollet, A.-J., et al. (2015). The anthropogenic nature of present-day low energy rivers in western France and implications for current restoration projects. *Geomorphology* 251: 64–76.

Light, A. (2000). Restoration, the value of participation, and the risks of professionalization. In: *Restoring Nature: Perspectives from the Social Sciences and Humanities*, (eds P.H. Gobster and R.B. Hull), 163–181. Washington DC: Island Press.

Linton, J. (2015). Introduction: water as a social opportunity. In: *Water as a Social Opportunity*, (eds S. Davidson, J. Linton, and W. Mabee), 1–13. Montreal, Canada: McGill-Queen's University Press.

Linton, J. and Budds, J. (2014). The hydrosocial cycle: defining and mobilizing a relational-dialectical approach to water. *Geoforum* 57: 170–180.

Molle, F. (2012). La gestion de l'eau et les apports d'une approche par la political ecology. In: *Environnement, discours et pouvoir: L'approche Political Ecology* (eds D. Gautier and T.A. Benjaminsen), 219–238. Versailles, France: Éditions Quae.

Molle, F. (2008). Nirvana concepts, narratives and policy models: insight from the water sector. *Water Alternatives* 1(1): 131–156.

Molle, F. (2009). River-basin planning and management: the social life of a concept. *Geoforum* 40: 484–494.

Morandi, B. (2014). *La restauration des cours d'eau en France et à l'étranger: de la définition du concept à l'évaluation de l'action*. Doctoral thesis. Université de Lyon. ENS de Lyon.

Morandi, B., Piégay, H., Johnstone, K., et al. (2016). Les Agences de l'eau et la restauration: 50 ans de tensions entre hydraulique et écologique. *VertigO – la revue électronique en sciences de l'environnement* 16(1). https://doi.org/10.4000/vertigo.17194.

Nones, M. (2016). River restoration: the need for a better monitoring agenda. Conference paper presented at the 13th Int. Symposium on River Sedimentation. Stuttgart, Germany (September 2016).

Perrault, T., Bridge, G., McCarthy, J. (eds) (2015). *The Routledge Handbook of Political Ecology*. London and New York: Routledge.

Perrin, J.-A. (2018). *Gouverner les cours d'eau par un concept: étude critique de la continuité écologique des cours d'eau et de ses traductions*. Doctoral thesis. Université de Limoges.

Perry, S.L. (2009). More than one river: local, place-based knowledge and the political ecology of restoration and remediation along the Lower Neponset River, Massachusetts. Doctoral thesis. University of Massachusetts at Amhurst.

Réseau d'Études et d'Échanges en Sciences Sociales sur l'Eau (2019). Université Paris 10, Nanterre. https://reseaux.parisnanterre.fr (accessed 2 July 2019).

Robbins, P. (2012a). *Political Ecology: A Critical Introduction* (second edition). Malden, MA: Wiley-Blackwell.

Robbins, P. (2012b) Qu'est-ce que la *political ecology*? In: *Environnement, discours et pouvoir: L'approche Political Ecology*, (ed. D. Gautier and T.A. Benjaminsen), 21–36. Paris: Éditions Quae.

Robertson, M.M. (2000). No net loss: wetland restoration and the incomplete capitalization of nature. *Antipode* 32(4): 463–493.

Sneddon, C., Barraud, R., and Germaine, M.-A. (2017). Dam removals and river restoration in international perspective. *Water Alternatives* 10(3): 648–654.

Swyngedouw, E. (2004). *Social Power and the Urbanization of Water: Flows of Power*. Oxford: Oxford University Press.

Swyngedouw, E. (2015). *Liquid Power: Water and Contested Modernities in Spain, 1898–2010*. Cambridge, MA: The MIT Press.

Tadaki, M., Brierley, G., and Fuller, I.C. (2014). Making rivers governable: ecological monitoring, power and scale. *New Zealand Geographer* 70: 7–21.

Trottier, J. (2008). Water crisis: political construction or physical reality? *Contemporary Politics* 14(2): 197–214.

Trottier, J. and Fernandez, S. (2010). Canals spawn dams? Exploring the filiation of hydraulic infrastructure. *Environment and History* 16(1): 97–123.

Weng, Y.-C. (2015). Contrasting visions of science in ecological restoration: expert–lay dynamics between professional practitioners and volunteers. *Geoforum* 65: 134–145.

Wheaton, J.M., Darby, S.E., and Sear, D. (2008). The scope of uncertainties in river restoration. In: *River Restoration: Managing the Uncertainty in Restoring Physical Habitat* (eds S.E. Darby and D. Sear), 21–39. Chichester: Wiley.

Wittfogel, K.A. (1957). *Oriental Despotism: A Comparative Study of Total Power*. New Haven, CT: Yale University Press.

Wohl, E.E., Lane, S.N., and Wilcox, A.C. (2015). The science and practice of river restoration. *Water Resources Research* 51(8): 5974–5997.

Worster, D. (1985). *Rivers of Empire: Water, Aridity, and the Growth of the American West*. New York: Pantheon Books.

Part III

Governance and Power Relationships Between Stakeholders

5

The Policy and Social Dimension of Restoration Thinking: Paying Greater Attention to "Interdependency" in Restoration Governing Practice

Caitriona Carter[1], Gabrielle Bouleau[2], and Sophie Le Floch[1]

[1] UR ETBX INRAE, Centre de Nouvelle Aquitaine Bordeaux, Cestas Cedex, France
[2] Université Gustave Eiffel, Laboratoire Interdisciplinaire Sciences Innovation Société (LISIS), CNRS, ESIEE, INRAE, UGE, Marne-la-Vallée, France

5.1 Introduction

5.1.1 Importance of the policy and social dimension of river restoration

Restoration has an important policy and social dimension influencing its achievement (Lejano et al. 2007). A central challenge is for rivers to meet all human needs sustainably, especially when climate change affects water abundance (Daniel et al. 2012). As a result, even though restoration has increasingly been taken up in policy documents (Suding 2011), it continues to compete with other policy and social objectives on a given territory (Perring et al. 2015). Indeed, environmental and sectoral policies governing rivers and water uses (Auerbach et al. 2014; Blackstock et al. 2012), along with cultural and territorial attachments to rivers (Le Floch 2014) and scientific and local knowledges about rivers and their infrastructures (Bouleau and Pont 2015), are all important elements contributing to effectiveness in restoration projects.

The importance of incorporating the policy and social dimension of restoration in project design and evaluation has been increasingly understood over the last decade (Drouineau et al. 2018). Ecologists working on the effectiveness of restoration have both come to identify the limitations of restoration evaluation focusing solely on ecological outcomes and to highlight evaluation-related knowledge gaps on the social and policy dimension. For example, in their literature review of evaluation studies of territorial restoration, published up until November 2012, Wortley et al. (2013, p. 539) found that only "3.5% of papers also included social and economic attributes." In a similar vein, Aronson et al. (2010), in their meta-analysis of papers (2000–2008) in *Restoration Ecology* (and 12 other scientific journals), highlighted the minimal amount of scholarly attention paid to the social and economic benefits of restoration and public policy. They summarize that "only 3% [of the papers they reviewed] devoted resources to interviewing people"; that there was a "gap between research on ecological restoration and the rest of society"; and finally that "80% of

River Restoration: Political, Social, and Economic Perspectives, First Edition. Edited by Bertrand Morandi, Marylise Cottet, and Hervé Piégay.
© 2022 John Wiley & Sons Ltd. Published 2022 by John Wiley & Sons Ltd.

the papers ... did not discuss or analyze direct policy impacts or implications of the restoration work" (Aronson et al. 2010, pp. 150–151). Their overall conclusion was that project evaluation at best underemphasized restoration's broader contribution to society, at times failing to address it at all.

These general observations have also been directed at river restoration studies. Although environmental historians have been writing river histories for decades (Mauch and Zeller 2008), for a long time, river restoration evaluation was limited to assessing ecological outcomes (Drouineau et al. 2018). It has only recently turned its attention to assessing social values (Morandi et al. 2014). For example, in their overview of the state of the art in assessments of "river environmental flow requirements," Pahl-Wostl et al. (2013) identified a lack of comparative research on the governance of environmental flows. Further, although they documented social factors (e.g. conflicts of interest, limited input from stakeholders, inappropriate governance structure) as the main obstacles to the successful implementation of river environmental flow requirements (Pahl-Wostl et al. 2013, p. 342), they nonetheless argued that there were major knowledge gaps on how these social factors caused failure in different settings.

Yet, even though there exists a comparatively low investment by social scientists in river restoration evaluation compared with natural scientists, social scientists have of course examined the social and policy dimension of river restoration from a number of angles (Sneddon et al. 2017). For example, McClenachan et al. (2015) produced results on social benefits arising from the completion of a fishery restoration project in the United States of America (USA) (coastal Maine). They interviewed stakeholders after restoration had taken place to determine different attitudes toward the project and local perceptions of social benefits arising from restoration practices. In another example, social scientists developed new methods for evaluating stakeholder engagement in river management plans toward restoration, for example in Scotland (Blackstock et al. 2012; Blackstock and Richards 2007). These produced new results on the social meaning of participation and contributed to establishing stakeholder participation as a research object in its own right, and not just as a variable measuring river restoration success. Whereas in both these examples the aim was to assess a restoration project or plan once in place, a slightly different focus has been taken by research carried out in environmental sociology (Barthélémy and Armani 2015), sociology of science (Lave et al. 2010), geography (Eden and Tunstall 2006), and political ecology (Fox et al. 2016). Here, restoration social and political processes contributing to project planning were examined, including identifying dominant actors and their alliances ultimately shaping outcomes (and hence social benefits incurred). Overall, these (and other) researches have centered upon the controversies of river restoration, including detailing conflicts and how these were ultimately resolved or not in project planning.

5.1.2 River restoration and "interdependency"

This increased investment by social science in river restoration has involved the mobilization of a "plurality of conceptual frames" (Sneddon et al. 2017, p. 650) and diverse analytical approaches, e.g. actor-network theory (Germaine and Lespez 2017) and political ecology (Barraud 2017; Jørgensen 2017). However, to date "none explicitly engages with state theory ... [or] how state power is exerted at both structural and political levels" (Sneddon 2017,

p. 653). Yet, changes in social and political relations governing river restoration practices (e.g. between public and private rights, between people and rivers, between expert and local knowledges) are not disconnected from wider processes of political transformation of the state – including the re-assignment of political authority which this transformation brings about (Carter and Smith 2008). Indeed, when considered through the lens of public policy analysis (Muller 2015), it is ever more apparent that tensions over river restoration are as much about the distribution of political authority as they are about water–society relations.

More specifically, drawing on Carter (2018), our starting point in this chapter is that river restoration takes place in a wider context of political transformation structured by three major processes of political change: (i) the rise of regulation and changing relations between states and markets and civil society; (ii) territorialization, globalization, and decentralization; and (iii) the democratization of knowledge use in policy-making. As documented by Carter (2018), these changes, stretching back over many years, have given rise to three types of interdependencies in addition to those traditionally associated with sustainability (Figure 5.1). These are: (i) regulatory interdependencies (e.g. between public policies and public/private rights), (ii) territorial interdependencies (e.g. between people, rivers, fish, plants, artifacts at different scales), and (iii) knowledge interdependencies (e.g. between different sciences and knowledge forms). Our working hypothesis is that, in the governing of river restoration, public and private actors have to grapple with these interdependencies and how they do so affects their restoration choices, including the choice not to restore the river.

Rather than treating these interdependencies as a mere background political and social context in which restoration takes place, the aim is to treat them instead as a key object of research. This has certain implications for research because the very idea of

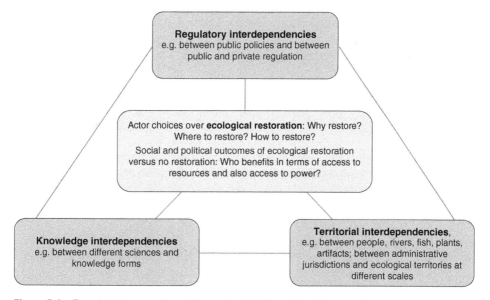

Figure 5.1 Template to grasp the policy and social dimension of restoration.

"interdependency" points to a specific understanding of relationships. According to Hay (2010), when a change in one component or variable may result in a change in another, we can talk in terms of "interdependence." Accordingly, in this definition of interdependency, there is no unique or unilinear causality: "A" can equally affect change in "B" and "C," as "C" can affect change in "A" and "B" (Hay 2010, p. 7). Consequently, studying interdependencies necessarily involves studying dynamic tensions between the different components which make up these interdependencies and, more specifically, grasping how actors arbitrate these tensions when governing river restoration projects (i.e. engage in a frontier politics of interdependency with Carter 2018).

Our argument is that through focusing upon interdependencies we can focus on specific relationships (and not all interactions) which enable us to explain many elements of restoration choices. This is because we view these different interdependencies as structuring of restoration choices, challenges, and blockages. However, our understanding of how interdependencies structure restoration is not a deterministic one. Among all possible existing interdependencies, in their political work actors may select and promote specific ones. Therefore, there is nothing inevitable about how interdependencies shape choices. For this very reason, an approach paying greater attention to "interdependency" operates on two levels. First, as a conceptual framework to redefine the social and policy dimension; second, as an epistemological method for examining how different components are socially constructed as interdependent.

Study of interdependency finds congruence with broader perspectives on river restoration which do not limit restoration to an administrative project (Fox et al. 2016): "ecological restoration is best seen not only as a technical task but as a social and political project" (Baker and Eckerberg 2013, p. 17). Even though national governments can frame restoration "as a 'win-win' situation, combining economic, ecological and safety improvements" (Buijs 2009, p. 2681), the different values underlying restoration projects are often more implicit than explicit, and objectives may not be clearly debated or expressed (Morandi and Piégay 2011). Diverging and fluctuating objectives among those concerned produce social conflict either over the very need for restoration in the first place or over the method chosen (Barthélémy and Souchon 2009; Jørgensen and Renöfält 2013). The trade-offs between disparate and changing social goals are replete with power and resource inequalities, asymmetric distribution of information, and unequal outcomes (Paavola and Hubacek 2013). For these reasons, too, thinking in terms of "interdependency" treats social and political processes as being at the heart of restoration (Baker and Eckerberg 2013), rather than as external potential obstacles to ecological outcomes. Importantly, it also allows us to make specific hypotheses about when conflicts or bottlenecks are likely to arise. For example, conflicts may arise within the interdependencies either when actors politicize tensions between components, or when actors consider that there has been a lack of a promotion of a particular interdependency, e.g. when habitants consider that restoration policies have not taken into account the relation between the river and the floodplain (Germaine and Lespez 2017).

Paying greater attention to interdependencies in restoration governing practice encourages a line of questioning along three complementary fronts. We ask:

1) How are policies and regulation on restoration brought together or made to compete with other policies governing rivers' futures; how are public policies, public/private

regulation, and common property rights put together or made to compete with river restoration projects?

2) How does restoration impact territorial construction processes led by different or competing actors at different scales; how do these renewed processes alter people's relations to rivers or meet their sometimes competing needs/expectations? How do actors promote interdependencies and/or manage tensions between territories at different scales?

3) How are different sciences and knowledges aggregated (e.g. ecological, socioeconomic, historical, cultural); which sciences and knowledges dominate over others? Which ecological, economic, and social facts are mobilized to drive or block restoration choices? Which knowledge conflicts can be identified?

In the rest of this chapter, we review published river restoration case studies, rereading results through the lens of our "interdependency" framework. We finish this review by presenting a blueprint for a case study on the politics of interdependencies governing restoration of the Ciron river in France (Box 5.1). We conclude by inviting reflection upon a number of points in relation to river restoration which emerge logically from our "interdependency" framework.

5.2 River restoration and the importance of interdependencies between public policies and between public and private rights

The main policy challenge for river restoration is not normally a lack of enabling legislation. Actually, in France, wetlands connectivity, river continuity, and biodiversity are all regulated through public policy. Instead, our first hypothesis is that a first set of challenges arise due to the interdependency of regulation. This interdependency can be found when actors mobilize around tensions either between river restoration policies or between restoration policies and other public policies, including other environmental policies promoting ecological transition of sectors operating on rivers (e.g. agriculture, fish farming, energy transition). For example, in the European Union (EU), biodiversity initiatives like Natura 2000 (EU Habitats Directive) can classify as areas of conservation lakes or wetlands resulting from dam construction, even though the dam causes river fragmentation affecting sediment-free water circulation. Indeed, ecologists working on lake restoration (in Scotland) have documented both positive and negative unintended consequences and "knock on effects" which restoration measures to address one problem (e.g. fish restocking) have had on other restoration goals (e.g. water quality) and vice versa (May and Spears 2012). Actually, in France, ecological continuity is considered one of the main objectives of the EU Water Framework Directive (WFD). Hence this objective is potentially interdependent with EU policy on green energy transition and renewable energy (Directive 2009/28/EC). In the case of rivers, limiting carbon emissions can result in dam construction to generate hydroelectricity. Yet, dams can lead to ecosystem fragmentation, a major threat to biodiversity adversely affecting amongst others migratory fish (e.g. eel, salmon, shad), which need sea, estuary, and river continuity to complete their life cycle (Drouineau et al. 2018). Removing dams in the name of ecological continuity and biodiversity has possible consequences beyond its potential interdependency with hydroelectricity production. This is because dam removal can also affect other river-related classifications, such as disease-free rivers under EU fish health directive (2006/88/EC). As argued by fish farmers in France, ecological continuity can facilitate the spread of diseases. Finally, restoring river continuity may be interdependent with other water usage regulations implemented by river basin committees, for example those within the framework of agriculture policy (Fernandez et al. 2014).

This first hypothesis that, in river restoration, important regulatory interdependencies are at stake is encouraged by a literature addressing river restoration from the perspective of interacting ecosystem services. Accordingly, conflicts over freshwater availability will become ever more pressing following climate and global change due to water shortages. In this light, one strand of scholarship has discussed how to minimize policy conflicts to produce positive synergies between interacting ecosystem services. For example, Bennett et al. (2009) highlight the need to provide water both for agriculture and for fish, and to reorient river managers' thinking around bundles of services. Meeting collective needs would require paying attention to the implementation of policies so as to maximize the number of services provided. This would entail developing management methods which would affect bundles of ecosystem services, rather than measures which would protect the provision of one service at the expense of others. This functionalist approach has also been advanced by Jager et al. (2015) examining riverscape design principles developing dam removal measures on rivers in the USA. In this case, the challenge was to calculate where best to remove some dams to maximize societal benefits. González del Tánago et al. (2012) also focus on the design of management measures of Spanish rivers. They argue that multiple public policies governing rivers in Spain provide increased institutional resources for those in favor of restoration in the face of societal resistance to change: managing tensions depend upon choices of measures and how these affect the implementation of other public policies.

Whereas in this research the politics of the regulatory interdependency is downplayed and rendered technical, in other research these political choices are highlighted. Following their study of rivers in the USA, Auerbach et al. (2014) argue against thinking about public policy trade-offs in terms of a balancing of services. Instead, they emphasize reasoning in terms of alternatives of river futures. They argue that, in their case study site, the alternative vision of a free-flowing river had been hidden behind policy discussions about interacting ecosystem services arising from an (already chosen) infrastructure river. In order to make informed choices on river futures, they argue, it is important to evaluate the alternative free-flowing river (and its accompanying bundle of services), rather than limiting evaluation and choices to interacting services arising from the default concrete river.

Value judgments inherent in the mediating of regulatory interdependencies become ever more visible when we consider the consequences of actors choosing public versus private rights governing river restoration choices. Neoliberalism in societal change has brought about a "shift in the style and goals of [state] intervention" (Goven and Pavone 2014, p. 311) and the redistribution of power from states to markets. Applied to rivers, Lave et al. (2010) illustrate how the rise of market-based intervention is altering how actors design restoration actions. In particular, it has encouraged the creation of new markets in ecosystem services. A possible consequence (also discussed by Lant et al. 2008) is that tensions between regulatory interdependencies will in future be adjudicated via regulated markets, prioritizing market values over public goods. This concern is also raised by Prévost et al.

(2016). They argue that the moral underpinnings of French public environmental law will be undermined by the contractualization and commercialization of ecosystem services, restoring nature for the sake of (some) human beings, rather than for its own sake. Whereas for these scholars, governing choices framed by ecosystem services automatically break with principles of equity, this is not the ultimate viewpoint of Lant et al. (2008). In order to protect principles of equity, these scholars make the case for developing new public policy on ecosystem services, which is currently absent. Ecosystem services public policy could protect common property regimes so that trade-offs could be managed within institutional frameworks respecting equal access to resources. Any such policy could take into consideration social disparities affecting under-represented groups' abilities to equally enjoy amenities resulting from river restoration (Goodling and Herrington 2014). Whereas for some scholars, therefore, managing public–private interdependencies through ecosystem services automatically resolves tensions toward regulated markets, for others new policies can be designed instead to foster joint custodianship and access to rivers.

5.3 River restoration and interdependencies in territorial construction

Important relationships exist between river restoration and territorial construction. Whereas the political modernization trend for many hundreds of years consolidated "political community ... institutions and policies ... on a single spatial scale: that of the nation-state" (Jeffery and Wincott 2010, p. 167), today the political meaning of territory can neither be reduced to state sovereignty (Sassen 2013) nor condensed to "linear notions of social progress" (Healey 2013, p. 1511). Rather, territorial construction is understood to be an emergent and contingent process (Healey 2013), at different times valorizing certain interdependencies between people and nature over others (Lussault 2007). Although in a common sense, "territory" refers to a (usually) bounded geographical space, more politically oriented definitions have been adopted. In this sense, territory is a spatial and practice-oriented construction (Gassiat and Zahm 2013) "actively formed and shaped through the political process" (Cochrane 2012, p. 104).

Although not framed using the language of "interdependency," case studies provide some initial insights for understanding how actors promote interdependencies through the building of particular links, for example between precise social groups and spatial areas (e.g. salmon anglers and salmon and the free-flowing river). In this vein, case studies stress the way the (expected) ecological output of the restoration project (e.g. the increase of fish populations) may reconnect local people to the river, offering new development opportunities for the territory and, hence, strengthening a sense of attachment to a particular place. Such imaginaries, whereby new interdependencies are viewed as boosting the territorial process, have a strong temporal dimension, in the sense that they invite the reactivation of links between different periods of time. For example, McClenachan et al. (2015) in their case study on the social benefits conferred by the restoration of habitat connectivity, fish populations, and local small-scale fisheries in Maine, USA, found that restoration was powerful in making connections between past and future relations between people and fish (McClenachan et al. 2015, p. 33). Through reinstating past cultural connections to the fish and, at the same time, building future local economies, the restoration project contributed to "the enhancement of personal connection to place and community that can come from restoring historically abundant fish populations" (McClenachan et al. 2015, p. 33). Indeed, both territory-building and an "increased sense of place and pride in the community" (McClenachan et al. 2015, p. 35) were a main social benefit and outcome of river restoration.

Other case studies provide more nuanced results on temporal linearity (i.e. between past, present, and future) and its capacity to feed a univocal territorial process. In contrast with McClenachan et al. (2015), the results of case study work on the removal of two hydroelectric dams in the Sélune valley (France) reveal territorial cleavages in relation to restoration which were not resolved (Germaine and Lespez 2014, 2017). Germaine and Lespez (2014) pose the question of territory from the outset and question whether river restoration in the name of biodiversity can be a driving force contributing to a collective sense of mobilization to manage natural resources. In the case studied, this did not happen. Actually, what they found was that conflicts emerged between different alliances of actors holding contrasting narratives of interdependencies between people and dams, between people and the river downstream, and between past, present, and future. Either people promoted positive interdependencies with dams seeing them as providing electricity, lakes for tourism and flood prevention, or they promoted negative ones, viewing dams instead as obstacles to migratory fish continuity which they valued and associated with the river run. Opponents to dam removal made links between a more or less recent past (e.g. their childhood) and the present. In so doing, they made connections between their experience and attachment to the lake as a popular tourist resort and ordinary biodiversity favored by the lake. Those in favor of dam removal focused instead on the river. They described this as "dynamic" and, in so doing, both reified the future whilst erasing the past. Opponents considered those in favor as being essentially external and elitist anglers giving priority to only emblematic species, such as salmon. These cleavages at the frontiers of interdependencies between people and dams halted collective community-building around shared river pasts and futures. In the end, the decision to remove the dams was taken by the French Minister of Ecology (commencing in 2018).

In yet another example in the USA, Fox et al. (2016) found common narratives of local people against dam removal. In this case, the dam had become part of local culture and symbolized historical landscape attachments and myriad interdependencies between people and rivers producing waterpower. As Fox et al. (2016, p. 102) argue, these deeper interdependencies and "landscape, historical and geographical contingencies" were more reliable factors explaining resistance to dam removal than glib comments such as "people don't like change" (Fox et al. 2016, p. 100). Drawing on two Catalonian case studies, Brummer et al. (2017), in line with Fox et al. (2016), insist on the importance of cultural and historical values in the understanding of the social tensions around restoration river projects and dam removals. Their examination of the spatial distribution of values associated with rivers, dams, and riverine forests can add to our understanding of territorial interdependencies, highlighting tensions. Cleavages emerged over different ways of promoting relationships. For example, for proponents of dam removal, the river was conceived as a "natural waterscape" and a "territorial backbone" articulating a wide set of interdependencies between forests and water and ecosystem services all along the river. For opponents, by contrast, bounded territories such as urbanized areas, or areas around a dam, were at the core of the territorial vision, interdependent with place-based ecosystem services which were historically connected.

Whereas these case studies produce interesting results illustrating the importance of a line of questioning on river restoration seizing the promotion of interdependencies between nature and people and hence contributing to local territorial construction, other research has highlighted the significance of examining how restoration practice can affect interdependencies between territories at different scales (up/downstream; local, regional, national). Taking the issue of longitudinal connectivity in respect of migratory fish who swim long distances and through many countries, Brevé et al. (2014, p. 207) argue that tackling dam removal only at the regional scale "lacks strategy and consistency at national and international levels." They demonstrate the problem through mapping fish trajectories and conducting a geospatial inventory of barriers in the Netherlands. What are required, they contend, are new governing arrangements for a "unified, multi-national and multi-river basin approach" to generate common understandings at a greater territorial scale over prioritization of barriers for removal and to address jurisdictional border problems. Territorial up/downstream interdependencies have also been the focus of research on rivers in southern Sweden (Tuvendal and Elmqvist 2011). Here researchers framed the issue more as a question of resilience of downstream people to upstream activities, rather than one of political action to re-scale governing practices (Friberg et al. 2016). In this case, the problem was how dissolved organic carbon upstream, brownificating the water, was experienced downstream by stakeholders. Results showed that stakeholders were more concerned about changing use values of the river (e.g. eel fishing and farming) than about changing biophysical properties. The authors concluded that, to engage people in restoration processes, it was important to specify the changing set of uses and well-being (people–river interdependencies) likely through nonaction at a regional scale.

These and other works thus raise the question of whether restoration can recompose and renew the meaning of territory at a larger scale. The question is particularly accurate, given that a recent argument made is that the "river basin" is the most appropriate scale for prioritizing restoration actions (Drouineau et al. 2018; Friberg et al. 2016). Accordingly, the "naturalness" of its boundaries makes it an unequivocal and a noncontroversial territory. Yet, other research can provide us with important insights on this point, even if they mostly deal with river management in general rather than restoration projects in particular. Some of this research has underlined that, whereas the fact of having a wider basin management plan in place does not necessarily create a river restoration project, any river restoration project set up when such a plan exists will nonetheless be impacted by the plan (Bernhardt et al. 2007). Other work has demonstrated how watershed lines are also a political construct, used by some actors at specific moments to discredit other ones (Molle 2009; Vogel 2012). Moreover, the ecological impacts of such projects designed at this natural scale are far from only being positive ones. Indeed, in some cases, they have even proved to wreak havoc on a river's ecosystem (Vogel 2012). These works reveal that naturalizing the scale may be critical, but they also underline the importance of not reducing the territorial dimension of "interdependency" to a matter of scale. Or, to put it differently, changing scales also changes the very (political) nature of the issues at stake (Massey 1999). Irrespective of the scale chosen, any new territorial frame proposed (or imposed) by actors will always come into interaction with other territorial processes, whether jurisdictional ones or territories as experienced by inhabitants (Di Méo 1996), or whether seemingly fixed administrative territories or more fluid and relational spatial imaginaries (Haughton and Allmendinger 2015). This is not, as is sometimes implied, merely a matter of coordination. As a brief focus on territorial interdependencies has shown, territorial conflicts over river futures are affected by deep and diverse interdependencies between people, rivers, fish, and artefacts. Moving upscale therefore necessitates the ability to command social legitimacy of any decisions taken at this scale. This means not only holding the ability to meet the underlying values and aspirations of the publics concerned but also having the capacity to foster common identities and political and social loyalty for choices made (knowing that there will always be winners and losers).

5.4 River restoration and interdependencies of sciences and knowledge forms

A third type of interdependencies potentially affecting successful river restoration is between knowledge types, including different approaches within disciplines (Dufour et al. 2017). Indeed, changes in environmental governance practice have accorded greater prominence to participatory forms of democracy to complement their representative counterparts (Blatrix 2009). Participatory governance has in turn endorsed uptake of multiple knowledge forms in decision-making. These have included: regulatory science, applied science, technical expertise, user knowledge, citizen science, and local knowledge. In this way, new interdependencies potentially come into play between knowledges and especially knowledge providers, potentially altering power relations between actors.

Actor mobilizations over knowledge interdependencies both have their origin in river restoration and can contribute to its contents. For example, in their analysis of the construction of ecological reference conditions for river restoration in Europe (within the framework of the WFD), Bouleau and Pont (2015) trace conflictual institutional positions of both scientists and their ecological concepts defining these conditions. As they show, after many months of exchange, reference conditions for river restoration were ultimately set when a "pristine model of nature" underpinned by an "old ecosystem paradigm" triumphed over a more inclusive biological, aesthetic and natural heritage paradigm (Bouleau and Pont 2015, pp. 35–36). But the story did not end there. Among scientists who lost out in these initial choices defining reference conditions, and who work instead within ecosystem services paradigms favoring participation, some have since sought to introduce their scientific understandings into the elaboration of sampling protocols to monitor rivers (Pont et al. 2007). In this way, and through mobilizing at the interface between knowledge interdependencies, these scientists have sought to regain their position as key knowledge providers on restoration practice – yet at the expense of nonscientific actors' participation (Bouleau 2017).

Whereas in the above case, tensions in hierarchies of power amongst scientists were connected to scientific definitions of the boundaries of the discipline of ecology and its authority for policy-making, in another example tensions across the interdependencies unfolded between scientists and other knowledge holders. In this example, it is emphasized that domination of certain kinds of science over other types of knowledge (e.g. stakeholder knowledge) can arise due to misunderstandings about public negative reactions to river restoration. In the case of four dam removals in rivers in Sweden, research results revealed that "public opposition is not based on knowledge deficiency, where more information will lead to better ecological decision-making, as is sometimes argued in dam removal science; it is instead a case of different understandings and valuation of the environment and the functions it provides" (Jørgensen and Renöfält 2013, p. 18). This has also been confirmed by case studies in Scotland. For example, in the case of the setting of river basin management plans, research revealed that, even though inclusive governing structures with stakeholders contributed to legitimizing outcomes, participation did not necessarily reduce conflict over knowledge uptake (Blackstock et al. 2012; Blackstock and Richards 2007). This was also the case for restoring urban rivers in England, where research has shown that even societal-knowledge-led approaches to restoration (as distinct from science-led ones) can face problems of social divisions (Eden and Tunstall 2006).

In other cases, the harnessing of different stakeholders and their knowledge has led to the redefinition of initial project objectives. In the example of the Rhone river restoration program, Barthélémy and Souchon (2009) found that scientists and national authorities had defined the initial project objectives in terms of restoring environmental flows, whereas local actors and authorities when associated to the implementation phase redefined the project in order to combine ecological improvement and flood prevention. This matters because, in the context of global change, leaving space for the redefinition of objectives is key for adaptation. Flitcroft et al. (2009, p. 1) argue that this may be addressed by designing a flexible social infrastructure which, in the case of watershed management, includes "mechanisms to design, carry out, evaluate, and modify plans for resource protection or restoration." Yet institutions can elude design (Cleaver and Franks 2005) even when they seek flexibility, and it is often left up to individual "entrepreneurs" to manage this interface (Richard-Ferroudji 2014).

Climate scientists now conceptualize our era as the Anthropocene (Crutzen 2002), an era where the scope of human impact on the environment is of a geological magnitude, even if not always noticed, understood, or managed. This change has also been documented for rivers (Kelly et al. 2017). In a discussion on competing anthropocentric values, Jørgensen (2017, p. 849) argues that there is an irreconcilable gap between those who oppose nature to culture and those who collapse these categories into a single category which she terms "natureculture." For the latter, there is room for debate on which categories of "socionature" (Braun and Castree 2001) or "waterscape" can and should be maintained, since most rivers will not be able to recover to a pristine state. Delineating and defining such categories is a political work (Bouleau 2014) which engages strongly with interdependencies between knowledge forms and, ultimately, goes beyond merely defining categories of knowledge to articulating these interdependencies with those in relation to both policy and territory.

Box 5.1   Blueprint for an empirical case study: restoring the Ciron river, south-west France

The Ciron river runs into the Garonne river, south of the city of Bordeaux. It is governed by a joint local authority committee which was first established in the 1980s. The river and its various tributaries are fragmented by around 70 river obstacles (dams, weirs, etc.). The vast majority of these are attached to privately owned mills which are no longer in use and have progressively been sold as private residences (often to second-home owners). Although for a long time the private use of these mills was not the object of extensive regulation, in 2010 this changed with the putting in place of a French-wide state river restoration policy of ecological continuity. In accordance with this policy, river obstacles must either be removed or modified (e.g. through installing fish passes). In the Ciron valley, state-wide policy especially targeted the lower stretch of the river and one of its tributaries, the Tursan.

In this blueprint, we draw on participant observation of a public meeting held in the Ciron valley and two exploratory interviews conducted by Arnaud Thomas, doctoral student ETBX INRAE, in 2017 with local officials working for the joint local authority committee charged with the implementation of this policy. Although investigative, already this material provides insights on the frontier politics of interdependency as a central feature of the political work of these public officials. To simultaneously establish their authority and influence local choices on river restoration, they have had to work politically on all three types of interdependencies discussed in this chapter. As they explain, for them, the scientific or technical act of ecological restoration is quite simple. What remains misunderstood and generates "deadlock situations" is "everything that surrounds it" (Interview, public official, 2017). Their work consists of managing tensions and bottlenecks. Because these tensions have not yet generated major conflicts, they are not immediately apparent. However, even at this exploratory stage, our "interdependency" framework can help us to realize their importance, both in terms of their content and their explanatory power for understanding restoration governing choices.

Regulatory interdependencies – During the implementation of river continuity regulation, many actors have promoted interdependencies between public policies and between public policy and private property rights. A first challenge faced by local public officials thus arises from this politicization by private actors of tensions between different public policies governing the river. For example, trout farmers have advanced political arguments claiming that river obstacles have permitted them to establish a health free zone on the river basin under EU rules, which they claim will be put in jeopardy by pursuing any restoration policy; private property owners are considering installing turbines in order to claim to produce hydroelectricity and hence escape from the objective of restoring continuity: or, internalizing policy inconsistencies, they have chosen to play the waiting game. A second challenge emerges from state-wide public action. Not only are the costs for dam and weirs removal not fully covered in state-wide funds, but also property owners can be subject to national cultural policy on archaeological heritage. In this case, playing the public-private interdependencies, state actors often opt for procedures which place costs on private property owners rather than on the state. Faced with both sets of arguments, local public officials have generally sought to de-politicize tensions. They have worked with property owners and trout farmers on a case-by-case basis and at a micro scale. In this way, they have sought to legitimize their own public authority over local restoration choices and methods, positioning themselves as indispensable for resolving inconsistencies between public policies and public and private responsibility.

Territorial interdependencies – Political work also consists of selecting and governing interdependencies promoted in the name of territory. In this case, public officials have taken an active role, seeking to carve out a new political space of action over which they can practice a legitimate governing authority – and this when viewed either from the perspective of the State or from any local actor. This space of action is geographically bounded to the river catchment area, which is reified by officials as an intermediary territory, capable of bringing together different jurisdictional competences and managing tensions both in relation to the French State and local authorities. In this political work, they are not only confronted with existing territorial jurisdictions, but also with a diverse range of territorial narratives put forward by a disparate group of private actors, some of whom are more organized than others (e.g. wine industry). At times, these territorial visions coincide with those of public officials. At others, they do not.

As local officials recognize that this depoliticized river catchment scale is not always the best place for debating territorial interdependencies choices, they have adopted three main governing strategies. First, as we saw for managing public–private interdependencies, they carry out restoration negotiations on a case-by-case basis, especially with, for example, mill owners. Second, they put trust in local politicians and technicians to carry out restoration in line with localized territorial visions. Third, they draw support from organized economic actors (e.g. Sauternes winegrowers) who find congruence between economic visions of valley–vineyard relations and restoration of ecological continuity. For example, when they highlight the quality of the Sauternes wine, winegrowers refer to the "physiology of the river," thus enhancing lateral connections between water, riverbanks, and cultivated lands. Support has also come from recreational fishers who were initially concerned that "their" trout would swim into the neighboring stretch of the river after weir removals, but have been pleasantly surprised by the overall restoration of fish stocks. These actors have introduced into debate other territorial interdependencies in the form of new up/downstream solidarities. However, despite these strategies, local officials can still find themselves at a disadvantage when faced with opponents, for example the local canoeing club (who initiated what became a state-wide petition against ecological continuity), some mill owners, and other local residents.

Knowledge interdependencies – Finally, political work has also consisted in promoting knowledge interdependencies. At a state-wide scale, the choice to restore the lower stretch of the Ciron and the Tursan rivers was based on generalized criteria shaped predominantly by regulatory science. However, local public officials have also acknowledged the importance of local knowledge. This can question the choice to restore certain stretches of the river, for example when rivers to be restored for lamprey do not contain any natural habitats for lamprey. Local users of these stretches of river can therefore have a more precise knowledge about these environments. In order to minimize public conflicts between regulatory and local knowledge, officials have avoided making generalized statements in public meetings knowing that these would be rejected out of hand. They have also carried out complementary studies on local habitats and aquatic species, co-constructing study protocols with local residents. These studies have ultimately questioned the pertinence of carrying out restoration projects on the Tursan, a river which naturally often runs dry and is consequently not favored by migratory fish. In the end, a local choice was taken collectively not to carry out any restoration actions on this river. This in turn consolidated the political authority of local public officials governing in the name of place-based knowledge.

In summary, therefore, we can observe strategies of depoliticization of the implementation of ecological continuity restoration and the political authority of these local officials. Local officials work at the interface of the three types of interdependencies to simultaneously shape their authority and river restoration choices. This has resulted in a protean and multifaceted restoration governing process, which is ongoing and cannot be reduced to individual restoration choices.

5.5 Conclusions

The chapter has proposed a conceptual framework highlighting interdependencies in order to renew our understanding of the governing and redistribution of political authority of river restoration. We have sought to illustrate the potential of this approach from a conceptual point of view. We have done this primarily through drawing upon already existing case study results of river restoration and rereading these through the lens of "interdependency." We have also offered a glimpse of an empirical demonstration through presenting a blueprint for a case study on the politics of interdependency governing restoration of the Ciron river in south-west France. Of course, the next step is to apply the framework to a set of comparative case studies and demonstrate it empirically through new research. And, although the chapter has taken a social science focus, ultimately we do not propose that this research take place in a silo, disconnected from research either in ecological economics or in ecology. To operationalize new social science research questions on interdependencies requires dialogue not only with stakeholders but also with other scientists (Drouineau et al. 2018; Jørgensen 2017).

We conclude by inviting reflection upon a number of points in relation to river restoration which emerge logically from our "interdependency" framework. In general terms, we consider that this approach strongly encourages letting go of certain conventional wisdoms about the policy and social dimension of restoration thinking. First, we need to let go of thinking about social factors solely in terms of obstacles to restoration. Instead, we need to consider in turn the different governing issues which are raised and how they might be addressed in the meeting of objectives. Second, we need to let go of the idea that a divergence of interests between stakeholders is a problem. Actually, it is a feature of politics and part of restoration stories, sometimes contributing to their success and sometimes contributing to their failure. This means that politics needs to be embraced and not feared. Politics is about conflict. The important question is not how to make it go away but how to harness divergent views to reach settlements commanding legitimacy. Third, we need to let go of the idea that restoration is only about the river and instead find ways to dialogue with those actors governing restoration, whether local public officials or stakeholders, on issues which extend above and beyond the substantive elements of restoration. Fourth, we need to let go of the idea that effectiveness in restoration can be achieved through a neat steering of stakeholders towards certain predefined goals. Actually, this way of thinking is part of the problem. Instead, we need to grasp how to interpret tensions at the heart of different types of interdependencies to support negotiation over common goals. Fifth, we need to let go of thinking about "natural" boundaries (i.e. the river basin) as automatically the most relevant territorial frame in which to integrate interdependencies of all kinds. On the one hand, natural boundaries need to be debated, defined, and designed. On the other hand, any essentializing discourses of this kind, whilst trying to avoid conflicts or to exclude some actors, can overlook democratic objectives as well as ecologically sound practices.

More specifically, an approach via "interdependency" invites new reflections. On regulatory interdependencies, these are first that the question of policy delivery is less about individual policies and their design and objectives but more about their interdependencies with other policies. Second, there may be compensation solutions which initially seem far away from river restoration, but nonetheless could contribute to the strengthening of certain values (e.g. new access rights to rivers for public use and/or recreational fishers, training of officials, equipping local actors with expertise and capacity to bring new arguments into policy arenas at wider scales). On territorial interdependencies, the question of the delivery is less about designing the best territorial frame for a restoration project to be both ecologically and socially relevant, and more about how to understand the complexity of intertwined territorial processes. These can both affect and be affected by the restoration project. A restoration project associated to a river, especially because it deals with continuity and borders, impacts already existing territorial processes which are already complex, mixing different scales, different objectives (economic regional development, habitats' preservation based on ecological units), and different spatial imaginaries. Moreover, there is a need to take into account the invisible territorial processes, which lie in the close-knit relations between inhabitants (and/or ordinary users) and their environment, as, in the end, their everyday practices matter in the success of the restoration project. On knowledge interdependencies, existing knowledge and laws produce descriptive categories that can be taken for granted and influence our way of seeing rivers and environmental problems. Yet, to solve river restoration problems, we sometimes need to question these categories, because they can carry specific visions of interdependencies between rivers and society which are themselves part of the problem. Consequently, lessons from knowledge interdependencies turn on the making available alternative categories and ways of seeing rivers and society relations in order to fully grasp the range of choices available and their potential consequences.

References

Aronson, J., Blignaut, J.N., Milton, S.J., et al. (2010). Are socioeconomic benefits of restoration adequately quantified? A meta-analysis of recent papers (2000–2008) in *Restoration Ecology* and 12 other scientific journals. *Restoration Ecology* 18:143–154.

Auerbach, D.A., Deisenroth, D.B., McShane, R.R., et al. (2014). Beyond the concrete: accounting for ecosystem services from free-flowing rivers. *Ecosystem Services* 10: 1–5.

Baker, S. and Eckerberg, K. (2013). A policy analysis perspective on ecological restoration. *Ecology and Society* 18(2): 17–27.

Barraud, R. (2017). Removing mill weirs in France: the structure and dynamics of an environmental controversy. *Water Alternatives* 10(3): 796–818.

Barthélémy, C. and Armani, G. (2015). A comparison of social processes at three sites of the French Rhône river subjected to ecological restoration. *Freshwater Biology* 60: 1208–1220.

Barthélémy, C. and Souchon, Y. (2009). La restauration écologique du fleuve Rhône sous le double regard du sociologue et de l'écologue. *Natures Sciences Sociétés* 17: 113–121.

Bennett, E.M., Peterson, G.D., and Gordon, L.J. (2009). Understanding relationships among multiple ecosystem services. *Ecology Letters* 12: 1394–1404.

Bernhardt, E., Sudduth, E., Palmer, M.A., et al. (2007). Restoring rivers one reach at a time: results from a survey of U.S. river restoration practitioners. *Restoration Ecology* 15(3): 482–493.

Blackstock, K.L. and Richards, C. (2007). Evaluating stakeholder involvement in river basin planning: a Scottish case study. *Water Policy* 9: 493.

Blackstock, K.L., Waylen, K.A., Dunglinson, J., et al. (2012). Linking process to outcomes: internal and external criteria for a stakeholder involvement in river basin management planning. *Ecological Economics* 77: 113–122.

Blatrix, C. (2009). La démocratie participative en représentation. *Sociétés contemporaines* 2(74): 97–119.

Bouleau, G. (2014). The co-production of science and waterscapes: the case of the Seine and the Rhône Rivers, France. *Geoforum* 57: 248–257.

Bouleau, G. (2017). Ecologisation de la politique européenne de l'eau, gouvernance par expérimentation et apprentissages. *Politique européenne* 55: 36–59.

Bouleau, G. and Pont, D. (2015). Did you say reference conditions? Ecological and socio-economic perspectives on the European Water Framework Directive. *Environmental Science & Policy* 47: 32–41.

Braun, B. and Castree, N. (2001). *Social Nature: Theory, Practice, Politics*. Basil Blackwell.

Brevé, N.W.P., Buijse, A.D., Kroes, M.J., et al. (2014). Supporting decision-making for improving longitudinal connectivity for diadromous and potamodromous fishes in complex catchments. *Science of the Total Environment* 496: 206–218.

Brummer, M., Rodriguez-Labajos, B., Thanh Nguyen, T., et al. (2017). "They have kidnapped our river": dam removal conflicts in Catalonia and their relation to ecosystem services perceptions. *Water Alternatives* 10(3): 744–768.

Buijs, A.E. (2009). Public support for river restoration: a mixed-method study into local residents' support for and framing of river management and ecological restoration in the Dutch floodplains. *Journal of Environmental Management* 90: 2680–2689.

Carter, C. (2018). *The Politics of Aquaculture: Sustainability Interdependence, Territory and Regulation in Fish Farming*. Abingdon: Routledge.

Carter, C. and Smith, A. (2008). Revitalizing public policy approaches to the EU: "territorial institutionalism," fisheries and wine. *Journal of European Public Policy* 15(2): 263–281.

Cleaver, F. and Franks, T. (2005). How institutions elude design: river basin management and sustainable livelihoods. Bradford Centre for International Development, Research Paper no. 12. http://core.kmi.open.ac.uk/display/5659 (accessed 24 May 2021).

Cochrane, A. (2012). Making up a region: the rise and fall of the "South East of England" as a political territory. *Environment and Planning C: Government and Policy* 30: 95–108.

Crutzen, P.J. (2002). Geology of mankind. *Nature* 415: 23.

Daniel, T., Muhar, A., Arnberger, A., et al. (2012). Contributions of cultural services to the ecosystem services agenda. *Proceedings from the National Academy of Sciences* 109(23): 8812–8819.

Di Méo, G. (1996). *Les territoires du quotidien*. Paris: L'Harmattan.

Drouineau, H., Carter, C., Rambonilaza, T., et al. (2018). River continuity restoration and diadromous fishes: much more than an ecological issue. *Environmental Management* 61(4): 671–686.

Dufour, S., Rollet, A.J., Chapuis, M., et al. (2017). On the political roles of freshwater science in studying dam and weir removal policies: a critical physical geography approach. *Water Alternatives* 10(3): 853–869.

Eden, S. and Tunstall, S. (2006). Ecological versus social restoration? How urban river restoration challenges but also fails to challenge the science–policy nexus in the United Kingdom. *Environment and Planning C: Government and Policy* 24: 661–680.

Fernandez, S., Bouleau, G., and Treyer, S. (2014). Bringing politics back into water planning scenarios in Europe. *Journal of Hydrology* 518: 17–27.

Flitcroft, R.L., Dedrick, D.C., Smith, C.L., et al. (2009). Social infrastructure to integrate science and practice: the experience of the Long Tom Watershed Council. *Ecology and Society* 14: 36.

Fox, C.A., Magilligan, F.J., and Sneddon, C.S. (2016). "You kill the dam, you are killing a part of me": dam removal and the environmental politics of river restoration. *Geoforum* 70: 93–104.

Friberg, N., Buijse, T., Carter, C., et al. (2016). Effective restoration of aquatic ecosystems: scaling the barriers. *WIREs Water* 4(1): e1190.

Gassiat, A. and Zahm, F. (2013). Améliorer la qualité de l'eau: quelle territorialisation? Exemple de MAE à « enjeu eau ». *Economie Rurale* 333: 85–104.

Germaine, M.-A. and Lespez, L. (2014). Le démantèlement des barrages de la Sélune (Manche): des réseaux d'acteurs au projet de territoire ? *Développement durable et territoires* 5(3). https://doi.org/10.4000/developpementdurable.10525.

Germaine, M.-A. and Lespez, L. (2017). The failure of the largest project to dismantle hydroelectric dams in Europe? (Sélune river, France, 2009–2017). *Water Alternatives* 10(3): 655–676.

Goodling, E. and Herrington, C. (2014). Reversing complete streets disparities: Portland's Community Watershed Stewardship Program. In: *Incomplete Streets: Processes, Practices, and Possibilities* (eds S. Zavestoski and J. Agyeman), 176–201. New York: Routledge.

González del Tánago, M., García de Jalón, D., and Román, M. (2012). River restoration in Spain: theoretical and practical approach in the context of the European Water Framework directive. *Environmental Management* 50(1): 123–139.

Goven, J. and Pavone, V. (2015). The bioeconomy as political project: a Polanyian analysis. *Science, Technology, & Human Values* 40(3): 302–337.

Haughton, G. and Allmendinger, P. (2015). Fluid spatial imaginaries: evolving estuarial city-regional spaces. *International Journal of Urban and Regional Research* 39(5): 857–873.

Hay, C. (2010). Introduction: political science in an age of acknowledged interdependence. In: *New Directions in Political Science: Responding to the Challenges of an Interdependent World* (ed. C. Hay), 1–24. Basingstoke: Palgrave.

Healey, P. (2013). Circuits of knowledge and techniques: the transnational flow of planning ideas and practices. *International Journal of Urban and Regional Research* 37(5): 1510–1526.

Jager, H., Efroymson, R.A., Opperman, J.J., et al. (2015). Spatial design principles for sustainable hydropower development in river basins. *Renewable and Sustainable Energy Reviews* 45: 808–816.

Jeffery, C. and Wincott, D. (2010). The challenge of territorial politics: beyond methodological nationalism. In: *New Directions in Political Science: Responding to the Challenges of an Interdependent World* (ed. C. Hay), 167–188. Basingstoke: Palgrave.

Jørgensen, D. (2017). Competing ideas of "natural" in a dam removal controversy. *Water Alternatives* 10(3): 840–852.

Jørgensen, D. and Renöfält, B.M. (2013). Damned if you do, dammed if you don't: debates on dam removal in the Swedish media. *Ecology and Society* 18(1): 18.

Kelly, J.M., Scarpino, P., Berry, H., et al. (2017). *Rivers of the Anthropocene*. Oakland, CA: University of California Press.

Lant, C., Ruhl, J., and Kraft, S. (2008). The tragedy of ecosystem services. *BioScience* 58(10): 969–974.

Lave, R., Doyle, M., and Robertson, M. (2010). Privatizing stream restoration in the US. *Social Studies of Science* 40: 677–703.

Le Floch, S. (2014). Les bords de Garonne et leurs nouveaux riverains. *Ethnologie française* 1: 165–172.

Lejano, R., Ingram, H., Whiteley, J., et al. (2007). The importance of context: integrating resource conservation with local institutions. *Society & Natural Resources* 20(2): 177–185.

Lussault, M. (2007). *L'homme spatial: La construction sociale de l'espace humain*. Paris: Le Seuil.

Massey, D. (1999). Space–time, "science" and the relationship between physical geography and human geography. *Transactions of the Institute of British Geographers* 24: 261–276.

Mauch, C. and Zeller, T. (2008). *Rivers in History: Perspectives on Waterways in Europe and North America*. Pittsburgh, PA: University of Pittsburgh Press.

May, L. and Spears, B.M. (2012). Managing ecosystem services at Loch Leven, Scotland, UK: actions, impacts and unintended consequences. *Hydrobiologia* 681(1): 117–130.

McClenachan, L., Lovell, S., and Keaveney, C. (2015). Social benefits of restoring historical ecosystems and fisheries: alewives in Maine. *Ecology and Society* 20(2): 31.

Molle, F. (2009). River-basin planning and management: the social life of a concept. *Geoforum* 40: 484–494.

Morandi, B. and Piegay, H. (2011). Les restaurations de rivières sur Internet: premier bilan. *Natures Sciences Sociétés* 19: 224–235.

Morandi, B., Piégay, H., Lamouroux, N., et al. (2014). How is success or failure in river restoration projects evaluated? Feedback from French restoration projects. *Journal of Environmental Management* 137: 178–188.

Muller, P. (2015). *La société de l'efficacité globale: Comment les sociétés se pensent et agissent sur elles-mêmes*. Paris: PUF.

Paavola, J. and Hubacek, K. (2013). Ecosystem services, governance, and stakeholder participation: an introduction. *Ecology and Society* 18(4): 42.

Pahl-Wostl, C., Arthington, A., Bogardi, J., et al. (2013). Environmental flows and water governance: managing sustainable water uses. *Current Opinion in Environmental Sustainability* 5: 341–351.

Perring, M.P., Standish, R.J., Price, J.N., et al. (2015). Advances in restoration ecology: rising to the challenges of the coming decades. *Ecosphere* 6(8): 131.

Pont, D., Hugueny, B., and Rogers, C. (2007). Development of a fish-based index for the assessment of river health in Europe: the European Fish Index. *Fisheries Management and Ecology* 14(6): 427–439.

Prévost, B., Rivaud, A., and Michelot, A. (2016). Économie politique des services écosystémiques: de l'analyse économique aux évolutions juridiques. *Revue de la régulation* 19: 1–32.

Richard-Ferroudji, A. (2014). Rare birds for fuzzy jobs: a new type of water professional at the watershed scale in France. *Journal of Hydrology* 519: 2468–2474.

Sassen, S. (2013). When territory deborders territoriality. *Territory, Politics, Governance* 1(1): 21–45.

Sneddon, C.S., Barraud, R., and Germaine, M.-A. (2017). Dam removals and river restoration in international perspective. *Water Alternatives* 10(3): 648–654.

Suding, K.N. (2011). Toward an era of restoration in ecology: successes, failures, and opportunities ahead. *Annual Review of Ecology, Evolution, and Systematics* 42: 465.

Tuvendal, M. and Elmqvist, T. (2011). Ecosystem services linking social and ecological systems: river brownification and the response of downstream stakeholders. *Ecology and Society* 16(4): 21.

Vogel, E. (2012). Parcelling out the watershed: the recurring consequences of organising Columbia River management within a basin-based territory. *Water Alternatives* 5(1): 161–190.

Wortley, L., Hero, J.-M., and Howes, M. (2013). Evaluating ecological restoration success: a review of the literature. *Restoration Ecology* 21(5): 537–543.

6

From Public Policies to Projects: Factors of Success and Diversity Through a Comparative Approach

Catherine Carré[1], Jean-Paul Haghe[2], and Pere Vall-Casas[3]

[1]*Université Paris 1 Panthéon-Sorbonne, Paris, France*
[2]*Rouen University, Rouen, France*
[3]*Universitat Internacional de Catalunya, School of Architecture, Barcelona, Spain*

6.1 Introduction

River restoration is relatively new, having started in the 1970s. At that time, it mainly concerned developed countries (Speed et al. 2016) and those with the same scientific culture and water management techniques (Mondragón-Monroy and Honey-Rosés 2016). Today, river restoration operations are gaining momentum in developing countries (Damanik and Patriwi 2017; Vollmer et al. 2015; Costa et al. 2010) and can be found worldwide, in a very wide range of contexts and at every level, from major rivers in rural areas to small streams in urban centers, involving a variety of stakeholders, both public and private. River restoration must be contextualized in the transition from an engineering-based to an ecosystem-based approach to river management (Hillman and Brierley 2005). This paradigm shift, still underway, started at 1980s and was formalized at 2000s in documents such as the European Union (EU) Water Framework Directive (WFD) that reflects the general trend toward holistic environmental management. The new approach moves beyond the command and control regulation focused on fluid mechanics and hydraulics, toward the integration of both the ecological and the social dimensions of river restoration. Regarding the ecological dimension, the river ecological integrity, including not only chemical and morphological but also biological components (Ormerod 2004; Pretty et al. 2003), is the center of the river restoration whose final goal is to heal the river through enhancing natural recovery mechanisms. Regarding the social dimension, emphasis is placed on multilevel and multi-actor river management. When stakeholders including local communities are sufficiently engaged and empowered in the decision-making process, the acceptance and effectiveness of the river restoration increase. Thus, linking social sciences with environmental sciences becomes paramount. Social sciences studies of restoration operations differ from those of environmental sciences in that they examine the way these operations are accepted by society, rather than their environmental objectives or material conditions of

River Restoration: Political, Social, and Economic Perspectives, First Edition. Edited by Bertrand Morandi, Marylise Cottet, and Hervé Piégay.

implementation. The factors leading to the success or failure of operations have been studied mainly by public action tools (regulatory framework, funding mechanism, and governance process). As observed by Baker and Eckerberg (2013, p.1), these operations "take place in the context of different distributions of power between the various public and private actors involved at the different stages of restoration policy making ... Ecological restoration is best seen not only as a technical task but as a social and political project."

The purpose of this chapter is to study how states set up policies to restore their rivers and streams and how restoration projects are (or are not) carried out in the context of the gradual paradigm shift previously defined, from an engineering approach to the river guided by utilitarian aims to a more ecological one. We refer to public policies designed at the national or federal level that are implemented in the form of local restoration projects. According to each country, the different facets of the eco-centered paradigm will be enhanced to different degrees and the impact of the river ecology on the projects will vary. On the other hand, as reported by authors such as Mollinga (2008, p. 12), "water policies, like other policies, are negotiated and re-negotiated in all phases or stages and at all levels, and are often transformed on their way from formulation to implementation, if not made only in the implementation process." Policies and projects are inseparable, and understanding the factors that allow project implementation is fundamental to assessing policy efficacy.

However, even if all the required factors are met, the project will not necessarily succeed. Beyond technical and financial considerations, one of the reasons for failure may be the opposition of local residents, despite good scientific and social reasons supporting the project. Local opposition may affect small projects of re-meandering on short sections of a river (Carré and Haghe 2013) as well as the rejection of dam removal (Brummer et al. 2017). Our last step will be to explore, by means of case studies, the acceptance or rejection of a restoration project. It is the realization of the operation that ultimately marks the degree of success of the new paradigm for river management.

With this in mind, we ask the following questions:

1) Do river restoration projects obey the same river management paradigm?
2) What are the necessary factors in implementing a project?
3) These factors being necessary but not sufficient, why do some projects succeed when others – especially dam removal – may be rejected and, ultimately, not carried out?

6.2   Toward a multifunctional, ecosystem-based paradigm for river restoration

River restoration today presents multiple objectives. In relation to the restoration of urban rivers, the European Environment Agency (2016, p. 9) states that "the use of the term restoration in this report is not limited to a management process striving to re-establish the structure and function of ecosystems as closely as possible to the pre-disturbance conditions and functions. The term restoration is used more broadly to refer to activities that aim to improve the status of degraded waters, by improving water quality or by changing hydromorphological conditions. Such activities, in addition, aim to serve other needs and preferences of the urban population, i.e. through multifunctional measures."

The multifunctional approach is not specific to the restoration of urban rivers; it is also found at larger scales, as in the Missouri River Ecosystem Restoration Plan, one of 16 programs presented by the Denver Bureau of Reclamation in 2011.[1] The main aim of this plan is to "restore the ecosystem, mitigate habitat loss, and recover fish and wildlife, and identify the tools needed for implementation." However, over time, a restored river could also contribute to "the revitalization of social, recreational, and business opportunities along the Missouri River, benefiting local citizens, local businesses, guides, outfitters, and recreational and commercial interests" (US Department of the Interior 2011, p. 8). These economic and social objectives are particularly sensitive in large urban centers as part of international city promotion, such as the Cheonggyecheon stream restoration project in Seoul, South Korea in 2005 or the Laojie River in Taoyuan, Taiwan in 2011: "Taoyuan has established national status by developing green infrastructure. The Laojie River, as the city's principal waterway, is the main focus of such development in the government plan … The two city-centre projects reintroduce landscape aesthetics and recreational value," while the improvement of the urban fabric acts as "a magnet for investment and tourists" (Chou 2016, p. 15).

This multifunctional approach is the result of a combination of specific objectives that have developed since the Seventies, reflecting changes in public action over time (Figure 6.1), from an instrumentalized approach concerned with human interests (maintaining fish stocks for fishing, quantity, and use of water) in the 1960s, to a more eco-centered approach in the 1980s and 1990s, concerned with ecological restoration, supported recently by economic justification of the net benefits of such projects (Cullinane et al. 2012). However, these trends are overlapping rather than linear.

These trends follow changes in national regulations, and explain the different ways the word "restoration" is used in public policies and projects. The main US federal laws relating to river restoration are the Wild and Scenic Rivers Act of 1968, the Clean Water Act (CWA) of 1972, and the Endangered Species Act (ESA) of 1973. The anthropocentric dimension is apparent in the title of the Wild and Scenic Rivers Act, which stipulates that "selected rivers … which possess outstandingly remarkable scenic, recreational, geologic, fish and

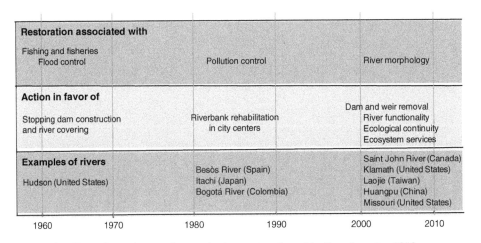

Restoration associated with			
Fishing and fisheries Flood control		Pollution control	River morphology
Action in favor of			
Stopping dam construction and river covering	Riverbank rehabilitation in city centers		Dam and weir removal River functionality Ecological continuity Ecosystem services
Examples of rivers			Saint John River (Canada) Klamath (United States) Laojie (Taiwan) Huangpu (China) Missouri (United States)
Hudson (United States)	Besòs River (Spain) Itachi (Japan) Bogotá River (Colombia)		
1960	1970	1980	1990	2000	2010

Figure 6.1 Chronology of some changes in river restoration objectives from the 1960s.

wildlife, historical, cultural or similar values shall be preserved in free-flowing condition and that they and their immediate environments shall be protected for the benefits and enjoyment of present and future generations" (Wild and Scenic River Act 16 USC 1273). The word "scenic" was specifically used by a group of citizens and local fishermen in 1963 when they formed the Scenic Hudson Preservation Conference to fight a power plant project on Storm King Mountain, which would have destroyed part of the cultural heritage of the Hudson River Valley.

The Federal Water Pollution Control Act of 1948, as amended in 1972 (the CWA), regulates discharge of municipal and industrial pollutants and makes provisions for funding sewage treatment plants, with the emphasis on chemical impairment to aquatic ecosystems. The ESA is the primary legal instrument used to protect stream flows for ecological benefits (Christian-Smith et al. 2012). All 16 programs presented by the Bureau of Reclamation of Denver in 2011 addressed endangered species issues and cited the ESA as one of the program authorities (US Department of the Interior 2011). Finally, the Coastal Zone Management Act of 1972 recognizes the relationship between retained dam sediments and coastal resilience, and also serves as a reference to initiate dam removal (Grabowski et al. 2017).

The increasing consideration given to the biological and morphological dimensions of rivers in addition to their physicochemical properties can also be seen in European directives. The earliest, such as that dated 16 June 1975, concerned the quality of groundwater for the production of drinking water. The following focused, like the CWA, on limiting the discharge of industrial and municipal pollutants, for example the EU Urban Waste Water Treatment Directive of 21 May 1991, and the Nitrates Directive of 12 December 1991 for the protection of waters from pollution by agricultural sources. The EU WFD of 23 October 2000 gave priority to ecosystems, with the aim of achieving "good ecological status" for all waters by 2015 (at the latest by 2027). One of the aims of the directive is to restore the ecological continuity of rivers, namely allowing the migration of aquatic organisms (in particular fish) and sediment flow, by encouraging EU Member States to remove or make passable any kind of obstacle to their circulation.

Alongside this change in priorities of public action (based on the idea that protecting ecosystem functioning ultimately protects water for human use), there was a change in flood prevention policies, with a shift from seeking total flood protection to the creation of controlled flood areas. River restoration operations can thus combine flood control, water uses, and ecological measures, as shown by policy changes in Europe (Wharton and Gilvear 2006) and in Japan (Nakamura et al. 2006).

The Itachi River restoration project (Japan, Yokohama), implemented in 1982, was one of the earliest projects in Japan designed to provide flood protection: "the lower channel was excavated, and sediments were placed along both banks to restore a meandering channel with alternating riffles and pools. It was an epoch-making project for a highly regulated urban river" (Nakamura 2008, p. 8). This was followed by further projects, instigated by the River Bureau of the Ministry of Land, Infrastructure, and Transport, the governmental authority for river management, which launched the "Nature-oriented River Works" in 1990, with the main aim of conserving and restoring river corridors and their biodiversity.

Restoration operations seem, then, to follow a slowly evolving trend. Indeed, case study review shows that a paradigm shift is underway in river restoration practice, marking a transition from a reductionist engineering-dominated approach with total disregard for ecosystem values, toward a multifunctional perspective and use of natural recovery mechanisms for improving rivers' ecological health (Hillman and Brierley 2005). The general acceptance of the multifunctional ecosystem-based paradigm can be explained in various ways. First, there is a convergence in the objectives of public policies, at least in the USA and the EU, as described above. Second, restoration operations tend to adopt the same technical methods throughout the world. This may be explained by the dissemination of successful operations; knowledge sharing between contractors and project managers from the same regulatory, scientific, and technical culture; and skills-sharing through international associations of engineers (International Water Association), scientific and professional networks (Society for Ecological Restoration, Asian River Restoration Network, European Centre for River Restoration), and non-governmental organizations (NGOs) (International Union for Conservation of Nature, American Rivers in the USA, Eau et Rivières de Bretagne in France, European Centre for River Restoration).

However, the change of paradigm and supporting technical methods has not yet been systematically matched by a change in stakeholder attitudes or professional culture. In Japan, while operations have adopted an ecological dimension ("New Age" river design) since the 1990s, bureaucratic inertia and the interests of politicians and construction industry lobbyists remain firmly entrenched. The priority continues to be major flood protection and navigation infrastructures. Even the objective of restoring rivers to their natural state is seen from a human-centered perspective (Chakraborty 2013). We are still a long way from the true meaning of the new paradigm, with its emphasis on the intrinsic values of rivers (functions maintained with the least possible human intervention).

6.3 Political factors that determine river restoration

High variability is observed in the implementation of the common paradigm in different countries. In particular, literature on EU WFD implementation shows that progress toward meeting the substantive objectives of this directive has been patchy (Jager et al. 2016). The very uneven results can be explained by the different contexts in which river restoration occurs. Concretely, the conditions of implementing the operation, which ensure that it will be carried out and determine the forms that it will take, depend on a combination of contextual factors (Figure 6.2) that are often specific to the country or to the state in the case of federations, as regards: (i) regulatory framework, (ii) funding mechanism, and (iii) governance process, the third being the most relevant to social sciences research.

6.3.1 Funding mechanism

According to Khorshed (2008, p. 593), restoring degraded rivers "is traditionally perceived as a public responsibility with very limited contribution from the private sector." The public funds available for carrying out restoration operations vary with regional context and political choices. In Europe, operations are financed by taxes on water levied on users (industry, farmers, consumers, residents) by local water authorities, on the principle that "water pays for water." The equipment for treating wastewater discharges into the river is subsidized by

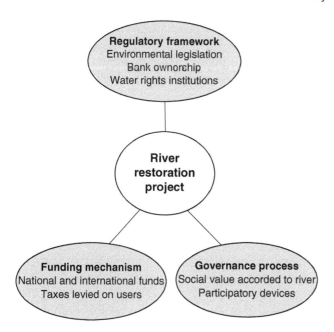

Figure 6.2 Political factors in river restoration projects.

local governments and ultimately depends on local savings and mobilization. The implementation of this principle is very uneven between countries, and the varying degree of public expenditure on sanitation remains linked to the political criterion.

However, even in so-called rich countries, local budgets are generally insufficient to provide the money needed to build a wastewater treatment plant. The complementary financing comes from public, national, or international funds. European countries can appeal to different organizations, such as the European Central Bank (ECB), which acts specifically by financing projects that contribute to economic and sustainable development with a direct impact on the daily life of inhabitants. In France, 1.6 billion euros have been spent by the ECB on the water and sanitation sector since 2008. Likewise, in the USA, financing is provided by federal funds (see the Clean Water Act, 1977), while the World Bank and international funds play a major role in China. "It is only with the support of the national government that New York and Shanghai were able to raise the capital needed to create wastewater treatment plants ... In the USA, the EPA provided over six billion US dollars for municipalities to construct and upgrade wastewater treatment plants. Seventy wastewater treatment plants were built in the Hudson River watershed. In China, there is a major involvement of international organisations in Shanghai, particularly in installing the necessary wastewater pollution control infrastructure. This is primarily due to the participation of organisations such as the Asian Development Bank and the World Bank" (Xiangrong et al. 2015, p. 8).

In developing countries, public funding is particularly limited because "environmental restoration activities compete with other priorities such as poverty alleviation, basic education or health care. The scarcity of financial resources in these countries, including the public domain, is often an insurmountable barrier to the restoration work aimed at improving the ecological health of rivers" (Khorshed 2008, p. 593). In order to reverse this situation that compromises sustainable development and expands the negative impacts of dying rivers, full economic assessment of the environmental benefits of river restoration is needed to demonstrate the desirability of redirecting public funding.

6.3.2 Regulatory framework

River restoration requires legal and organizational capacities disposed by states according to their particular traditions with regard to water management. In this respect, the inertia of preexisting legal and institutional structures has been relevant in the case of EU WFD implementation (Jager et al. 2016). The authors observed how, given the leeway afforded by the Directive, many states have opted to retain existing structures and procedures as far as possible, and this has meant a variety of forms of implementation of the Directive at national and subnational levels according to differing degrees of legal and institutional adaptation. Regarding legal capacity, the possibility of initiating a restoration operation at local level will depend to a large extent on the legal status of the river, and ownership of the riverbed and banks; if they are privately owned, negotiation with the landowners may be required (Heldt et al. 2016). It is generally ownership rights that are applied and taken into consideration by the courts in the event of opposition to a project. Regarding organizational capacity, river restoration is a question of multilevel, multidomain coordination, and is only possible when the actions of agencies with different responsibilities and jurisdictions are synchronized and mutually reinforcing. With this in mind, Warner et al. (2013) identify river restoration as being based on four strategies: (i) formal (national) legislation in countries, such as France, with a strongly centralized governmental system; (ii) policy programs stimulating actors to jointly formulate project proposals in countries such as the Netherlands, with a strong consensus-oriented culture; (iii) bottom-up network governance, including authorities from different institutional levels and actors from different policy domains, in countries such as the United Kingdom, with an absence of powerful state actors and major institutional fragmentation; and (iv) self-organized informal networks of enthusiastic actors in states such as Hungary and Romania with formal bureaucracies that lack the necessary capacity to implement river restoration.

The lack of a legal framework may compromise the inclusiveness of the organizational structure. The right of private parties (environmental organizations and residents) to defend the collective interest on an equal footing with public stakeholders differs depending on the political culture of the country. For example, in the USA, when the association formed in 1963 – the Scenic Hudson Preservation Conference – decided to initiate legal action against the Federal Power Commission[2] in 1965, the court ruled that the association had legal standing on the basis that the cost of a project is only one of several factors to be considered, and that these should also include the preservation of natural beauty and national historic sites. This decision established a precedent and played a significant role in subsequent land-use and environmental battles (Lifset 2014). In order for environmental interest groups to become involved in such cases, they need to have an organizational structure that is recognized in the legal system of the country and approved by a public authority such as the Hudson River Valley Commission formed by the State of New York in 1965, the first time that such a body was created by the state.

In a study comparing the Hudson Valley case with that of the River Huangpu in China, the authors note that "public activism and NGO environmental organisations do not play the strong role in Shanghai that they did in the cleaning of the Hudson … It was only after the economic reforms of the late 1970s that nongovernmental groups had the option to be involved in such things as environmental protection … Still, there are signs that they could eventually have an influence on Shanghai's water politics" (Xiangrong et al. 2015, p. 12).

Similarly, local residents and activists can also come together to oppose restoration operations outside formal institutional channels, typified by environmental legislation and organized social movements (including the work of NGOs). For example, Fox et al. (2016, p. 99) described a grassroots effort by the Friends of the Upper Roberts Meadow Reservoir to preserve a dam on the Mill River in Northampton, Massachusetts in order to "protect an important wildlife habitat and protect a rare water vista at this site," which they understand as "a beautiful spot where even the casual passer-by might glimpse a river otter or blue heron, Merganser ducks, beaver, or even a kingfisher. The wildlife living here depends upon this established habitat." This case and similar examples of dam removal opposition illustrate the key role played by people's engagement with their riparian landscape, in disregard of existing regulatory frameworks.

6.3.3 Governance process

Finally, the existence of laws and institutions is indispensable to be able to initiate river restoration operations, but it is not sufficient to ensure that they are carried out. Inclusive decision-making processes based on governance systems that involve all the public and private actors are also essential for these operations' acceptability and feasibility. The case of the removal of four large dams along the Klamath River (Oregon and California, USA) described by Gosnell and Kelly (2010, p. 365) demonstrates that laws alone are not enough to remove dams and overcome opposition by local stakeholders: "While the laws [particularly the ESA, the Federal Power Act[3] (FPA) and a range of federal-tribal trust responsibilities] were necessary to force change and, to some degree, forestall further ecological impacts, they were inadequate by themselves for resolving complex social and ecological problems without the right forum to explore socially sustainable ways to implement them." The interests of all the stakeholders in the river must be brought into play. The authors explained that multiple factors contributed to the agreement to remove the dams. One critical aspect was the improved social relation between formerly antagonistic Indian tribes and nontribal farmers and ranchers, which came about due to a number of local collaborative processes during the early 2000s. The authors described three developments that "catalysed constructive problem-solving: (1) investment in novel approaches that move from a focus on prevention of loss to investment in recovery; (2) investment in more collaborative, 'horizontal' forums and institutions that support decentralised, flexible approaches; and (3) more explicit attention to the tribal trust responsibility" (Gosnell and Kelly 2010, p. 364).

However, public acceptance and consensus achievement do not result from all participatory endeavors. Many authors have highlighted the importance of the procedural features of participatory process, such as incorporation of participants' knowledge and two-way interaction among interested parties (Kochskämper et al. 2016), while others have focused on the importance of the context in which participatory process occurs. With regard to the latter, Ballester and Mott (2016), in a comparative study of effectiveness of public participation in the Ebro River Basin (Spain) and the Tucson Basin (Arizona), observed that participatory legal framework (existence of detailed legal requirements for public participation), political leadership (involvement of the competent authority in providing the resources necessary for effective participation), and social awareness (existence of social movements specifically devoted to the reason for the participatory process), determine the capacity of public participation to improve the environmental standards of plans. Authors reported that "Despite the more developed participatory legal framework for water management in Europe, the impact of the participatory process is similar in both case studies. This may indicate that higher democratic culture in Tucson case is compensating for its less developed participatory legal framework. On the other hand, the mere existence of a participatory legal framework guarantees the development of a participatory process, and therefore the beginning of a more or less productive interaction between the public and regulators, which otherwise would not have taken place in a context of a less democratic culture, such as in the Ebro Basin" (Ballester and Mott 2016, p. 15).

What emerges from the scientific literature is a convergence on the fact that, in order to carry out a restoration operation, the acceptance of the stakeholders must be obtained by demonstrating the gain in quality for the area and for all occupants. To ensure that all stakeholders agree to the removal of a weir or dam, the perimeter of the gains (and losses) must be extended beyond the immediate perimeter of the river, and all the stakeholders must be involved and participate in the decision. A study conducted by Buijs (2009) is based on a Dutch example. It describes how residents base their opposition or support of a restoration operation on three distinct frames that give meaning to river restoration projects: "(i) an attachment frame, focusing on cultural heritage and place attachment, (ii) an attractive nature frame, focusing on nature as attractive living space and the intrinsic value of nature, (iii) a rurality frame, focusing on rural values, agriculture and cultural heritage. Resistance to river restoration plans stems from the attachment and rurality frames. People using these frames challenge safety arguments for river restoration and highlight potential threats to sense of place and to agriculture" (Buijs 2009, p. 2680). The author concludes with a warning for future projects: "Positive attitudes towards restoration are mainly related to the enhancement of scenic beauty. However, as the most recent trends in Dutch river management policy are to focus more on safety measures and less on ecological restoration, this support may turn into resistance in future projects. If project initiators promote new river restoration projects only as enhancing safety, many residents may be susceptible to arguments expressed by opponents of river restoration, who argue that such plans are ineffective or inefficient to diminish the risk of flooding" (Buijs 2009, p. 2689).

Likewise, if the gains expected from an operation are to be extended to a larger perimeter with more issues, and hence more stakeholders, the negative impacts on the functioning of the river must be considered more extensively, and treatment of pollution at a different scale must also be considered. Thus, the treatment of pollutants and their prevention at source remains one of the main objectives of restoration operations. However, they encounter the problem of a spatial mismatch between water resource management and land-use policy and planning. It is increasingly important to be able to take action against the landowners (farmers, residents), particularly in the case of non-point-source pollution, as stressed by Moss (2004, p. 85): "On the one hand water and river authorities generally possess limited means of influencing uses of land and water both up-stream (longitudinally) and across the river catchment or basin (laterally). On the other hand, they exercise little leverage over other policy fields which have a direct bearing on water quality and quantity issues." This situation is not resolved by setting up river basin authorities, because "the replacement of existing institutional units by institutions oriented around biophysical systems will inevitably create new boundary problems and fresh mismatches" (p. 87).

6.4 Field-testing the river restoration new paradigm: from operation acceptance to rejection

Consensus achievement in river restoration takes variable forms depending on the (i) actors, (ii) objectives, and (iii) conflicts associated with the operations. Regarding the actors, river restoration is carried out by different agents according to the local or regional and state levels of decision involved in the operation, ranging from individuals, local communities, and cities to river unions, basin committees, and states. These actors may be private actors, owners of the land to be developed or the hydraulic structure (mill, dam), and/or public actors. We can distinguish, for example, European countries where these operations are conducted by platforms specializing in water management – such as water committees, bringing together users of water by watershed, or river unions – and interacting with other public actors (state services, local authorities). In other places, such as North American states, restoration operations involve federal agencies, states (with their own restoration programs), tribal governments with their sovereign rights over fisheries, and owners of hydraulic works (in the USA, 64% of dams are privately owned, as reported by the American Society of Civil Engineers in 2017).

Regarding the objectives, restoration operations show varying degrees of ambition in recovering the natural condition of riverine ecosystems, ranging from partial improvements that modify riverbank geomorphology using bioengineering techniques to complete transformations that include dam dismantling. The most radical step concerns the removal of large dams (usually hydroelectric). These interventions have a biological dimension (to restore ecological continuity) as well as a hydromorphological dimension (to enhance sediment transfer). However, they are essentially presented to stakeholders for the return of fish communities in the river to achieve good biological quality and proper functioning of the stream. Conversely, the natural transfer of sediments is rarely evoked in presentation of the projects or debates.

Finally, regarding the conflicts, disagreements in restoration operations show different degrees of complexity depending on the functional dimensions involved. We can distinguish less-complex conflicts (proximity conflicts) related to the social use of the river, and more-complex conflicts that include social, economic, and environmental dimensions. Examples of the latter are found in projects to dismantle dams that are not necessarily met with the approval of local residents, who even raise opposition leading to the abandonment of projects (Figure 6.3). In the case of small dams, resistance is frequently a response to being told what to do by outsiders, rather than opposition to dam removal itself, which highlights the importance of the participation of local residents in the success or failure of a restoration operation (Fox et al. 2016).

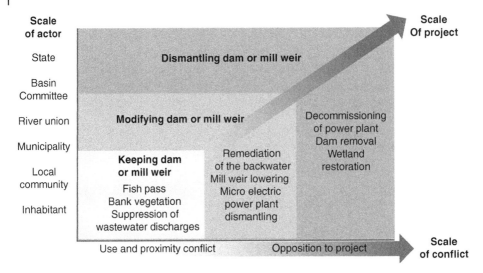

Figure 6.3 River restoration projects according to scales of actors, objectives and conflicts: the example of mill weirs or dams dismantling.

The latter examples presented illustrate the paradigm shift and varying degrees of controversy in local reception. Two different patterns are seen, according to the main goal of the restoration projects: (i) partial restoration in urban contexts aimed at reconnecting cities and dwellers (e.g. Colombia and Spain) and (ii) complete restoration in rural contexts including dam removal (e.g. Canada). Examples of the first group can be found worldwide, in developed and developing countries. In these cases, cities have taken advantage of the rivers to boost ambitious urban renewal operations. These projects have changed the general perception of rivers from polluted, dangerous environments to leisure and meeting places, thereby raising the land value of the riverbanks. Despite some concerns as to gentrification and banalization of the river landscape, there is a general consensus on the benefits of these projects that are widely accepted by local public and private actors (Hartig and Wallace 2015; Chang and Huang 2011). Examples of the second group are more recent and limited to developed countries that show higher standards of rivers' ecological performance and greater awareness of the advantages of adaptive environmental governance. The improvement of the ecological health of rivers lies at the center of these projects that are more controversial due to contrasting perceptions among local actors of the potential benefits deriving from dam and channel removal.

6.4.1 River restoration projects led by public actors in developing countries: the case of the Bogotá River (Bogotá, Colombia)

The Bogotá River is one of the most contaminated bodies of water in Colombia, and the world. It originates in páramo de Guacheneque (Villapinzón, Cundinamarca) in the center of the country and runs 336 km, through 41 municipalities with 1.3 million inhabitants and the Bogotá District with 6.763 million inhabitants, before joining the Magdalena River. Along its course, the river receives the sewage of approximately 20.9% of the Columbian population. The restoration of this river is cited as a successful case that can serve as a model for emerging countries and all of Latin America: "it is important to highlight that the project was catalogued as a model at continental level at the Integrated Management Seminar in Buenos Aires and at national level as the best environmental project in Colombia in 2014 (*Noticias Colombianas*, 2015)" (Mondragón-Monroy and Honey-Rosés 2016, p. 16). The project is called "Megaproyecto del río Bogotá" and is directed by the Regional Autonomous Corporation of Cundinamarca.[4] It is valued at over 500 million US dollars and financed by city of Bogotá property taxes and a World Bank loan of a further $250 million.

This project took 50 years to complete, from the first study in 1974 for the treatment plant, in 1989 the first wastewater treatment plants project, a general sanitation plan in 2004 and, in 2010, an agreement with the World Bank (Costa Posada and Urazán Bonells 2015). However, the ruling of the Colombian Federal Court of Justice in 2014 on disagreements between regions and municipalities is still pending. A Council of State judgment ordered that basic sanitation and water quality improvements be the top priorities for recuperating the Bogotá River. It declared that 23 national bodies and 43 municipal administrations were responsible for the environmental catastrophe in the river and needed to trace strategies for its sanitation and restoration, and made the Colombian state responsible for the river.

The scientific literature describes the collaboration between public actors and, for the restoration project, the river transformed into a stable ecosystem capable of sustaining life. Depollution is achieved with the construction of three wastewater treatment plants, flood protection by means of conventional hydraulic methods (containment, digging of the riverbed), and the evacuation of the residents of the banks to restore a floodplain, habitats, and wetlands, with recreational spaces for city dwellers. "The project includes the construction of a linear ecological corridor with pedestrian and bike paths, the construction of passive and active recreational spaces, the plantation of 500 000 native trees" (Mondragón-Monroy and Honey-Rosés 2016, p. 16). What is not mentioned is the consultation of the local residents. The evacuation of riverside homes was only addressed by the relocation of families, partly financed by the World Bank. "Plus, the relocation of 188 families established in the riverbank area is planned by giving them three subsidies covering their property's total value" (Mondragón-Monroy and Honey-Rosés 2016, p. 16).

6.4.2 River restoration projects led by public actors in developed countries: the case of the Besòs River (Barcelona, Spain)

The Besòs River Basin (BRB) lies to the north of Barcelona, has a surface area of approximately 1000 km^2, and is a representative example of a Mediterranean basin with occasional intense floods and scarce flow during much of the year, especially in dry seasons. In the 1960s and 1970s, its six main streams, formerly devoted to providing irrigation, energy, and raw materials in a rich agrarian landscape, were subject to overexploitation of water resources, contamination of ecosystems, and flooding due to the uncontrolled development of riverbanks. The Besòs River and floodplains, with over a million inhabitants living around them, became: (i) areas of growth; (ii) spaces for road and railway communications between Barcelona's urban core and its successive metropolitan rings; and (iii) channels for drainage, waste, and energy supply. In the process, major stretches of the riverbanks were neglected and left to decline, leading to the dwindling of local riverside residents' feeling of belonging that had been characteristic of the first half of the twentieth century. However, this condition underwent a reverse in the Eighties as a result of significant efforts to reappraise the river's social and environmental assets (Gordi 2005).

The BRB, one of the most urbanized European basins, belongs to the River Basin District of Catalonia (RBDC), part of the decentralized Spanish water management system. The RBDC is administered by the Catalan Water Agency, according to the multifunctional basin-based approach of the EU WFD. The Water Agency is responsible for developing the River Basin Management Plan of the RBDC and providing financing for water environment projects through water user tax and subsidies. Within this regional regulatory framework, a specific agency, the Consortium for the Defense of the Besòs River Basin, was created in 1988 for the purpose of coordinating all the local and regional administrations involved in water management and river restoration in the BRB. During the last thirty years, significant efforts have been made to provide water treatment plants, prevent flood risk, restore riparian ecosystems, and make riverbanks accessible. Likewise, resources have been invested in education about the river environment, combining information and environmental volunteering with local culture and sports promotion (Benages-Albert and Vall-Casas 2014).

Special restoration efforts have centered on river stretches closest to major urban centers, as a precondition for promoting urban renewal. The transformation of the lower stretch of the Besòs River where it meets the city of Barcelona exemplifies the significant social contribution of the restored river to a deprived urban area with a shortage of public spaces. In the 6 km of this stretch alone, renewed in the late 1990s as part of an emblematic urban project based on the parameters of smart water management and European funding, it has been estimated that 300 000 people who live within sight of the river benefit from its proximity in their everyday lives. The most significant contributions of this project were: floodplain recovery as public space (the Besòs River Park), ecological restoration of the river mouth, wastewater treatment improvement using constructed wetlands, removal of energy supply lines, and upgrading of riparian neighborhoods (Diputació de Barcelona 2017). For this purpose, in 1995, the Besòs River Consortium, a new governance platform at intermediate level between basin and municipalities was created.

Since 2000, public participation in the BRB has been undertaken according to the WFD rationale and deliberately promoted by the public administration. The BRB has its own basin committee to identify public perception of relevant water issues and elicit proposals to be included in the River Basin Management Plan. Previous participatory processes led by the Catalan Water Agency support planning cycles of six years. The first process (2006–2010) involved 152 people representing 110 bodies (Catalan Water Agency 2010). Although open to the general public, influential and organized stakeholders have a leading role in the committee, and lay public participation remains a challenge. Interestingly, the lay public is emerging as a relevant actor as the riverbanks recovered during recent decades become increasingly overused. The Besòs River Park, with two million users a year, manifests growing pressure derived from the river's social success, such as damage to the natural environment and conflicting overlap of leisure practices. In particular, repeated accidents between cyclists and walkers have raised social alarm, and planning for conflict prevention is currently a central concern of the Besòs River Park's management endeavors (Diputació de Barcelona 2017).

6.4.3 River restoration projects rejected by local communities: the case of the Mactaquac Dam (Canada)

In New Brunswick, Canada, the restoration project described here is associated with the renewal of hydropower concession and the decision to remove the Mactaquac Dam, a hydroelectric dam on the Saint John River (Sherren et al. 2017). With a height of 42 m, the Mactaquac Dam is one of the biggest dams that was planned to be dismantled in North America. Here it is the owner of the dam, the company New Brunswick Power, which conducted the public consultation. The company proposed in 2014 to the residents of New Brunswick that they decide on one of the following three choices: (i) the removal of the dam and a return to the original valley, (ii) a technical renovation of the dam and an extension of its operating life, (iii) or maintenance of the dam and restoration of the water spillways. Respondents were uncertain as to the environmental improvements associated with the removal of the dam. The operation costs were taken into account; the removal was seen as a waste, with negative aspects on the local economy (tourism related to the presence of the lake). In the end, New Brunswick Power decided to restore the dam despite the protests of regional Native American organizations and environmental associations (the local natives having voted for conservation).

This case study shows fundamental divergences around the functions of dams, opponents to their deletion favoring the production of green energy and, implicitly, the nonuse of nuclear energy. This situation is not specific to the case study but is found for other dams (Barraud 2017). Time also appears as one of the fundamental variables explaining the failure or success of dam removal. Opponents reject the immediate loss of scenery, aesthetics, and fun linked to the disappearance of the lake reservoir associated with a dam. They have little understanding of future gains from the restoration of ecological continuities for which they suspect significant development costs and delays (Fox et al. 2016).

In these cases, successful operation "typically depends upon achieving co-operation between actors operating in both the public and private spheres and that may hold different and not necessarily compatible interests" (Baker et al. 2014, p. 517). It is worth mentioning that in 2017, for the first time, the USA appears to have an annual net loss of dams, being the US Federal Energy Regulatory Commission (FERC)[5] licensing process for power generation "an important window of opportunity for institutionalising adaptive environmental governance toward the renegotiation of social and ecological values associated with rivers" (Chaffin and Gosnell 2017, p. 819).

6.4.4 Comparative reading of the case studies on restoration acceptance

The first and second case studies show partial river restoration projects in highly urbanized areas. The case of Bogotá involves a larger scale with a time delay in relation to the case of Barcelona. All the components of the ecosystem-based paradigm are present, even though the pollution and flood control part remain central. Both examples illustrate the importance of public international funds for financing the green infrastructure, as well as the coordination between public actors and the time needed to change the ways of doing things. The complex political adhesion of the public actors at the state, regional, and municipal levels of government, and the process of negotiations, ultimately ensures the project is passed. These projects represent a significant enhancement of metropolitan dweller's quality of life and, in general terms, are perceived as beneficial. Differences between cases, associated with economic and political specificities, can be found. They include the affluence of the society that conditions the time of project implementation and the speed at which the ecosystem-based paradigm permeates technical and political circles; and the level of public control of the urban transformations that leads toward more public administration-driven (Barcelona) or speculative market-driven (Bogotá) scenarios.

As for the social dimension, both cases coincide in the lack of consultation in the restoration projects that are deployed autonomously by the public administration for the sake of the public good. Only collateral social implications in the implementation phase, such as the relocation of riverbank inhabitants or changes in polluter practices, are worth mentioning. This residual role of local actors in the inception of the projects contrasts with recent operations to de-culvert rivers in densely populated areas that also form part of this group. They have previously required compensation to local landowners and reassuring residents that flood prevention has been taken into consideration. For example, in Taoyuan (Taiwan), the local authority has held over 50 open forums since 2011 with the aim of building public participation, consensus, and support for de-culverting and subsequent improvements. Certainly, the Barcelona, Bogotá, and Taoyuan examples meet social and environmental demands with the support of regional and local actors, and are characterized by wide public acceptance. However, over time these operations may present malfunctions derived from their social success. In the case of Barcelona, conflicts among users and poor maintenance are emergent issues that must be solved. In Taoyuan, the role of local residents has gone beyond just accepting the project; it has also involved maintenance, which is a crucial factor in the long-term feasibility and success of river space design. In this case, where local authorities usually allocate relatively small budgets for the maintenance of river projects, "local NGOs and communities are widely regarded as a solid, bottom-up force supporting urban river space as a multifunctional natural resource by regular cleaning-up, disposing of rubbish and environmental monitoring" (Chou 2016, p. 16).

Conversely, dam removal for achieving complete river restoration in rural contexts, illustrated by the third case study in Canada, is more controversial because it entails complex scenarios dominated by the political treatment of the mill's energy use and its implications on the environment (Barraud 2017) and local economy. In these cases, restoration challenges current river governance and is prone to generate rejection on the part of residents. Although decommissioning can bear economic, environmental, and cultural benefits, dam removal is unlikely to occur if local actors perceive that decommissioning brings more losses than opportunities (Brummer et al. 2017; Pejchar and Warner 2001). This points to the need to understand the stakeholders' contrasting perceptions in landscape change, and to develop participatory decision-making processes aimed at solving the conflicts. Consensus achievement becomes unavoidable in scenarios where local opposition may compromise river transformations. In these cases, the importance of the social dimension for the project's feasibility is paving the way for a prevalent role of social sciences research on river restoration.

6.5 Conclusions

The principles of river restoration have now permeated most public policies, in national and international institutions. Analysis of the implementation of restoration operations reveals a relative similarity between these operations across continents due to their shared conceptual and technical approach. It includes the shift of restoration objectives (improvement of water chemical quality, fight against floods, and, since 2000, the biological and morphological restoration of rivers).

Differences between countries in the way operations are carried out can be explained by contextual political factors such as regulatory framework, funding mechanisms, and governance processes. Although flood control is practiced almost everywhere, ecological restoration remains the preserve of Anglo-Saxon and European countries, with a recent extension to Asian cities (China, Korea, Japan, Taiwan). As for the recalibration of rivers for navigation, it may be presented in some countries as restoration operations – in Colombia, the Magdalena River, and in Peru, the Rímac River (Mondragón-Monroy and Honey-Rosés 2016) – which does not seem to fit the principles of restoration at all. However, river artificialization continues, whether it is in the destruction of agricultural streams, culverting in cities, dam construction to supply water and energy, or setting up wide gauge for navigation. In Europe, the WFD classification "Heavily Modified Water Body" and "Artificial Water Body" make it possible to continue the artificialization of rivers, such as the project in progress of setting up a wide gauge of 17 km of the Seine in France (between Bray and Nogent-sur-Seine).

As for opposition to restoration operations, it is generally related to the maintenance of the water supply status quo for hydroelectric production. Rejection of restoration is also observed when operations are imposed from outside, thus depriving local residents of their own management of the river. Against this backdrop, participatory schemes that foster convergence of interests are identified as a key indicator for successful river restoration (see Palmer and Allan 2006). In this respect, an exploration of knowledge and perspectives for supporting public involvement and consensus has become the prevalent aim of current social science research in river restoration. The success of restoration seems ultimately to depend on the capacity of public and private actors to achieve consensus on the projects, or the capacity of public actors to deploy them autonomously.

Notes

1 This federal agency under the US Department of the Interior has overseen, since 1902, water resource management to advance settlement in the West through the construction of large dams, reservoirs, and canals. Today, it assures the oversight and operation of projects built throughout the western United States for irrigation, water supply, and hydroelectric power generation to industries and municipalities.

2 It is an independent commission of the US government, which operated since 1930 to 1977 to license hydroelectric projects on the land or navigable water owned by the federal government.

3 Enacted in 1920 and amended several times since, its purpose is to coordinate the development of hydroelectric projects in the United States. This act created the Federal Power Commission.
4 This company operates mainly in the water and waste sectors, under the Colombian environmental government authority, and develops projects for a population of 7.3 million people, covering an area of 1.8 million hectares of seven river basins.
5 The Federal Energy Regulatory Commission is an independent agency that regulates the interstate transmission of electricity, natural gas, and oil.

References

American Society of Civil Engineers (2017). Infrastructure Report Card: Dams report. https://2017.infrastructurereportcard.org/wp-content/uploads/2017/01/Dams-Final.pdf (accessed 18 May 2021).

Baker, S. and Eckerberg, K. (2013). A policy analysis perspective on ecological restoration. *Ecology and Society* 18(2): 17.

Baker, S., Eckerberg, K., and Zachrisson, A. (2014). Political science and ecological restoration. *Environmental Politics* 23(3): 509–524.

Ballester, A. and Mott, K.E. (2016). Public participation in water planning in the Ebro River Basin (Spain) and Tucson Basin (US, Arizona): impact on water policy and adaptive capacity building. *Water* 8: 273.

Barraud, R. (2017). Removing mill weirs in France: the structure and dynamics of an environmental controversy. *Water Alternatives* 10(3): 796–818.

Benages-Albert, M. and Vall-Casas, P. (2014). Vers la recuperació dels corredors fluvials metropolitans: el cas de la conca del Besòs a la regió metropolitana de Barcelona. *Documents d'Anàlisi Geogràfica* 60(1): 5–30.

Brummer, M., Rodríguez-Labajos, B., Thanh Nguyen, T., et al. (2017). "They have kidnapped our river": dam removal conflicts in Catalonia and their relation to ecosystem services perceptions. *Water Alternatives* 10(3): 744–768.

Buijs, A.E. (2009). Public support for river restoration: a mixed-method study into local residents' support for and framing of river management and ecological restoration in the Dutch floodplains. *Journal of Environmental Management* 90: 2680–2689.

Carré, C. and Haghe, J.-P. (2013). Spatialization of political action applied to waterways management. *Urban Environment* 7: 1–17.

Catalan Water Agency (2010). Memòria dels processos participatius de l'Aigua. Directiva Marc de l'Aigua 2006. Barcelona: Generalitat de Catalunya.

Chaffin, B.C. and Gosnell, H. (2017). Beyond mandatory fishways: federal hydropower relicensing as a window of opportunity for dam removal and adaptive governance of riverine landscapes in the United States. *Water Alternatives* 10(3): 819–839.

Chakraborty, A. (2013). Developing rivers: how strong state and bureaucracy continue to suffocate environment-oriented river governance in Japan. *SAGE Open* 3(4). https://doi.org/10.1177%2F2158244013501329.

Chang, T.C. and Huang, S. (2011). Reclaiming the city: waterfront development in Singapore. *Urban Studies* 48(10): 2085–2100.

Chou, R.J. (2016). Achieving successful river restoration in dense urban areas: lessons from Taiwan. *Sustainability* 8(11): 1159.

Christian-Smith, J., Gleick, P.H., Cooley, H., et al. (2012). *A Twenty-First Century US Water Policy*. Oxford: Oxford University Press.

Costa, L.M., Vescina, L., and Barcellos, D. (2010). Environmental restoration of urban rivers in the metropolitan region of Rio de Janeiro, Brazil. *Urban Environment* 4: 13–26.

Costa Posada, C. and Urazán Bonells, C. (2015). The Río Bogotá environmental recuperation and flood control project. In: *Water and Cities in Latin America Challenges for Sustainable Development* (eds I. Aguilar-Barajas, J. Mahlknecht, J. Kaledin, et al.), 166–181. London: Routledge.

Cullinane T., Skrabis, K.E., and Gascoigne, W. (2012). Ecosystem restoration. In: *The Department of the Interior's Economic Contributions*, FY 2011. US Department of the Interior. https://www.doi.gov/sites/doi.gov/files/migrated/ppa/upload/Chapter-4.pdf (accessed 9 December 2020).

Damanik, F.K. and Pratiwi, W.D. (2017). Consideration of tourism riverfront development elements for Pekanbaru City transformation. *Journal of Regional And City Planning* 28(2): 140–150.

Diputació de Barcelona (2017). Parc fluvial del Besòs. Retrieved November 2017. http://parcs.diba.cat/es/web/fluvial/el-parc-fluvial (accessed 09 December 2020).

European Environment Agency (2016). *Rivers and Lakes in European Cities: Past and Future Challenges*. Report no. 26, Luxembourg: Publication Office of the EEA.

Fox, C.A., Magilligan, F.J., and Sneddon, C.S. (2016). "You kill the dam, you are killing a part of me": dam removal and the environmental politics of river restoration. *Geoforum* 70: 93–104.

Gordi, J. (2005). El paisatge fluvial a la conca del Besòs. Ahir, avui…, i demà? *Lauro* 30: 82.

Gosnell, H. and Kelly, E.C. (2010). Peace on the river? Social–ecological restoration and large dam removal in the Klamath basin, USA. *Water Alternatives* 3(2): 361–383.

Grabowski, Z.J., Denton, A., Rozance, M.A., et al. (2017). Removing dams, constructing science: coproduction of undammed riverscapes by politics, finance, environment, society and technology. *Water Alternatives* 10(3): 769–795.

Hartig, J.H. and Wallace, M.C. (2015). Creating world-class gathering places for people and wildlife along the Detroit riverfront, Michigan, USA. *Sustainability* 7(11): 15073–15098.

Heldt, S., Budryte, P., Ingensiep, H.W., et al. (2016). Social pitfalls for river restoration: how public participation uncovers problems with public acceptance. *Environmental Earth Sciences* 75: 1053.

Hillman, M. and Brierley, G. (2005). A critical review of catchment-scale stream rehabilitation programmes. *Progress in Physical Geography* 29(1): 50–70.

Jager, N.W., Challies, E., Kochskämper, E., et al. (2016). Transforming European water governance? Participation and river basin management under the EU Water Framework Directive in 13 Member States. *Water* 8: 156.

Khorshed, A. (2008). Cost–benefit analysis of restoring Buriganga River, Bangladesh. *Water Resources Development* 24(4): 593–607.

Kochskämper, E., Challies, E., Newig, J., et al. (2016). Participation for effective environmental governance? Evidence from Water Framework Directive implementation in Germany, Spain and the United Kingdom. *Journal of Environmental Management* 181: 737–748.

Lifset, R. (2014). *Power on the Hudson: Storm King Mountain and the Emergence of Modern American Environmentalism*. Pittsburgh, PA: University of Pittsburgh Press.

Mollinga, P. (2008). Water, politics and development: framing a political sociology of water resources management. *Water Alternatives* 1(1): 7–23.

Mondragón-Monroy, R. and Honey-Rosés, J. (2016). *Urban River Restoration and Planning in Latin America: A Systematic Review*. Vancouver, Canada: University of British Columbia Faculty Research and Publications, Open Collections.

Moss, T. (2004). The governance of land use in river basins: prospects for overcoming problems of institutional interplay with the EU Water Framework Directive. *Land Use Policy* 21: 85–94.

Nakamura, K. (2008). River restoration efforts in Japan: overview and perspective. *The 4th International Workshop on River Environment*, Korea (3 June 2008).

Nakamura, K., Tockner, K., and Amano, K. (2006). River and wetland restoration: lessons from Japan. *Bioscience* 56: 419–429.

Ormerod, S.J. (2004). A golden age of river restoration science? *Aquatic Conservation-Marine and Freshwater Ecosystems* 14(6): 543–549.

Palmer, M.A. and Allan, J.D. (2006). Restoring rivers. *Issues in Science and Technology* 22(2): 40–48.

Pejchar, L. and Warner, K. (2001). A river might run through it again: criteria for consideration of dam removal and interim lessons from California. *Environmental Management* 28(5): 561–575.

Pretty, J.L., Harrison, S.S.C., Shepherd, D.J., et al. (2003). River rehabilitation and fish populations: assessing the benefit of instream structures. *Journal of Applied Ecology* 40: 251–265.

Sherren, K., Beckley, T., Greenland-Smith, S., et al. (2017). How provincial and local discourses aligned against the prospect of dam removal in New Brunswick, Canada. *Water Alternatives* 10(3): 697–723

Speed, R., Li, Y., Tickner, D., et al. (2016). *River Restoration: A Strategic Approach to Planning and Management*. Paris: UNESCO.

US Department of the Interior (2011). A Summary of 16 Programs and Shared Institutional Challenges. Report for River Restoration: Exploring Institutional Challenges and Opportunities. Albuquerque, New Mexico (14–15 September 2011).

Vollmer, D., Prescott, M.F., Padawangi, R., et al. (2015). Understanding the value of urban riparian corridors: considerations in planning for cultural services along an Indonesian river. *Landscape and Urban Planning* 138: 144–154.

Warner, J.F., van Buuren, A., and Edelenbos, J. (2013). *Making Space for the River: Governance Experiences with Multifunctional River Flood Management in the US and Europe*. London: IWA Publishing.

Wharton, G. and Gilvear, D. (2006). River restoration in the UK: meeting the dual needs of the European Union Water Framework Directive and flood defence? *International Journal of River Basin Management* 5(2): 143–154.

Xiangrong, W., Shixiong, W., Guotao, P., et al. (2015). Ecological restoration for river ecosystems: comparing the Huangpu River in Shanghai and the Hudson River in New York. *Ecosystem Health and Sustainability* 1(7): 1–14.

7

How to Better Involve Stakeholders in River Restoration Projects: The Case of Small Dam Removals

Marie-Anne Germaine[1], Ludovic Drapier[2], Laurent Lespez[2], and Beth Styler-Barry[3]

[1] Université Paris Nanterre, Nanterre, UMR, 7218, CNRS LAVUE, France
[2] Université Paris Est Créteil, UMR 8591 CNRS Laboratoire de Géographie physique, Meudon, France
[3] The Nature Conservancy – New Jersey Field Office, Chester, NJ, USA

7.1 Introduction

In the Western world, the removal of dams is one of the most emblematic of all ecological river restoration projects (Sneddon et al. 2017). Removals often target large dams but, above all, the numerous mill dams, many of which are now unused, and also fisheries or agricultural dams. These small dams (height between 0.5 and 8 m) may obstruct the movement of migratory fish (Kemp and O'Hanley 2010; Brown et al. 2013), disturb the sediment transport, and create artificial water bodies behind the dam, leading to ecological habitat modification, eutrophication, etc. (Nilsson and Berggren 2000; Heinz Center 2002; Garcia de Leaniz 2008). Their removal aims to restore natural channel processes and fish habitats (Hart et al. 2002; Doyle et al. 2003). Their implementation remains problematic, however, because the large majority of small dams are privately owned (for the most part by riparian landowners but also companies). Furthermore, these dams are located on small rivers (<6 Strahler stream order) along which a wide variety of stakeholders are present: river users (anglers, kayakers, nature conservation associations, etc.) but also riparian landowners, inhabitants, and elected officials who are more or less involved in, or affected by, the dam removal process. Projects for the removal of small dams often generate local conflict.

Social sciences are concerned with the conflictual nature of dam removal projects. Whilst most dams are small and not used for economic purposes, plans to remove them have nonetheless often been conflictual, both in North America (Born et al. 1998; Fox et al. 2016; Magilligan et al. 2017; Sneddon et al. 2017) and in Europe (Eden et al. 2000; Lejon et al. 2009; Jähnig et al. 2011; Emery et al. 2013; Germaine and Barraud 2013a, 2013b; Germaine and Lespez 2017). These studies highlight the importance of cultural and historical dimensions in anti-dam removal protests. Dams constitute symbols of local history which communities identify with (Lejon et al. 2009; Fox et al. 2016; Howard et al. 2016;

River Restoration: Political, Social, and Economic Perspectives, First Edition. Edited by Bertrand Morandi, Marylise Cottet, and Hervé Piégay.

Brummer et al. 2017), family or private legacy which riparian landowners have invested heavily in rehabilitating buildings (mills converted into main or secondary homes) or hydraulic structures (Le Calvez 2017), while the reservoir provides a scenic landscape for populations established upstream (Barraud and Germaine 2017).

Conflicts also arise from the fact that users do not share the same vision as dam removal advocates regarding the river and its associated natural environment (Germaine and Barraud 2013a; Fox et al. 2016; Jørgensen 2017). User interviews reveal not only the attachment of riverside populations to the landscape of the modified river but also the fact that the landscape is perceived as a fully fledged ecosystem which has been in place long enough to support attested biodiversity (Barraud 2017; Jørgensen 2017). The species inhabitants highlights are not only migratory fish but fish living permanently in the water bodies, birds and other animals, and also plant species. All of which are identified as important by local inhabitants and anglers who use these areas on a daily basis. They describe these artificial environments, such as the water bodies created upstream of the Vezins and Roche Qui Boit dams in the Sélune valley in France as "natural" or "wild" (Germaine and Lespez 2017). The production of renewable energy is also put forward by dam removal opponents as an ecological and economic argument which justifies maintaining such hydraulic works, especially in the case of public facilities as for the Sélune dams in Normandy or in the Kennebunk dam removal project in Maine (USA) (Drapier et al. 2017; Germaine and Lespez 2017).

Studies focused on the governance of dam removal projects, and more generally river management, reveal the difficulty of defining a shared project which takes into account ecological stakes and the needs of local communities (Vogel 2008). Conflicts can be the result of governance that does not comply with local expectations (Germaine and Lespez 2017; Perrin 2018). We propose to focus here on a preliminary point: identification of stakeholders. Our hypothesis is that this identification is an essential prerequisite for taking into account and involving everyone in the construction of such projects. While this is one of the conditions for the success of such projects, this identification is not easy, especially for small rivers where stakeholders are not well structured. The difficulty is largely related to the multiplicity of uses associated with a dammed river. River managers take into account the needs and expectations of easily identified direct users (fishermen, kayakers, energy production, etc.) and dam owners, whereas other users (often not generating any economic benefits) like walkers (leisure, aesthetic appreciation, nature observation, etc.) are neglected often for purposes of efficiency (Reed 2008; Barraud and Germaine 2013; Lespez et al. 2016). Furthermore, it seems crucial to take into account local inhabitants because they express distinctive problems concerning property issues and risk (flood or erosion, for example). This raises the issue of defining who the stakeholders are and how managers could involve all of them (Grimble and Wellard 1997, Duram and Brown 1999, Junker et al. 2007) since social studies demonstrate the conflict comes from users but also from local inhabitants, who are deeply attached to their river.

This chapter focuses on stakeholders but not on conflict or governance processes. Our objective is to demonstrate that the involvement of all stakeholders is a key element of a project's success. The aim is to examine how projects can be enriched by drawing on the expectations and reactions of all stakeholders concerned. The objective is also to help managers ensure that all stakeholders are involved. For this reason, we chose to start from the example of two small dam removal projects where the promoters have tried to take into account all the people concerned to build the dam removal project. Conflicts have therefore been prevented, and although there may be people dissatisfied with the final draft, no objections have been publicly expressed. This study is based on two case studies, one in France, the other in the United States of America (USA). These regions are geographically similar but have a very different institutional frameworks (Drapier et al. 2018) as explained in the case studies presentation. This choice is intended to examine the weight of stakeholders in project development and not the effects related to distinct governance processes.

7.2 The role of stakeholders in dam removal in two different institutional contexts

The two case studies are located in similar geographic and geomorphologic contexts (Normandy in north-western France and New Jersey in north-eastern USA) but in very different regulatory contexts.

7.2.1 Regulatory frameworks in France and USA

In Europe, the WFD (2000) requires member states to achieve "good ecological status" of water bodies. Ecological continuity is one of the parameters required to achieve good ecological status. France is the country which has gone furthest in this direction by tightening regulations: Article L214-17 of the Environmental Code imposes barriers located on important rivers to be modified or removed. France has adopted a regulatory arsenal that obliges the maintenance or restoration of ecological continuity on the part of the hydrographic network (approximately 90 000 dams) (SANDRE 2014). It concerns especially dams located on rivers classified under Lists 1 and 2 by the French Environmental Code, which means that no new barriers can be built and that existing ones must be modified in order to allow the movement of sediment and of migratory fish. In the USA, there is no regulatory obligation to maintain ecological continuity. Hydroelectric facilities located on rivers which are deemed navigable under federal legislation are subject to control by the Federal Energy Regulatory Commission (FERC). The Commission grants operating licenses (with a duration of 30 to 50 years) and decides on their renewal. Since 1986, the Electric Consumers Protection Act has specified that the FERC must give equal consideration to the various aspects: energy use, recreation (navigability, for example), wildlife protection, and environmental quality considerations (Richardson 2000; Pohl 2002). The management of nonproductive dams, like the Finesville dam, however, falls under state legislation and is therefore highly heterogeneous. It is often during the renewal of operating permits for hydroelectric facilities that removal projects are initiated (Doyle et al. 2003; Germaine and Lespez 2017). While they aimed primarily at achieving ecological quality (and particularly healthy fish stocks), safety and the loss of economic profitability play a key role in triggering such projects (Drapier et al. 2018).

7.2.2 Interviews methods to analyze stakeholders involvement

We conducted semistructured interviews in both sites once dams were removed (six years after in Finesville, two years after for the Bateau dam). First, we met dam removal promoters (project managers, financial and technical partners) and local officials, and analyzed project reports and documents to rehearse the history of each project. This method demonstrates the way the dialog can be enriched by increasing the number of people involved in it. The aim was also to clarify the different stages of opening: when, for what reason, and by what means, and through which stakeholder new protagonists enter the scene? Then the objective was to reconstitute in detail the dam removal process.

At the same time, we met people directly involved in the participation process (official or more informal meetings) or identified as river users, but also waterfront inhabitants usually not taken into account (20 people in Finesville and 13 people in the Bateau dam). Users are represented by associations' leaders (fishing, paddling, for example), whereas inhabitants were contacted using a door-to-door method within a 500 m radius of the restored sites (Tapsell 1995; Tunstall et al. 1999; Åberg and Tapsell 2013). These interviews aimed to understand inhabitants' feedback: the aim was not to assess their satisfaction with regard to the final project but rather concerning its building. What is inhabitants' appreciation of the concertation process? Do they feel they have been able to express themselves? Have they been listened to and taken into account?

7.2.3 Dam removal case studies in France and the USA

Located on two low-energy rivers, the two dams were minor barriers typical of the types of small dams which are often targeted to restore free movement of migratory fish (Figures 7.1 and 7.2).

- The Bateau dam is in Normandy on the Orne River, which flows into the English Channel after having run its 170 km long course. The dam is 65 m long and 1.9 m high. It was built in 1830 and served as a mill until 1970. Then it produced hydroelectric energy until 2012. Despite being equipped with a fish pass, the dam still hindered fish migration (Atlantic salmon, sea trout, eel). The water impoundment was also damaging the ecosystem (FDPPMA Calvados 2015).
- The Finesville dam is located in New Jersey, on the Musconetcong River, which runs 68 km to the Delaware River. The dam is 33 m long and 2.7 m high. It had been put to several different purposes after its construction in the second half of the eighteenth century (first a forge, then a sawmill, a spinning mill, and a cutlery factory) before being converted into a paper mill in the twentieth century and finally closing in 1990. Recreational fishing on the Musconetcong River targeted trout and bass. Federal agencies and local associations hoped that dam removal would allow shad to migrate up the river.

Both sites are located in rural areas surrounded by agricultural land with low population density (Figures 7.1. and 7.2). The sites, especially the disused dam water reservoirs, are nonetheless bordered by residential properties located in the immediate vicinity of the river, which is appreciated by inhabitants for well-being and aesthetic purposes. They also use the river for leisure purposes, such as fishing and canoeing (see Figure 7.3).

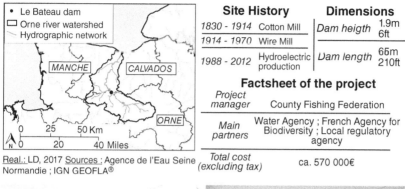

- Le Bateau dam
- ☐ Orne river watershed
- Hydrographic network

MANCHE   CALVADOS

ORNE

0   25   50 Km
N 0   20   40 Miles

Real.: LD, 2017 <u>Sources :</u> Agence de l'Eau Seine
Normandie ; IGN GEOFLA®

Site History		Dimensions	
1830 - 1914	Cotton Mill	Dam heigth	1.9m 6ft
1914 - 1970	Wire Mill		
1988 - 2012	Hydroelectric production	Dam length	65m 210ft

Factsheet of the project

Project manager	County Fishing Federation
Main partners	Water Agency ; French Agency for Biodiversity ; Local regulatory agency
Total cost (excluding tax)	ca. 570 000€

Pre removal   AESN, FR, October 2012

Post removal   LD, November 2016

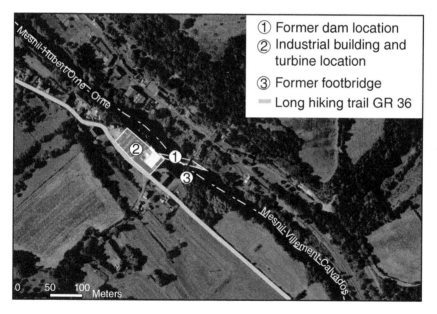

① Former dam location
② Industrial building and turbine location
③ Former footbridge
▬ Long hiking trail GR 36

Figure 7.1 Location and characteristics of the Bateau dam removal project, Orne River (France).
Source: IGN Geoportail-LD, 11/2018.

Site history

1751 - 1789	Iron forge
1790 - ca. 1850	Saw mill Grist mill
ca. 1850 - ?	Cutlery
1950 - 1990	Paper mill

Dimensions

Dam heigth 2.7m
9ft

Dam length 33m
109ft

Real.: LD, 2017
Sources : USGS

Factsheet of the project

Project manager	Musconetcong Watershed Association
Main partners	Natural Resources Conservation Service ; NOAA ; US Fish and Wildlife Service ; American Rivers
Total cost	$ 265000

Pre removal   USDA NRCS, May 2008

Post removal   LD, April 2017

① Former dam site

② Old mill active until the early XXth century

③ Former turbine premisses

④ Production facility until 1990

Figure 7.2 Location and characteristics of the Finesville dam removal project, Musconetcong River (New Jersey, USA). *Source:* US Geological Survey-LD, 11/2018.

(a)

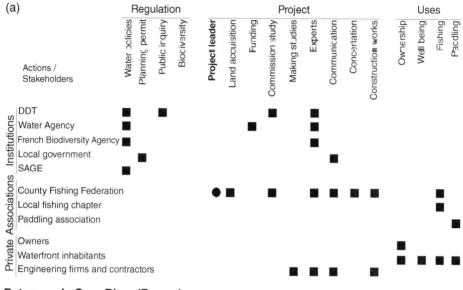

Bateau weir, Orne River (France)

(b)

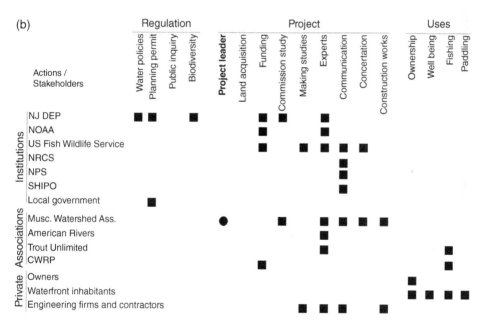

Finesville dam, Musconetcong River (USA)

Figure 7.3 Involvement of stakeholders in the Bateau (a) and Finesville (b) dams removal projects.

7.3   From the involvement of the stakeholders to the enrichment of the project

The analysis of these two dam removal projects reveals the multiplicity of stakeholders, and how they were identified and consulted, in a more or less satisfactory fashion, during the inception stage of the project.

7.3.1   A project initially builds by ecologically driven stakeholders

The projects were launched at a time when the dams were no longer in use and their owners were seeking to free themselves from the associated liabilities (in 2007 for Finesville, and in 2012 for the Bateau dam). The two dam removal projects were driven by local associations which advocated the ecological restoration of rivers: the Musconetcong Watershed Association for the Finesville dam and the County Fishing Federation[1] for the Bateau dam. Figure 7.3 summarizes the characteristics of all the stakeholders who took part in the definition of the project according to their status – institutions, associations, and individuals – but also their role in the development of the project. Some have a regulatory role: they check that the project complies with water and biodiversity regulations, deliver building permits, or monitor the conduct of any public inquiry. Others participate directly in the building of the project which requires leadership, land acquisition, funding, commissioning, and carrying out studies, monitoring by experts, and finally carrying out work. Both projects also include a communication and concertation component. Finally, a third category of actions refers to uses that connect local stakeholders to the project: they are owners of the land affected directly or indirectly, users (fishermen and kayakers here) or whether they are attached to the site by the well-being it provides them.

Created in 1992, the Musconetcong Watershed Association brings together local stakeholders (individuals, families, and associations) with the objective of improving the quality of natural and cultural resources in the watershed area through education and awareness programs and through the promotion of environmentally friendly practices. Amongst its actions, the association promotes the ecological restoration of rivers. Toward the end of 2007, the mill owner contacted the association after hearing about previous projects it carried out. The association then spearheaded the Finesville dam removal project in conjunction with its permanent partnership (the Musconetcong River Restoration Partnership) which facilitates the organization of river continuity restoration projects by bringing together experts and potential backers: federal agencies: Natural Resources Conservation Service (NRCS), National Oceanic and Atmospheric Administration (NOAA), US Fish and Wildlife Service (FWS), National Park Service (NPS); state agencies: New Jersey Department of Environmental Protection (NJDEP) Division of Fish and Wildlife; associations: American Rivers, North Jersey Resources Conservation and Development, Trout Unlimited; and private owners. The Corporate Wetlands Restoration Partnership (CWRP) also participated in the project funding. It is not part of the permanent partnership (the Musconetcong River Restoration Partnership) and brings together public (federal and state agencies) and private (companies and associations) stakeholders to fund wetlands restoration projects. The first public event regarding the Finesville dam removal was a meeting held in late 2008 to receive comments and questions from the public. Then, a feasibility study was launched at the beginning of 2009 followed by an impact study in 2010 under the supervision of the NRCS. The official decision to remove the dam was taken in August 2010. Work began in November 2011. The total cost of the project was $265 000.

In France, legal provisions required that the owner of the Bateau dam brings it into conformity with the law. Since 2012, the Orne River has been register on Lists 1 and 2 by the French Environmental Code, which means that no new dams can be built and that existing ones must be modified in order to allow the movement of sediment and of migratory fish. The river is a priority area under the French "National eel management plan" and the dam was also designated as a priority during the French multiparty environment debate (Grenelle de l'environnement) initiated in 2007. While there are other technical options (fish passes or river diversion structures), removal is usually the most favored option since the Seine-Normandy Water Agency, in charge of water management of the Seine river basin, subsidizes between 80 and 100% of removal costs. The owner of the Bateau dam was officially notified of their responsibility by the local regulatory agency (DDT[2]) which represents French State authority at the Orne county level). Since they considered the cost of the works excessive, and based on advice from both the County Fishing Federation and the Water Agency, the owner chose to sell the site (both the hydraulic dam and the adjoining mill). The County Fishing Federation acquired the entire site in August 2013 and has taken charge of the project in accordance with its mission to protect aquatic environments. It is assisted by a steering committee made up of the Water Agency, which funded the project (the cost of studies and works including the purchase of the site and the demolition of the mill), institutions providing technical expertise (French Agency for Biodiversity[3]), and the local regulatory agency (DDT). Whereas the Musconetcong Watershed Association steered the removal project, the institution in charge of the Water Development and Management Plan on the Orne River (SAGE[4]) was, on the contrary, much less heavily involved. Indeed, SAGE did not participate in the planning of the removal project, even though part of its remit is to bring together elected officials, state authorities, and users responsible for implementing action programs to improve water and environment quality. A study was commissioned in 2013 to estimate the cost of works and the feasibility of demolishing the adjacent mill, which required an asbestos removal project. Works eventually took place from June to November 2016 at a total cost of almost €600 000.

For both projects, the same types of stakeholders were represented and played roughly the same roles. Institutions not only regulated but also participated in the development of the project in each case. Both the Musconetcong Watershed Association on the one hand and the County Fishing Federation on the other centralized project management. However, some elements illustrated the top-down approach that dominates in France and the weakness of local environmental associations there compared to the USA (Drapier et al. 2017). In the USA, a large array of stakeholders (public and private) work together to achieve a common goal, and Musconetcong Watershed Association was the project leader because it was a grassroots institution. The range of stakeholders is narrower in France, and the County Fishing Federation held a central position in the project because of its historic bonds with numerous French environmental institutions.

7.3.2   A project enriched by considering a rising demand expressed by other stakeholders

During the implementation of the two projects (Figure 7.4), the project leaders – County Fishing Federation and Musconetcong Watershed Association – received additional information and requests from other participants. The latter expressed their views during compulsory procedures (public inquiry, public meetings, and the canvassing of opinions) or during the setting-up of a voluntary consultation process. In the case of the Orne River, only one public meeting was held (September 2015). Interestingly enough, it was organized two years after the County Fishing Federation became the property owner, thus making the removal inevitable. On the Musconetcong River, two public meetings in December 2008 and 2009 were held by the project leader before the decision to remove the dam was finally taken in 2010. Figure 7.5 illustrates the different way stakeholders' expectations could be heard by project leaders. The heart of the flowers (black round) represented the initial objective of both projects which consisted of removing the dam. The two bigger circles represent the stages of the project expansion by studies and public meetings. Progressively new protagonists entered the scene and shared their points of view (expectations or fears), leading to a more complex project. They contributed to build a more inclusive approach to river restoration.

The project managers personally contributed to the development of the projects by including additional actions, mostly aimed at even better restoration of fish and sedimentary continuity (habitat improvement, work on the river course, etc.). The steering committee anticipated a number of questions which concerned stakeholders. They were contacted directly by the project leader (bilateral meetings). The findings of the studies also led to new recommendations. Some additional issues were revealed by surveys. Finally, others issues were raised by inhabitants and users during public meetings.

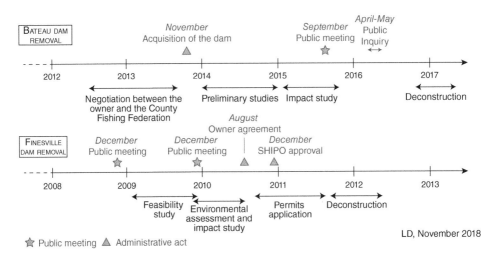

Figure 7.4   Chronologies of the Finesville and the Bateau dams removal projects figuring the main administrative steps and the public events related to. *Source:* Based on Drapier, L., Germaine, M.-A. and Lespez, L. (2018). Politique environnementale et territoire : Le démantèlement des ouvrages hydrauliques en France à l'épreuve du modèle nord-américain., *Annales de Géographie,* (722): 339–368.

Stakeholders

Expectations and demands (dotted lines: not taken into account in the final project)

➡ from steering committee or bilateral dialogue

➡ from studies

➡ from public meetings

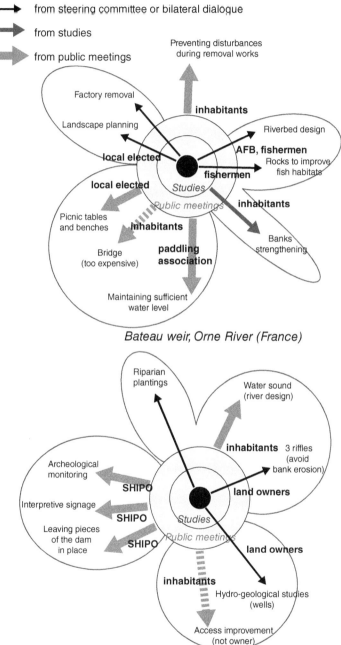

Bateau weir, Orne River (France)

Finesville Dam, Musconetcong River (USA, NJ)

Figure 7.5   The enrichment of the Bateau and Finesville dams removal projects by the different stakeholders.

On the Orne River the County Fishing Federation aimed to remove the dam for regulatory and ecological reasons. Once it had acquired the site (November 2014), the Federation included the demolition of the factory, and the subsequent need for landscaping work (Figure 7.6a), from a very early stage in the project. Discussions with elected officials and the proximity of a walking trail along the Orne River led to the decision to set up benches and picnic tables on the newly landscaped site (Figures 7.5 and 7.6c). In agreement with the local fishing chapter, the layout of the river was redesigned and boulders were placed in the riverbed in order to improve local fish habitat (Figure 7.5). Following the conclusions of the feasibility study, an agreement was signed with the riparian landowners for construction work to strengthen riverbanks using geo-grid and plant textiles in order to protect buildings and gardens. A public meeting was held on September 2015 in a riverine village (Le Mesnil-Villement: around 300 inhabitants). The County Fishing Federation presented the project once the decision to remove the dam had been taken (summer 2012) and prior to applying for the permits (2014) (Figure 7.4). Inhabitants of the municipalities concerned by the projects as well as local user associations and elected officials were invited and some 40 people attended the meeting. At this occasion, the director of the local canoeing and kayaking association called for the creation of a navigable stretch of water at the dam. This request was reiterated by email and supported by local elected officials who were also anxious to maintain such recreational activities. The public meeting also allowed inhabitants to express their fears about the works and led the owner to take further

Figure 7.6 The works carried out on the Orne River: (a) landscaping on the former factory site; (b) rocks for ecological habitat improvement; (c) picnic tables and benches. *Source:* Ludovic Drapier 2017.

precautionary measures during the works. The local inhabitants' request to install a footbridge to connect the two banks separated by the removal of the dam, which also served as a pedestrian bridge, was not met, since the cost was considered prohibitive. At this meeting, the setting-up of a mandatory public inquiry was announced (Figure 7.4).

On the Musconetcong River, the Watershed Association assisted the owner in their project to remove the dam in order to enable the return of shad fish stocks and, more generally, to improve the quality of the aquatic environment. Between the first discussions, in late 2007, and the beginning of the works, in late 2011, two public meetings were organized in December 2008 and 2009. Meetings took place before the official decision to remove the dam was taken in August 2010. The exchanges between the project owner and local stakeholders also helped to broaden the project's objectives. Even before these meetings, bilateral meetings with landowners raised the issue of how to ensure continued water supplies for wells on riverside properties. The director of the Musconetcong Watershed Association met all the potentially affected riparian landowners after a hydrogeological study had been carried out in order to reassure them about the sustainability of water use. At the end of the first public meeting, inhabitants highlighted their attachment to the noise of the waterfall. The noise was recreated through modification of the riverbed (using gravel) in order to maintain a similar sound to the previous waterfall. The local inhabitants' request for improved access to the river was not met, since the owner did not have property rights over the riverbanks. After the second public meeting, a memorandum of understanding was signed with the State Historic Preservation Office (SHIPO). This memorandum requested that an archeologist be present for the duration of the works, and also that the project included heritage enhancement (preservation of the remains of the previous dam and installation of an information panel outlining the site's history and the objectives of the work carried out). The outcome of the project assessment conducted by the NJDEP (the relevant regulatory authority), convinced the project owner to install three grade controls to reduce erosion. In addition, the owner decided to plant trees to replenish a riparian forest along the riverside in order to improve habitat conditions for fish (thus providing increased shade, etc.).

The whole process thus contributed to further enhance both projects. These two examples illustrate how the consultation process, including mandatory but also voluntary meetings between users, experts, and external institutional participants, may broaden the initial project. Once completed, the two projects went beyond the simple dam removal and thus met more ambitious environmental challenges or complementary concerns (uses, well-being). Beyond a purely technical operation, the two projects were enhanced by taking into account the requests and points of view of the various stakeholders. This input influenced both the final objectives of the project and the way in which it was carried out (Figure 7.4). Whilst the extent of the challenges and the breadth of participants involved in a dismantling project appeared empirically over the course of each of these two projects, they clearly do need to be planned for and anticipated beforehand. The project design benefited from being done with stakeholders in a comprehensive manner rather than in opposition, and was designed in this way from the beginning.

However the dialogue between the stakeholders is organized, which was different in France and the USA (Drapier et al. 2017), the identification of the stakeholders concerned is a prerequisite for their participation. Neglecting, or failing to help organize, the expression of stakeholders' opinions can encourage conflict. Either way, it is vital to consider the opinions of stakeholders, users and local inhabitants when considering a project. The heritage dimension occupies a stronger place in the USA, thanks to SHIPO, while cultural services are just beginning to take up this issue in France and are struggling to intervene in the debates on the removal of dams.

7.4   Some key points to improve stakeholders' involvement

According to these experiences, several key points may be highlighted to improve the potential success of such actions. One of the difficulties in implementing river restoration policies is the inadequacy of the formal jurisdictions compared to the true extent of the challenge. Developing these projects on an operational scale makes it possible to take into account all the participants affected by the project. Once this has been done it is possible to set up procedures which ensure genuine consultation of all the stakeholders involved whilst also taking into account the existing balance of power.

7.4.1   Stakeholders and scales: uses and issues, a new geography of stakeholders

The definition of the relevant project parameters, and hence of the stakeholders, requires that the dam be located on a broader scale than that of the barrier running across the river. In general, ecological issues are considered at different scales, particularly when it comes to restoring the ecological continuity of migratory fish, and the hydrosystem is the scale of reference. However, in the conduct of the project at the local level, it is first of all the dam to be removed that is at the center of all attention even if the spatial scale arrangement is an issue at the local level. For example, as local boundaries are often based on the hydrographic network, many dams are located between two administrative areas. In our case studies, both dams are actually located on the border between two administrative areas (French departments for the Bateau dam and counties for Finesville). Especially in the case of Finesville and to a lesser extent for the Bateau dam, it significantly increased the number of stakeholders involved in the process. More generally, the project required de facto creation of space for ad hoc dialogue. Above all, the notion of hydraulic site (Lespez and Germaine 2016; Barraud and Germaine 2017) is preferable in order to take into account the context in a wider perspective of whether it is to be considered as a barrier from a morphological or fish point of view by the advocates of ecological restoration. It is important to take into account all the elements which make up the surrounding landscape: the hydraulic infrastructures, including the dam, but also the inflow and evacuation channels, gates, and other water regulation devices; the associated buildings (old mills or factories, lock houses, and more recent constructions); the access roads to the site; and lastly the pattern of the surrounding landscape (land status, land use, etc.). In both case studies, inhabitants and users sought access to the river. The creation of paths along the river allows inhabitants to experience the continuity of the river for themselves. This widening of scale implies considering the whole valley: taking into account the valley bottom and slopes makes it possible to consider the hydraulic structures in their immediate geographical context and thus to identify their landscape and heritage characteristics. There is a wide variety of situations: some dams are neither visible nor accessible, while others benefit from a location which attracts local inhabitants (residents, anglers, hikers, etc.) who appropriate the site. The affected area is therefore flexible and specific to each situation.

If stakeholders' inventories were not planned at the beginning of the two studied cases, the analysis post hoc demonstrates the interest of anticipating and formalizing this stage while adapting to the specificities of each case. These processes teach us that the definition of stakeholders must be based on an inventory of the uses of the river and of the hydraulic site (Orr et al. 2007). These uses are not confined to the narrow river corridor. They are usually dependent on the upstream dam water body, which can stretch for several hundred meters (Le Calvez 2017). Downstream users also have a stake in dam removals and may be concerned about flooding or enthusiastic about fish migration. It is therefore necessary to define all the possible consequences of the removal from the early stages of the project (Reed 2008). These may be directly linked to the removal: the production of energy ceases or the possibility to cross the river thanks to a gateway or bridge installed on the dam disappears. The removal of the upstream dam water body also generates disturbances. Activities which involve drawing water from the river may be jeopardized, thus affecting farmers (irrigation and watering livestock) but also companies which may need water reserves against fire, for example. In the case of the Musconetcong, private wells were a concern for riparian landowners and this concern could be expressed, and subsequently addressed, thanks to the local involvement of the project manager at public meetings but also by using a door-to-door approach. Recreational uses such as swimming or certain kinds of boating may be threatened with the lowering of the water level. The removal of water bodies may deprive users of facilities such as a kayak polo court or a beginners' area for canoeing and kayaking as in the case of the Orne River. Anglers are also affected, since the fish population is modified (disappearance of certain fish species) and some anglers, accustomed to static fishing from a pontoon or banks, like upstream from the Bateau dam, find their environment modified. The two case studies demonstrate that dam removal can also affect riparian landowners. Indeed, properties may be weakened by bank erosion.

The removal of a barrier does not only affect commercial uses or material goods. Disturbances also involve immaterial and sensorial aspects. Because it involves a more or less significant transformation of the river landscape, removal projects affect not only the enjoyment so dear to property owners but also the heritage value attributed to many dams. This also involves aesthetic choices which emphasize a preference for maintaining a river which is "full" right up to its banks (Germaine and Barraud 2013b). These uses can be exclusively private, as is the case when only the owner benefits from the amenities offered by the hydraulic site (the noise of the waterfall from the dam, the presence of water, the mirror effect of the water, etc.). In other cases, when the site is more widely visible and accessible, like the Bateau dam (close to a hiking trail, kayak route, or located in a village, etc.), or when it is open to the public, all local inhabitants and walkers may benefit from its features. In the latter case, the removal may threaten the local history and vernacular heritage to which elected officials and inhabitants are attached. Removing traces of past activities, whether they were old grain mills or more recent textile mills, may be perceived as an attack on the cultural and traditional heritage which is emblematic of the history of a region (Fox et al. 2016). In some cases, these intangible aspects are linked to economic interests. Some sites are, in fact, the object of economic development (holiday cottages, campsites, local museums) or heritage enhancement by local authorities (site development in a town or village, or along a hiking trail), thus contributing to the attractiveness of the local area. In such cases removal appears unthinkable.

It is therefore only through an inventory of the various interests associated with maintaining or removing a dam that the stakeholders can be identified. All those who may be affected by the dismantling process are considered legitimate participants in the debate. It is therefore essential that they be involved in the conception of the project and that they take part in its wider development. This implies widening the affected area both upstream and downstream of the dam. The stakeholder group is also variable and may differ from one project to another. Thus, the definition of the people concerned and the reference scale must be the first step in the work at the local level for the acceptance of the ecological project.

7.4.2 Convincing or consulting?

The studies that precede projects to restore ecological continuity usually focus on technical and environmental aspects. In addition to feasibility and impact studies, which are mandatory stages, the project managers rely on expert assessment by specialists in fish populations and in morphological river functioning, etc., in order to assess the feasibility of the project. Immaterial benefits provided by ecosystems (whether they are recreational, cultural, aesthetic, educational, etc.) are, on the other hand, more rarely taken into account (Morandi et al. 2014). Such benefits are rarely assessed financially and are, in fact, rarely discussed at all, except for some projects, mainly big dams, studied by economists[5] (Loomis 1996, 2002; Gowan et al. 2006; Robbins and Lewis 2008). It is nevertheless crucial to identify all the environmental services (including ecological services and amenities) actually provided by rivers (Lespez et al. 2016) but the use of standardized river management practices tends to call into question the multifunctional nature of rivers. Moreover, these projects are rarely accompanied by the dissemination of knowledge and information, thus tending to exacerbate the hostility of inhabitants worried about the fate of their river (Lejon et al. 2009; Barraud and Germaine 2017; Germaine and Chilou 2017; Le Calvez and Hellier 2017).

State authorities have become aware of the importance of the consultation issue. Methodological guidelines have been published in France, for example, but many of them are in fact more akin to tools for implementing communication exercises aimed at getting local inhabitants to support river restoration projects (Barraud and Germaine 2017). It is also significant that this phase is only envisaged once the guidelines have already been defined. When dialogue is established, other problems arise, such as vocabulary (Eden and Tunstall 2006; Notte and Salles 2011; Lespez et al. 2016). Language is a bone of contention. For example, inhabitants are disturbed by the issue of mud (after dam removal), while river managers seek to reassure them by asserting that they can handle sediment management. Users express their attachment to lakes, while removal project managers prefer the term "water bodies," thus emphasizing their artificial nature (Germaine and Lespez 2017). Moreover, the various stakeholders do not all have the same understanding of key concepts such as the notion of ecological continuity which is rarely fully understood by inhabitants who are unaware of its wider signification beyond their local experience.

Consultation aims to share knowledge and spread information. This implies a two-way exchange (Eden 2017). Project managers must be able to listen to, and take into account, the expectations of local inhabitants (Figures 7.4 and 7.5) and they must also answer their questions. They must be able to adapt to unforeseen issues and provide solutions to allay fears (e.g. water availability for wells). The examples of the Finesville and Bateau dams testify to the importance of taking into account different types of knowledge in order to enhance a project. The expertise of inhabitants and users helps strengthen the project. In France, however, the increasingly conflictual nature of the removal issue has made this a highly charged topic. On the one hand, removal opponents, for the most part made up of mill heritage supporters and advocates of hydroelectricity, have set up networks and attend consultation meetings where they express their opposition to removal projects (Barraud 2017). On the other hand, state agencies and environmental associations also sometimes radicalize their discourse, making it very difficult to establish any form of meaningful dialogue. These types of participants, who refuse meaningful discussion and entrench themselves in their positions, still have to be consulted in the project development process. It is probably not necessary to try to convince them, but informing them is a necessary action for them to not feel left out of the process because they disagree (Reed 2008).

7.4.3 Inevitable power relations

When there is disagreement on hydraulic dams, removal advocates tend to refer to the natural functioning of the river and highlight the services rendered by "natural" river. Dam removal aims to restore (or repair) ecological functions. On the contrary, mill supporters view rivers as heritage resulting from the age-old interaction between nature and human society. They highlight the services rendered by these dams (renewable energy, water oxygenation, anthropogenic but ancient and specific ecosystems, etc.). This divergence of opinion reveals that technical and scientific objectives also represent social choices, and that management decisions are above all a matter of power relations (Kothari 2001; Eden 2017). This bears witness to the difficulty in reconciling wider concerns, (which are viewed from a global, planetary point of view in order to justify top-down decisions), while local expectations tend to be anchored in the socio-economic reality of local areas. The implementation of ecological restoration projects is marked by an imbalance in power relations, with different stakeholders having different levels of influence. Removal advocates enjoy the power which accrues from possession of the expert knowledge that is difficult to replace, even if they are less well informed about local practices and customs than local inhabitants. River managers are in a privileged position due to their control over information, as well as over formal and hierarchical rules (regulations, laws, financing plans, etc.). Since water and the environment continue to be perceived as purely technical subjects the inhabitants and users, and even elected officials, find themselves powerless to intervene in the debate (Eden 2017). Removal advocates included environmental protection associations and representatives of fishing associations, but also experts and scientists are often viewed as outsiders by inhabitants (Fox et al. 2016). Stakeholders sometimes consider them to be illegitimate because they are not local. While it is important for inhabitants to understand that there are wider issues at stake in such collective projects, it is also crucial that the proponents of ecological projects realize that these are in fact political issues and that they must shoulder their political responsibility since they too are stakeholders in the final analysis.

The ecological restoration of rivers implies change. Such change involves collective issues and requires public support. Mermet and Berlan-Darqué (2009) distinguish two major models of concerted change: co-construction and negotiation. In the first instance, change emerges from the collective co-learning of the participants by developing a common understanding through the construction of a common vision of the problems which lead to the definition of a collective strategy. Conversely, a negotiation-based situation involves a project manager (e.g. the watershed association in the USA), or a politician (e.g. government authorities in France), facing their opponents. The latter model integrates the asymmetry of the collective context in which the objective is to reach a decision. This lends itself to the French situation, where the restoration of ecological continuity is enshrined in regulation. It is thus a question of identifying the people and organizations who can lead the debate in each case. It would appear that watershed associations and institutions (e.g. SAGE) are the best structures for this purpose since they are able to play a mediating role when bringing the various stakeholders into discussions.

7.5 Conclusions

According to Habermas (1987), we have to keep in mind that the implementation of a participatory process requires that the stakeholders be free to join the process and have the same power and must be sincere. If the objective is to co-construct a project, all of the participants must recognize their roles in the process and sincerely assume their position. This is the case both for the riparian landowners, who are defending private property which they cannot transform into collective property if their hydraulic site is not of heritage interest, but also for removal advocates who have to accept the contradiction inherent in the definition of the public interest. There is frequent disagreement as to the definition of what constitutes public interest. The ecological argument oscillates between protecting biodiversity (concerning fish especially) and promoting renewable energies in the fight against global warming as against the social demand for the preservation of cultural heritage and landscapes for the sake of quality of life and well-being. This raises the question of the roles of the expert and of the citizen in the decision-making process and suggests that it is necessary to broaden both the geographical and thematic scope of studies and the array of stakeholders involved in the construction of such projects. It is believed that involving the greatest possible number of people in an ecological project increases the chance of convincing the largest number of inhabitants of its merits. This can be complicated and risky, but it is probably the best way to sustainably increase public support for a river restoration project.

Notes

1 In French: Fédération Départementale pour la Pêche et la Protection du Milieu Aquatique (FDPPMA).
2 In French: Direction Départementale des Territoires.
3 In French: Agence française de la biodiversité.

4 In French: Schéma d'Aménagement et de Gestion des Eaux.
5 We should stress, though, that these economic studies are mainly about bigger projects which greatly differ from the small dam removal considered here.

References

Åberg, E.U. and Tapsell, S. (2013). Revisiting the River Skerne: the long-term social benefits of river rehabilitation. *Landscape and Urban Planning* 113: 94–103.

Barraud, R. (2017). Removing mill weirs in France: the structure and dynamics of an environmental controversy. *Water Alternatives* 10(3): 796–818.

Barraud, R. and Germaine, M.-A. (2013). Defining and achieving good water status: expert rule versus local participation: case studies on dam removal in western France. In: *European Continental Hydrosystems under Changing Water Policy* (eds G. Arnaud-Fassetta, E. Masson, and E. Reynard), 233–246. Munich, Germany: Friedrich Pfeil Verlag.

Barraud, R. and Germaine, M.-A. (eds) (2017). Démanteler les barrages pour restaurer les cours d'eau: Controverses et représentations. Versailles, France: Éditions Quae.

Born, S.M., Genskow, K.D., Filbert, T.L., et al. (1998). Socioeconomic and institutional dimensions of dam removals: the Wisconsin experience. *Environmental Management* 22(3): 359–370.

Brown, J.J., Limburg, K.E., Waldman, J.R., et al. (2013). Fish and hydropower on the US Atlantic coast: failed fisheries policies from half-way technologies. *Conservation Letters* 6(4): 280–286.

Brummer, M., Rodríguez-Labajos, B., Nguyen, T.T., et al. (2017). "They have kidnapped our river": dam removal conflicts in Catalonia and their relation to ecosystem services perceptions. *Water Alternatives* 10(3): 744–768.

Doyle, M.W., Stanley, E.H., Harbor, J.M., et al. (2003). Dam removal in the United States: emerging needs for science and policy. *Eos Transactions, American Geophysical Union* 84: 29–36.

Drapier, L., Germaine, M.-A., and Lespez, L. (2017). Comparaison de projets de démantèlement d'obstacles en travers en Europe et aux Etats-Unis. In: *Démanteler les barrages pour restaurer les cours d'eau: controverses et représentations* (eds R. Barraud and M.-A. Germaine), 195–211. Versailles, France: Éditions Quae.

Drapier, L., Germaine, M.-A., and Lespez, L. (2018). Politique environnementale et territoire : Le démantèlement des ouvrages hydrauliques en France à l'épreuve du modèle nord-américain. *Annales de Géographie* (722): 339–368.

Duram, L. and Brown, K.G. (1999). Insights and applications assessing public participation in U.S. watershed planning initiatives. *Society and Natural Resources* 12(12): 455–467.

Eden, S. (2017). *Environmental Publics*. Oxford: Routledge.

Eden, S. and Tunstall, S. (2006). Ecological versus social restoration? How urban river restoration challenges but also fails to challenge the science policy nexus in the United Kingdom. *Environment and Planning C: Government and Policy* 24(5): 661–680.

Eden, S., Tunstall, S.M., and Tapsell, S.M. (2000). Translating nature: river restoration as nature-culture. *Environment and Planning: Society and Space* 18(2): 258–273.

Emery, S.B., Perks, M.T., and Bracken, L.J. (2013). Negotiating river restoration: the role of divergent reframing in environmental decision-making. *Geoforum* 47(1): 167–177.

FDPPMA Calvados (2015). Effacement de seuils en rivières sur le cours de l'Orne: le Bateau, la Fouillerie et Danet sur les communes de Le Mesnil-Villement (14), Menil-Hubert-sur-Orne

(61), Saint-Philbert-sur-Orne (14), Rapilly (14) et Les Isles Bardel (14). Dossier de demande d'autorisation unique.

Fox, C., Sneddon, C., and Magilligan, F. (2016). "You kill the dam, you are killing a part of me": the environmental politics of dam removal. *Geoforum* 70: 93–104.

Garcia de Leaniz, C. (2008). Weir removal in salmonid streams: implications, challenges and practicalities. *Hydrobiologia* 609(1): 83–96.

Germaine, M.-A. and Barraud, R. (2013a). Les rivières de l'ouest de la France sont-elles seulement des infrastructures naturelles? Les modèles de gestion à l'épreuve de la directive-cadre sur l'eau. *Natures Sciences Sociétés* 21(3): 373–384.

Germaine, M.-A. and Barraud, R. (2013b). Restauration écologique et processus de patrimonialisation des rivières dans l'ouest de la France. *VertigO* (16). https://doi.org/10.4000/vertigo.13583.

Germaine, M.-A. and Chilou, A. (2017). Construire un diagnostic élargi et partagé des sites hydrauliques ? Intérêts et limites d'une grille multicritère. In: *Démanteler les barrages pour restaurer les cours d'eau. Controverses et représentations* (eds R. Barraud and M.-A. Germaine), 145–162. Versailles, France: Éditions Quae.

Germaine, M.-A. and Lespez, L. (2017). The failure of the largest project to dismantle hydroelectric dams in Europe? (Sélune River, France, 2009–2017). *Water Alternatives* 10(3): 655–676.

Gowan, C., Stephenson, K., and Shabman, L. (2006). The role of ecosystem valuation in environmental decision making: hydropower relicensing and dam removal on the Elwha River. *Ecological Economics* 56(4): 508523.

Grimble, R. and Wellard, K. (1997). Stakeholder methodologies in natural resource management: a review of principles, contexts, experiences and opportunities. *Agricultural Systems* 55(2): 173–193.

Habermas, J. (1987). *Theory of Communicative Action*. Cambridge: Polity Press.

Hart, D.D., Johnson, T.E., Bushaw-Newton, K.L., et al. (2002). Dam removal: challenges and opportunities for ecological research and river restoration: we develop a risk assessment framework for understanding how potential responses to dam removal vary with dam and watershed characteristics, which can lead to more effective use of this restoration method. AIBS Bulletin 52(8): 669–682.

Heinz Center (2002). *Dam Removal: Science and Decision Making*. Washington DC: The Heinz Center.

Howard, A., Knight, D., Coulthard, T., et al. (2016). Assessing riverine threats to heritage assets posed by future climate change through a geomorphological approach and predictive modelling in the Derwent Valley Mills WHS, UK. *Journal of Cultural Heritage* 19: 387–394.

Jähnig, S.C., Lorenz, A.W., Hering, D., et al. (2011). River restoration success: a question of perception. *Ecological Applications* 21(6): 2007–2015.

Jørgensen, D. (2017). Competing ideas of "natural" in a dam removal controversy. *Water Alternatives* 10(3): 840–852.

Junker, B., Buchecker, M., and Müller-Böker, U. (2007). Objectives of public participation: which actors should be involved in the decision making for river restorations? *Water Resources Research* 43(10). https://doi.org/10.1029/2006WR005584.

Kemp, P.S. and O'Hanley, J.R. (2010). Procedures for evaluating and prioritising the removal of fish passage barriers: a synthesis. *Fisheries Management and Ecology* 17(4): 297–322.

Kothari, U. (2001). Power, knowledge and social control in participatory development. In: *Participation: The New Tyranny* (eds B. Cooke and U. Kothari), 139–152. London and New York: Zed Books.

Le Calvez, C. (2017). Les usagers confrontés à la restauration de la continuité écologique des cours d'eau: approche en région Bretagne. PhD dissertation. Université Rennes 2 – Haute Bretagne.

Le Calvez, C. and Hellier, E. (2017). Expérimenter la continuité écologique sur une masse d'eau fortement modifiée, ou la mise au jour des tensions entre les représentations de l'Aulne canalisée. In: Démanteler les barrages pour restaurer les cours d'eau: Controverses et représentations (eds R. Barraud and M.-A. Germaine), 115–128. Versailles, France: Éditions Quae.

Lejon, A.G.C., Renöfält, B.M., and Nilsson, C. (2009). Conflicts associated with dam removal in Sweden. *Ecology and Society* 14(4): 19.

Lespez, L. and Germaine, M.A. (2016). La rivière désaménagée? Les paysages fluviaux et l'effacement des seuils et des barrages en Europe de l'Ouest et en Amérique du Nord-Est. *Bulletin de la Société Géographique de Liège* 67: 223–254.

Lespez, L., Germaine, M.-A., and Barraud, R. (2016). L'évaluation par les services écosystémiques des rivières ordinaires est-elle durable ? *VertigO* 25. https://doi.org/10.4000/vertigo.17443.

Loomis, J. (1996). Measuring the economic benefits of removing dams and restoring the Elwha River: results of a contingent valuation survey. *Water Resources Research* 32(2): 441–447.

Loomis, J. (2002). Quantifying recreation use values from removing dams and restoring free-flowing rivers: a contingent behavior travel cost demand model for the Lower Snake River. *Water Resources Research* 38(6). https://doi.org/10.1029/2000WR000136.

Magilligan, F.J., Sneddon, C.S., and Fox, C.A. (2017). The social, historical, and institutional contingencies of dam removal. *Environmental Management* 59(6): 982–994.

Mermet, L. and Berlan-Darqué, M. (eds) (2009). *Environnement: décider autrement, nouvelles pratiques et nouveau enjeux de la concertation*. Paris: L'Harmattan.

Morandi B., Piégay H., Lamouroux N., et al. (2014). How is success or failure in river restoration projects evaluated? Feedback from French restoration projects. *Journal of Environmental Management* 137: 178–188.

Nilsson, C. and Berggren, K. (2000). Alterations of riparian ecosystems caused by river regulation: dam operations have caused global-scale ecological changes in riparian ecosystems: how to protect river environments and human needs of rivers remains one of the most important questions of our time. *BioScience* 50(9): 783–792.

Notte, O. and Salles, D. (2011). La prise à témoin du public dans la politique de l'eau: la consultation directive-cadre européenne sur l'eau en Adour-Garonne. *Politique européenne*, 1(33): 3762.

Orr, P., Colvin, J., and King, D. (2007). Involving stakeholders in integrated river basin planning in England and Wales. *Water Resources Management* 21(1): 331–349.

Perrin, J.A. (2018). Gouverner les cours d'eau par un concept: Etude critique de la continuité écologique des cours d'eau et de ses traductions. PhD dissertation, Université de Limoges.

Pohl, M.M. (2002). Bringing down our dams: trends in American dam removal rationales. *Journal of the American Water Resources Association* 38: 1511–1519.

Reed, M.S. (2008). Stakeholder participation for environmental management: a literature review. *Biological Conservation* 141(10): 2417–2431.

Richardson, S.C. (2000). Changing political landscape of hydropower project relicensing. *William & Mary Environmental Law and Policy Review* 25: 499–531.

Robbins, J.L. and Lewis, L.Y. (2008). Demolish it and they will come: estimating the economic impacts of restoring a recreational fishery. *Journal of the American Water Resources Association* 44(6): 1488–1499.

SANDRE (2014). Description des ouvrages faisant obstacle à l'écoulement: dictionnaire des données. https://www.sandre.eaufrance.fr/urn.php?urn=urn:sandre:dictionnaire:obs:FRA:::ressource:latest:::pdf&pk_vid=57dfaa8a0863a3d816200249100af79b (accessed 18 May 2021).

Sneddon, C., Barraud, R., and Germaine, M.-A. (2017). Dam removals and river restoration in international perspective. *Water Alternatives* 10(3): 648.

Tapsell, S.M. (1995). River restoration: what are we restoring to? A case study of the Ravensbourne River, London. *Landscape Research* 20(3): 98–111.

Tunstall, S.M., Tapsell, S.M., and Eden, S. (1999). How stable are public responses to changing local environments? A "before" and "after" case study of river restoration. *Journal of Environmental Planning and Management* 42(4): 527–545.

Vogel, E. (2008). Regional power and the power of the region: resisting dam removal in the Pacific Northwest. In: *Contentious Geographies: Environmental Knowledge, Meaning, Scale* (eds K.T. Evered, M.T. Boykoff, and M.K. Goodman), 165–86. Aldershot: Ashgate.

8

Letting the Political Dimension of Participation in River Restoration have its Space

Nora S. Buletti, Franziska E. Ruef, and Olivier Ejderyan

USYS TdLab, ETH, Zürich, Switzerland

8.1 Introduction

Participatory decision-making has become a central feature of river restoration practices (Pahl-Wostl et al. 2011; Sivapalan et al. 2012). This is part of a global trend to open decision-making to the public, which gained momentum with the United Nations' Agenda 21 (UN 1992) and was later reinforced by international agreements such as the Aarhus Convention (UNECE 1998). The European Union's Water Framework Directive (EC 2000), as well as the endorsement of approaches such as integrated river basin management by international organizations (World Bank 2006), have contributed to the quick adoption of participation in river management.

River restoration practitioners have tended to adopting participatory decision-making procedures early on (Nijland and Cals 2000; Niehuis and Leuven 2001; Smith et al. 2014). The need to acquire land to ensure enough space for rivers, coordination with other activities such as farming or leisure, and the opportunity to inform the public about what restoration entails have all supported the rapid adoption of participation in river restoration practice (Eden et al. 2000; Pahl-Wostl 2006; Rohde et al. 2006; Petts 2007; Wohl et al. 2015). There is now a breadth of literature illustrating how participation can contribute to the implementation of successful restoration, flood protection, and water-use projects (Mees et al. 2016; Angelopoulos et al. 2017; Zingraff-Hamed et al. 2017).

The underlying idea behind what has been labelled a "paradigm change" (Pahl-Wostl et al. 2011) is that early participation of individuals and organizations affected by the project (so-called stakeholders) makes river restorations more acceptable because their views can be integrated into the projects. Another widely shared belief is that project developers will be able to better communicate their goals and hence convince stakeholders about potential benefits. Yet, despite such a positive depiction of participation being largely shared among practitioners, the implementation of river restoration projects in a participatory way is still considered a challenge (Pahl-Wostl et al. 2011; Maynard 2013; Buletti et al. 2014; Heldt et al. 2016).

River Restoration: Political, Social, and Economic Perspectives, First Edition. Edited by Bertrand Morandi, Marylise Cottet, and Hervé Piégay.
© 2022 John Wiley & Sons Ltd. Published 2022 by John Wiley & Sons Ltd.

Including stakeholders or the public in river restoration projects implies departing from ideal river restoration defined in purely scientific terms. It recognizes that "river restoration is not just a biophysical process, but also a social, political and economic process" (McDonald et al. 2004, p. 279). Some scientists and practitioners may believe that such an approach leads to suboptimal projects and not wish to acknowledge the political factors brought by participation (Latour 2004; Donaldson et al. 2013).

It is not so much a lack of tools and guidelines that leads to the difficult implementation of participatory river restoration projects but rather the lack of engagement with the political dimension of participation on the part of project developers (Tsouvalis and Waterton 2012; Donaldson et al. 2013; Anderson et al. 2016). Too often, they consider participation a technical instrument that should enable the implementation of successful projects in a linear way. In such approaches, participation becomes depoliticized: discussions or conflicts about a project's purpose and goals are avoided at all costs and the search for a solution is reduced to an administrative and technical matter. In such cases, participation becomes unable to address issues that matter to the participants. Depoliticization is linked to the fundamental tension between the opening of the decisional process to stakeholders and the necessity for project managers to reach a decision by closing down participation at some point (Chilvers 2008; Stirling 2008).

The goal of this chapter is to assess how the depoliticization of participation in river restoration projects operates and its potential negative consequences. We propose ways for practitioners to acknowledge and integrate this political dimension. We discuss why such a move can benefit the goals of river restoration and argue that it is important to consider the political consequences of participation, as a way to revitalize public engagement in river restoration.

Building on a review of the literature from the social sciences and humanities on the politics of participation in environmental policy and river management and on three case studies from Switzerland, we show common ways through which the political dimension of participation is overlooked – and sometimes deliberately neglected – by practitioners in river management. We discuss the implications of this neglect of the politics of participation for river restoration and offer ways to acknowledge it within existing participatory practices.

8.2 Participation and river restorations

"Participation" generally refers to the "practice of consulting and involving members of the public in the agenda-setting, decision-making, and policy-forming activities of organizations or institutions responsible for policy development" (Rowe and Frewer 2004, p. 512). Such an understanding of participation includes a spectrum of modes of involvement, ranging from informing the public about project content to fully integrating citizens in constructive collaborative effort during the decision-making process (Thaler and Levin-Keitel 2016). It acknowledges that there are various appropriate ways to have stakeholders or the public participating in decision-making, depending on the project context. Owing to this variety of understanding, the very notion of participation has come under academic scrutiny and debate, and its definition in both the restoration practice and academic debates is far from definitive (Chilvers 2008; Chilvers and Kearnes 2015).

An important strand of the literature on participation in river restoration has dealt with questions of design of participatory processes to implement projects ranging from issues such as stakeholder identification to communicating technical aspects in an appropriate way (Junker et al. 2007; Reichert et al. 2007; Luyet et al. 2012; Norton et al. 2012; Hassenforder et al. 2019). This body of work principally focuses on more practical recommendations, strategies, and tools to include the point of view of different stakeholders.

Another significant body of work on participation in river restoration focuses on stakeholders' perceptions and representations of river restorations and how these can be integrated through participatory measures (Tunstall et al. 1999; Buchecker et al. 2013; Le Lay et al. 2013; Heldt et al. 2016). This body of work focuses on the strength of taking into account the perception of stakeholders for a better design and acceptance of the project and often supports their argument for the potential of social learning in such processes.

Other studies argue that difficulties in implementing river restoration projects do not necessarily come from a lack of tools and are not determined by actors' perception but rather result from a lack of engagement with the political dimensions of river restoration and participation (Tsouvalis and Waterton 2012; Donaldson et al. 2013; Anderson et al. 2016). These works acknowledge that there are politics inherent in the practice of river restoration due to the need to coordinate different interests and the diverse values of stakeholders involved. Furthermore, there are various reasons why decision-makers implement participatory processes. Fiorino (1990) identifies three rationales for participation: a normative one based on the assumption that participation is the right thing to do, a substantive one that considers that participation can improve the quality of projects by integrating inputs from participants, and an instrumental one that sees participation as a way to reach predefined goals.

Further, authors have underlined that a specific aspect of the politics of participation in river restoration relates to defining the natural state that needs to be restored. This leads to negotiations about which natural aspects should be prioritized from a sociocultural perspective but can also spark controversies among experts, such as defining the reference period or prioritizing knowledge from different scientific disciplines involved in a project (Light and Higgs 1995; Eden et al. 2000; Ejderyan 2014). Indeed, the way expert knowledge is presented in a participatory process frames the process and, as such, contributes to defining what is negotiable and what is not (Wynne 2008). Such findings have brought social scientists working on participation to shift their attention to the politics of expertise.

Work on post-politics has focused on participation as a means for experts to tame opposition (Swyngedouw 2011; Allmendinger and Haughton 2012). In post-political settings, those defined as "competent" settle struggles or negotiations to handle a situation or tackle a problem, either because of their knowledge or because of their ability to comply with formalized rules and decision-making procedures. Some authors have criticized the generalization of participatory processes for river restorations through top-down regulations, as they turn participation into administrative processes conducted in bureaucratic and technocratic ways (Graefe 2011; Parés 2011; Anderson et al. 2016).

Influenced by science and technology studies (STS), a further strand of literature focuses on the political role of experts by looking at their practices within controversies related to river restoration projects (Whatmore 2009; Tsouvalis and Waterton 2012). This literature also pays attention to how experts in river restoration frame participatory processes.

However, it does not see expertise as a structurally determined position but rather as the result of expert practices that might be questioned and contested. Uncertainties and inter-disciplinary controversies, but also divergences in what matters to local populations and to experts, can open room for negotiation within participatory processes (Landström et al. 2011; Lane et al. 2011).

While these two theoretical bodies of work have much in common in terms of their diagnosis of expertise and the way they implement participation as a depoliticizing force, they differ considerably in their view of possibilities for challenging depoliticization (Tsouvalis 2016; Buletti and Ejderyan 2021). Hence, while scholars of the post-political regard public engagement activities with suspicion, scholars influenced by STS perceive them as opportunities for controversies and generative of the political (Donaldson et al. 2013).

In Section 8.3, we illustrate how such depoliticization through participation operates in actual river restoration projects by looking at cases from Switzerland.

8.3 Participation in Swiss river management

Switzerland is exemplary in its changes in river management over the past decades. The federal government introduced policies to restore rivers in the late 1990s (BWG 2000) and encouraged participatory decision-making in river management (OFEFP and OFEG 2003). In 2008, the federal government introduced an additional subsidy to support cantonal authorities (the regional level for decision-making) who implement flood protection projects in a participatory way (OFEV 2008). As such, the Swiss case allows us the opportunity to examine the politics of participation at various levels. In Section 8.3.1, we present the Swiss context, our three case studies, and some results from the analysis of the national level and the case studies.

8.3.1 Participation from a national perspective

In Switzerland, the federal government formulates the general legal framework for river management and guidelines for its implementation (OFEV 2019), but cantons – the subnational level – are in charge of its implementation. The Swiss federal law on river works states that all interventions on rivers must contribute to maintaining or restoring their natural or near-natural status. Moreover, participatory decision-making is strongly encouraged (OFEV 2018, 2019).

In federal guidelines, participation is mainly presented as a way to increase the acceptance of river management projects (OFEFP and OFEG 2003). At the federal level, participation has only recently been recognized in its complexity and as a process that needs to be built with precise scrutiny of the context, transparent design, and trust between practitioners and participants (OFEV 2019). Two surveys, during which cantonal officers of river management were interviewed on the topic of participation, have shown that the consideration of participation as a tool to build acceptance has also been widely shared by cantonal authorities in charge of river management (Zaugg et al. 2004; Buletti et al. 2014). The two surveys, carried out in 2004 and 2014, had the principal goal of exploring the different considerations of participation for those responsible for river works at the local level. The more recent study by Buletti et al. (2014) highlights the tension between instrumental and more collaborative understandings of participation among cantonal officers in charge of rivers. Other studies have also shown that cantonal officers in charge of river restoration in local governments see participation in an ambiguous way. They consider it as a very important factor in the success of river restoration projects and for their legitimacy among the population, but at the same time complain of how time consuming and unpredictable it is. In addition, a significant number of them see participation as a necessary evil they would prefer to avoid, as they report that they systematically have to reformulate less ambitious environmental goals for their projects (Utz et al. 2017).

8.3.2   Feedback from river restoration projects

Looking at specific cases provides a more detailed picture of the different ways participation is implemented in Swiss river restoration projects. We present insights from three case studies, each located in different parts of the country (Figure 8.1). The case studies were selected through maximum variation sampling, which brings to light shared patterns cutting across the cases in distinct settings as well as drastic differences between them. Each of the authors of this chapter has worked on one of the three cases, and insights from each of these contexts have been shared along the way (Ejderyan 2014; Ruef 2015; Buletti 2019; Buletti and Ejderyan 2021). For all three cases, we conducted interviews with stakeholders

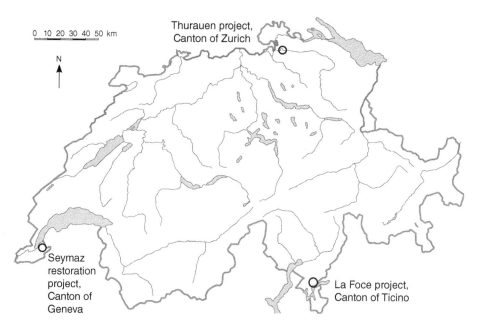

Figure 8.1   Map locating the Thurauen, Seymaz, and La Foce case studies. *Source:* Based on Buletti, N., Utz, S., Ejderyan, O., et al. (2014). Définitions et mise en œuvre des processus participatifs dans l'aménagement des cours d'eau en Suisse. Résultats d'une enquête auprès des services cantonaux responsables de l'aménagement des cours d'eau Projet. Fribourg, Lausanne.

and managers of the participative processes and/or the river restoration projects. We analyzed the interview data using content analysis, focusing on the discourse of the different categories of actors. Table 8.1 provides an overview of the case studies.

8.3.2.1 The Thurauen project

The Thurauen project is located in the north of the Canton of Zurich in the German-speaking part of the country and concerns the eastern part of the Thur river. Between 1983 and 2005, the first sectors of the Thur were restored. The second sector, between Andelfingen and the inflow into the Rhine river, started in 2008, and the project ended in 2017. The restoration of the Thur river is executed by the Cantonal authority of Zurich under the supervision of the federal government.

Concerning the stakeholders' and citizens' participation, several meetings with concerned parties were conducted with the help of an external moderator. The principal points of concern expressed during the meeting were related to the fear of mosquitoes, property rights, property acquisition, and the financial aspect of the project. Over the entire course of the participation process, no single appeal was lodged against the project, and consensus over several discussed issues was reached with the accompaniment group.

This *positive* outcome of the participation process for the project managers is directly linked to the way the contents of the discussion were framed during the meetings. The main pillars of the project were not up for discussion and were interlinked in a way that made it difficult to call into question individual elements of it. As the project manager put it, it was a "complete package."

This case study highlights how those responsible for the project in question considered participation as instrumental, as a certain order was maintained in the planning and decisional structure of the project and no modifications were made to the initial planning of the project. In this sense, participation was used as a tool to ensure consensus building.

8.3.2.2 The River Seymaz restoration project

In 2002, after three years of discussions within a participatory group called "Charte Seymaz," the planners in charge of designing the project presented a project plan. The restoration project of the river Seymaz was initiated in December 1998 in the Canton of Geneva in the French-speaking part of the country. In 2002, after three years of discussions with a participatory group, the planners in charge of designing the project presented a project plan. Farmers agreed with the principle of restoring the river but expressed dissatisfaction with the participatory process. Their critique was that project developers did not consider the farmers' perspective and that technical solutions that were proposed were presented as the only possible ones. They refused to accept the proposed plan. They required direct contact with the engineers in charge of the project implementation. A restricted "accompanying group" was constituted. This group elaborated a new project between August 2003 and December 2004.

Within the accompanying group, engineers and other experts took into account the farmers' local knowledge to modify the restoration project. They translated the farmers' local sense of identity (expressed, for instance, in requests that the project respect the place's history) into technical options (proposing to recreate marshes, instead of widening the river as a way to respect the place's history). Paradoxically, whereas this second alternative

Table 8.1 Summary of Thurauen, Seymaz, and La Foce case studies' principal characteristics.

	Thurauen Project, Thur River, Canton of Zurich	River Seymaz Restoration Project, Canton of Geneva	La Foce Project, Cassarate River, Canton of Ticino
Date	2008–2017	1998–2005	2004–2014
Geographical context	Rural	Rural and urban	Urban
Aim of the project	More flood security, more space for nature, and more leisure spaces	Restore the Seymaz river to improve environmental quality in the upstream rural part and provide flood protection to the downstream urban part	Increase accessibility to the river; improve the ecological system of the delta, for the flora and fauna; ensure flood security
Participatory process	Approach through representation Several meetings organized for the accompaniment group Stakeholders selected by the project managers	Approach through representation Participatory process was based on the selection of individuals or organizations representing groups of interest Phase 1: a participatory group was set up Phase 2: after complaints, a restricted "accompanying group" was set up	No participatory process organized at any phase of the project, even though cantonal officers advised the municipal authorities in charge to do so Project was contested after publication on the official bulletin leading to a referendum Participation reduced to institutionalized forms of contestation offered by Swiss semidirect democracy
Manager of the participatory process	External moderator mandated by the project managers	Phase 1: participatory process managed by public officers Phase 2: participatory process managed by the coordinator of the engineering companies	No participatory process
Participants	25 representatives of different interest groups: community representatives, farmers, hunters, nature conservationists, cantonal deputies, representatives of the electric power plant, foresters	Phase 1: authorities, farmers, environmental nongovernmental organizations, and residents Phase 2: engineers, some farmers, municipal officers, and an environmentalist	Citizens participated in the referendum However, only binary logic of a yes/no vote – no possibility to truly express themselves about the project

(Continued)

Table 8.1 (Continued)

	Thurauen Project, Thur River, Canton of Zurich	River Seymaz Restoration Project, Canton of Geneva	La Foce Project, Cassarate River, Canton of Ticino
Output	Controversy took place, but controlled by the exclusionary dynamic of the representational model Consensus after discussion and "fighting well" No appeal was lodged	Controversy took place thanks to opening up the planning process to include local knowledge Dissatisfaction with inclusion process in the beginning (Phase 1), but acceptance later, even though restoration project demanded more agricultural land to be frequently flooded Identity claims of the farmers were taken into account	Fundamental controversy could not take place Design dominated by experts Participation was disregarded The referendum reduced the citizens' possibility to express themselves to a popular vote

demanded more agricultural land to be permanently or frequently flooded, the farmers accepted this project. They saw it as more respectful of the area's history and their social identity as farmers, as extensive farming is still possible in some parts of the marshes, and as creating an aesthetically pleasing landscape.

8.3.2.3 The La Foce project

The La Foce project, located in the city of Lugano in the Italian-speaking canton of Ticino, planned to restore the Cassarate river delta. Concerning participation, cantonal officers urged the municipal authorities in charge of the project to inform the public and involve stakeholders in the decisional process. However, neither the team in charge of designing the project nor the local authorities organized a participatory process at any phase of the decisional process. And so it departed from standard practice in terms of participation.

Shortly after the publication of the project on the official bulletin, which nobody appealed against, the project became an object of contestation that came to a head with a referendum against the project launched by a transversal political coalition of municipal council members. After a heated campaign, from January to June 2011, citizens voted on the referendum on 5 June 2011, and the project was accepted by 50.55% of voters.

The work carried out, starting in August 2012, followed the initial plan drafted by the project team to the letter; the experts took no heed of the concerns expressed during the campaign, refusing to modify their original plan in any way. This case study offers an opportunity to focus on how experts and their discourses on participation contributed to maintaining the political outcome of the planning of the project, as participation was disregarded and the popular vote reduced the citizens' possibilities to express themselves to a binary choice.

8.4   Processes of depoliticization at work

We now build on the analysis of the three case studies introduced in Section 8.3 and the results of the more recent national survey conducted with cantonal officers in charge of river management (Buletti et al. 2014). In order to discuss the politics of participation at work. This analysis illustrates three common topics that highlight the experts' preferences on participation and the consequences of the practices of participation in river restoration projects. A first common pattern shows dynamics of inclusion and exclusion when it comes to participation in the decision-making process of such projects. Secondly, the way in which participation is implemented is justified with assumptions that disqualify the knowledge of some actors as inappropriate for participation. Thirdly, experts often favor a consensus-building model that negates conflicts and consequently silences the political dimension of participation. In this sense, the analysis highlights common patterns among the three case studies and the national survey. However, the three case studies also complement each other, as they exemplify in different ways the possibility for political potential to emerge.

8.4.1   Dynamics of inclusion and exclusion

The dynamics of inclusion and exclusion are common patterns when implementing any form of participation in the management of environmental issues (Chilvers et al. 2018). This relates to the principles through which a decision process is opened up to participants.

The La Foce project illustrates the fundamental tension between opening the decisional process to the public and the fact that certain phases of the decisional process remain inaccessible to public scrutiny. This tension appeared when the cantonal officers advised the municipal authorities responsible for the river works to involve local constituents, stakeholders, and citizens in the decision-making process as a way to convene people around the project and build consensus. In the La Foce and Thurauen projects, stakeholder participation was considered a necessary step for a smooth implementation of the river restoration project. In the La Foce project, cantonal officers wanted the municipality and the project team to implement participation in order to open up the planning and at the same time to control those who would participate as well as the processes' output. As stated by an interviewed officer, "You have to identify the *right people* and groups and present them things in the right way." Similarly, the Thurauen project team describes one of the steps to implement participation as the invitation of selected stakeholders to form an accompaniment group. However, both show an implicit instrumental consideration of participation, as it is a one-way relationship in which authorities "inform the people and present them things in the right way." They indirectly consider that the "right" way is the one that the authorities apply and consequently have control over who participates, entering in an avoidable dynamic of exclusion typical of stakeholder selection.

In the Seymaz project, the participatory process was based initially on the selection of individuals or organizations representing groups of interest. The refusal of the farmers to accept the project after a first phase of negotiations illustrates the limitations of such an approach through representation. First, the individuals were restricted to their roles of representatives of interests. It was thus not expected that the farmers would also express themselves as local inhabitants who want to enjoy the landscape, for instance. The second limitation was that it was not always clear what the representatives were supposed to represent. Again, in the case of farmers, it was not clear if the individuals sitting at the table represented the economic branch "agriculture," or if they were representing the specific interests of the farmers whose land would be directly affected by the project. In this case study, the farmers, even if included, felt at first excluded from some fundamental aspects of the decisional process.

8.4.2 Questioning the prevalence of expertise

As highlighted in the literature review, expertise can operate as a powerful instrument of depoliticization when it appears as unquestionable truth. The three case studies presented here highlight different mechanisms through which expertise can depoliticize.

A first and common mechanism that appeared in the three case studies consists of putting expertise beyond discussion. In the La Foce project, this was done very explicitly by arguing that there was no need to involve stakeholders, because the project had been designed by experts who "knew better," thus excluding any possibilities for debate. In the two other cases, expertise was placed beyond discussion, less explicitly, by making it a pre-established frame for participation (Wynne 2008), that is by letting expertise define what was or was not in the realm of discussion.

In the Seymaz project, in the first phase, an expert group from commissioned engineering companies and the administration designed the pre-project. The public officials then presented the project to be discussed in the participatory setting, in the absence of the technical experts who had designed it. The project was presented as based on facts that determined what should be done. In the participatory meeting, public officers would then negotiate with stakeholders what was acceptable or not, as well as potential compensations. In the case of the Thurauen project, the pre-framing by the experts was more explicit. There, the project manager indicated in an interview that they intentionally designed the project to create as many dependencies as possible between restoration and flood protection measures. This way of framing the project was done with the intention to sell the project as a "complete package," whose single elements could not be discussed separately.

The second mechanism through which expertise appears as a depoliticizing force is the disqualification of other categories of actors as not having adequate knowledge to make claims about the project. In the case of the La Foce project, experts judged other participants "too emotional" to make rational decisions or unable to "understand" the real stakes of the project. In the first phase of the Seymaz project, stakeholders were also presented as unable to decide rationally because they were directed by "particular interests" and thus unable to take the holistic view necessary to develop such projects.

These types of accounts tend to construct an opposition between experts and so-called lay actors and ignore the fact that even experts do not always agree on the best option for developing a project. This is well illustrated by the case of the river Seymaz, where there were strong disagreements between hydrologists and biologists from the beginning on how best to restore the Seymaz ecosystem. In the first phase, these disagreements were handled while hidden from the stakeholders, who would discuss only what apparently was agreed upon in the project. In the second phase, it became more obvious that some farmers' requests were not irrational but corresponded to possible options that the experts could have pursued.

In this case, making controversies between experts visible proved a way to re-politicize expertise, in the sense that it became visible that some technical choices were directed by decisions from the administration but were still open to discussion from a technical point of view.

8.4.3   Negating the conflict

Stakeholder participation – when used as a tool by the Swiss government to avoid conflicts through consensus building – reflects an instrumental form of participation (Fiorino 1990) and becomes a management tool for authorities of different administrative levels and experts to smoothly implement the policies of river restoration. The analysis of the three case studies highlights how the consensus-building model is differently interpreted and activated in the projects. This has consequences regarding the possibility for unruly publics (de Saille 2015) who question the purpose of participation and commitments to consensus as a prerequisite to engage with the project and for the political dimension of participation to be explored.

In the La Foce project, when the local authorities decided not to implement participation at all, a tension arose between them and the cantonal officer who suggested implementing participation. The cantonal officer had a clear consideration of participation that implicitly referred to the federal consensus-building model. Experts of the project team and local authorities considered participation more as a threat to the experts' authority and to the project than as a possibility to build consensus around the project. For the experts responsible for this project, it appeared more important to keep their authority than to ensure the fulfilment of the project, as they accepted that the project was put at risk with the referendum. However, the format of the referendum campaign prevented this controversy from fully unfolding and alternatives from being explored. Despite the citizens' participation in the popular vote, which brings into play a political aspect, the project had to be appraised through the narrow binary logic of a yes/no vote as a formal step of the decisional process of the project. Finally, the La Foce project is a clear exclusionary example in which stakeholder groups and citizens could not truly express themselves about the project.

The representative of the Thurauen project described the work done with the accompaniment group as being about "fighting well with each other," thereby reaching a common consensus among stakeholders. The project leader's interests were the good execution of the project and a smooth realization based on a participatory process. Thus, "fighting well" means that the participatory process is ideal, as the different parties can bring up their interests and doubts, with the final goal to find a common-consensual "solution." This gives the impression that the critical points were on the table and the group was able to resolve them through discussion. "Fighting well" makes a more robust impression than simply "discussing." Despite the unavoidable exclusion when selecting representatives for the accompaniment group composition, it appears that a discussion took place and that it included different points of view in reaching a final solution: consensus.

In the Seymaz restoration project, consensus building and the rise of controversy took a different turn and had a deeper meaning for the involved stakeholders and for the outcome of the project. The interest groups, with their representatives, were first selected and formally engaged in the project with the Charte Seymaz. After the first phase of stakeholder involvement and consultation, there was a fundamental shift in how an interest group – in this case the local farmers – considered their inclusion in the decisional process. They initially accepted their role in in the decision-making process. However, once the engineers translated their observations into technical terms on the project plan, the farmers openly questioned the representative model of stakeholder participation. The point of interest of this case study is the project team's reactions to this controversy. They fully considered the farmers' dissatisfaction and collectively called into question the whole planning of the project. Finally, what was different from the first phase of the involvement phase and consequently was the main factor in gaining the farmers' support for the project was that their identity claims were taken into account. This could be done because the engineers modified the restoration project by translating the identity claim in technological measures.

8.5 Recommendations for taking into account the political dimension of participation

As previously discussed in the literature review, the institutionalization of participation can play an important role in depoliticizing decision-making (Swyngedouw 2011; Allmendinger and Haughton 2012). The high density of conflicts of interest in the river space, and the way in which experts implement participation as a depoliticizing force, limits the potentiality of controversies to re-politicize participation even within its institutional context. The analysis of the three case studies and of the national survey reflect such observations. The previously presented findings illustrate the main dynamics that highlight a depoliticization of participation. In this section, we illustrate recommendations that invite practitioners and academics to explore a possible reconsideration of the political dimension of participation (Figure 8.2).

8.5.1 Allowing the controversy to emerge

When confronted with river restorations, stakeholders often express opposition, such as through claims about land ownership when the restored river is planned to take over farming lands along the river. This opposition differs depending on the unique historical and geographical contexts in which the project takes place. However, if experts allow the emerging controversy to unfold, it may suggest or underline aspects of the project that were not visible at the beginning of the decisional process (Cuppen 2018; Tsouvalis and Waterton 2012).

Our analysis shows how a positive and constructive view of opposition, as taken by engineers in the Seymaz restoration project, allows for a debate between stakeholders and practitioners that can be generative of new possibilities. The Seymaz project is characterized by two specific moments that are fundamental to the unfolding of the controversy: (i) the challenge to the representational system of participation that the farmers argued about and (ii) including the farmers' concerns about the project by the engineers, with the modifications of their plans. It is with these dynamics that the farmers feel that they can engage in

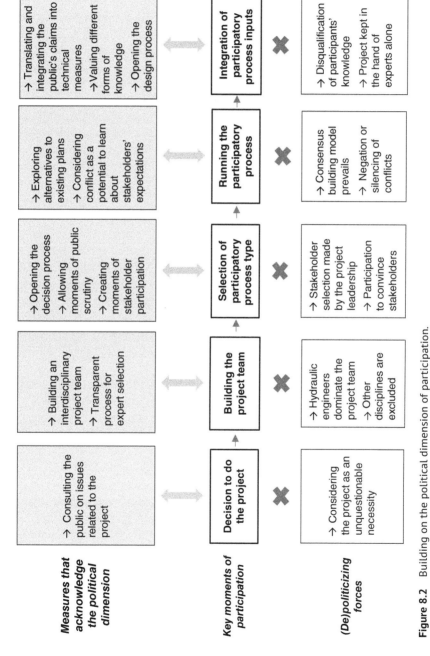

Figure 8.2 Building on the political dimension of participation.

project planning. Contrary to the Seymaz project, in the La Foce project, the decisional process was kept within the expert domain, and this led to the referendum and the popular vote, which were limited to a binary logic and did not allow fundamental controversy to take place.

On the contrary, when controversy is considered and a real dialogue takes place, participation can have political consequences, as the controversy destabilizes the course of events, placing the experts outside their comfort zone. Waterton et al. (2015) were involved in a cross-disciplinary participatory initiative around a problem of lake water management (Loweswater, English Lake District) to explore important issues in public engagement. The authors argue that the accompaniment group created from an early stage of the decisional process suggested an opening of spaces where controversy could emerge and all knowledge and expertise be debated, referring to different experiences of the lake water management of Loweswater. In this perspective, they argue that these "new collectives for local, bottom-up governance of water bodies can reframe problems in ways which both bind lay and professional people to place, and also recast the meaning of 'problem solving'" (Waterton et al. 2015, p. 1). Such reframing can happen at any stage in the process and needs to be done participatively to be most effective, as illustrated by the Seymaz case (Pearce and Ejderyan 2020).

8.5.2 An open design process

When the design of a project is forced, basing it and justifying it on the prevalence of expert knowledge, there is no space for controversy to emerge (Chilvers 2008). In the Seymaz project, engineers propose a new plan for the project, in which the farmers' propositions were considered. The modification of technical measures shows an open design process in which local knowledge is taken into account. Despite the unavoidable exclusionary dynamic of the representational model applied in the Seymaz project, its outcome results were highly inclusionary, as the farmers felt that their identity was respected and reflected in the project's final plan.

As a counterexample, the La Foce case study is one in which the design of the project was forced and maintained within the expertise domain. Experts responsible for this project related to argumentation to contest the political legitimacy of participatory procedures by contrasting them with other institutionalized forms of contestation offered by Swiss semi-direct democracy, such as the local referendum.

If spaces for debate are opened and experts agree to modify technical plans, the diversity of stakeholders' perspectives is considered and not silenced. Henze and her colleagues analyze the interests of the relevant actors in a case study for sustainable river landscape development in Germany and argue for the importance of capturing the diversity of points of view that characterize the river landscape development to enhance more sustainable ways of implementing river restoration (Henze et al. 2018). Similarly, Euler bases his study on case studies from Germany, England, and Spain and argues for taking into account the political consequences of participation as an opportunity to explore ways in which participation helps social and ecological sustainability (Euler and Heldt 2018).

Other examples of opening the design process to the public are present in Britain and are mostly considered under the umbrella of the co-production approach. Some academics and practitioners of flood risk management experimented with co-production of knowledge in implementing participatory processes during the early stages of the design process. For example, co-modeling between certified experts and lay people was conducted experimentally in the United Kingdom after heavy flooding in 2007 (Landström et al. 2011; Lane et al. 2011; Maynard 2013). In these projects, the knowledge of lay people was included from an early stage of the decisional process and valued as essential for an appropriate implementation of the project, thereby taking into consideration local perspectives and points of view regarding flood events.

8.5.3 Acknowledging the production of new political subjects

Experiences and practices that accompany the new regulatory regime, associated with plural, stakeholder-led, participatory governance, facilitate the shifting consciousness and the emergence of new political subjects. In order to deal with the controversies present in river restoration projects, it is essential to recognize the emergence of new political subjects – both positively and negatively engaged with the environment (Agrawal 2005).

The fact that there is a new law, a new paradigm of river restoration, unavoidably creates subjects that differently relate to the environment and are differently engaged in politics. The Seymaz case study is a clear example of how individuals – in this case farmers – are engaged in the decisional processes and with this engagement change their perspective of the project. It is only with consideration of the emerging political subjectivities that a more inclusive, locally based, and creative participation can be implemented.

In this regard, the La Foce and Thurauen case studies are examples in which experts consider their knowledge the only one legitimate for the planning of technical projects; they are uneasy about exposing this knowledge to detailed scrutiny. In one of the two case studies, this confrontation is avoided by keeping the project within expert boundaries. In the case of Thurauen, the confrontation is kept under control with the creation of an accompaniment group, here used as a management tool to smoothly implement the river restoration project. Employing different means, the experts of these two projects avoid dealing with newly formed political subjects that have different visions, other forms of knowledge, and other practices for the project. Within this dynamic, individuals are not considered political subjects but left in the sphere of procedural choices constrained within predetermined boundaries.

8.6 Conclusions

In this chapter, we argue that at the base of a full consideration of participation lies the inclusion of the stakeholders' points of view. We suggest that the whole process of selection and inclusion of various stakeholders should be more reflexive and explicit. For example, in the recently published Swiss guidelines about participatory processes in river works, most of the recommendations are focused on the idea of transparency and openness between the experts and the public, such as clarifying from the beginning the maneuver margin that the public has regarding the project (OFEV 2019). Being transparent about the whole process will help participants to see the gaps and weaknesses of the inclusionary process and identify the potentiality of the political consequences of the process. Saying this, we do not idealize inclusion, as "inclusion is always exclusion" (Chilvers 2008), and we recognize the risk of falling into a superficial and naive inclusion of everyone.

Our aim with presenting and discussing the three case studies is to turn attention away from the question of "Should there be participation or not?" to that of "What happens when a participatory process and controversies take place in a river restoration project, and how does it influence the political dimension of participation?" The reason for this shift is that even if there are examples in the literature that indicate that participation can be a depoliticizing force, many others show how participation and controversies are generative of politics. Therefore, our goal is to identify some aspects of participation wherein its political potential may lie and suggest recommendations to enhance this dimension of participation.

We identified three dynamics that illustrate a depoliticizing force within participatory processes: exclusionary dynamics when selecting the stakeholders to include in the decisional process, a prevalence of expert knowledge over other sources of knowledge, and the negation of conflicts by silencing them. However, when exploring the details of what happened in the case studies, we identify moments in which this depoliticization can be challenged and participation is re-politicized, suggesting that participation is a space where politics might or might not unfold (Whatmore 2009). In relation to the literature, our contribution suggests that, rather than entering into a recurrent critique of participation as an inherently depoliticizing force, we can explore ways to address its political nature considering the proposed principles in this chapter, ranging from a more vivid consideration of controversies to the acknowledgement of newly formed political subjects.

The proposed principles are addressed to practitioners in river restoration – principally governmental officers or experts in river restoration – who also have to apply recommended practices of organizing stakeholders and citizen participation in decisional processes. We address these principles for a more explicit and transparent conduct of participation in river restoration projects in order to revive the political dimensions of participation. We argue that opening the possibility of re-politicization enables a richer and more dynamic decisional process that can result in a better project, one in which stakeholders and citizens can identify themselves.

References

Agrawal, A. (2005). *Environmentality: Technologies of Government and the Making of Subjects.* Durham, NC: Duke University Press.

Allmendinger, P. and Haughton, G. (2012). Post-political spatial planning in England: a crisis of consensus? *Transactions of the Institute of British Geographers* 37(1): 89–103.

Anderson, M.B., Hall, D.M., McEvoy, J., et al. (2016). Defending dissensus: participatory governance and the politics of water measurement in Montana's Yellowstone River Basin. *Environmental Politics* 25(6): 991–1012.

Angelopoulos, N.V., Cowx, I.G., and Buijse, A.D. (2017). Integrated planning framework for successful river restoration projects: upscaling lessons learnt from European case studies. *Environmental Science & Policy* 76: 12–22.

Buchecker, M., Salvini, G., Di Baldassarre, G., et al. (2013). The role of risk perception in making flood risk management more effective. *Natural Hazards and Earth System Sciences* 13(11): 3013–3030.

Buletti, N. (2019). *Destabilizing Participation: Exercises of Power in Swiss River Management.* Doctoral thesis. University of Fribourg.

Buletti, N. and Ejderyan, O. (2021). When experts feel threatened: strategies of depoliticisation in participatory river restoration projects. *Area* 53(1): 151–160.

Buletti, N., Utz, S., Ejderyan, O., et al. (2014). *Définitions et mise en œuvre des processus participatifs dans l'aménagement des cours d'eau en Suisse: Résultats d'une enquête auprès des services cantonaux responsables de l'aménagement des cours d'eau Projet.* Fribourg, Lausanne, Switzerland: University of Fribourg and University of Lausanne.

BWG (Bundesamt für Wasser und Geologie) (2000). *Raum den Fliessgewässern!* Bern, Switzerland: Bundesamt für Geologie.

Chilvers, J. (2008). Environmental risk, uncertainty, and participation: mapping an emergent epistemic community. *Environment and Planning A: Economy and Space* 40(12): 2990–3008.

Chilvers, J. and Kearnes, M. (2015). *Remaking Participation: Science, Environment and Emergent Publics.* London: Routledge.

Chilvers, J., Pallett, H., and Hargreaves, T. (2018). Ecologies of participation in socio-technical change: the case of energy system transitions. *Energy Research and Social Science* 42: 199–210.

Cuppen, E. (2018). The value of social conflicts: critiquing invited participation in energy projects. *Energy Research and Social Science* 38: 28–32.

de Saille, S. (2015). Dis-inviting the unruly public. *Science as Culture* 24(1): 99–107.

Donaldson, A., Lane, S., Ward, N., et al. (2013). Overflowing with issues: following the political trajectories of flooding. *Environment and Planning C: Government and Policy* 31(4): 603–618.

EC (European Commission) (2000). Directive 2000/60/EC of the European Parliament and of the Council of 23 October 2000 establishing a framework for community action in the field of water policy.

Eden, S., Tunstall, S.M., and Tapsell, S.M. (2000). Translating nature: river restoration as nature-culture. *Environment and Planning D: Society and Space* 18(2): 257–273.

Ejderyan, O. (2014). Quels aménagements pour quelle nature? Hydrologie, patrimoine et biodiversité dans le projet de renaturation de la Haute-Seymaz à Genève. In: *Urbanités et biodiversité: Entre villes fertiles et campagnes urbaines, quelle place pour la biodiversité?* (ed. V. Bradel), 262–275. Saint-Étienne, France: Presses de l'Université de Saint-Étienne.

Euler, J. and Heldt, S. (2018). From information to participation and self-organization: visions for European river basin management. *Science of the Total Environment* 621: 905–914.

Fiorino, D.J. (1990). Citizen participation and environmental risk: a survey of institutional mechanisms. *Science, Technology, & Human Values* 15(2): 226–243.

Graefe, O. (2011). River basins as new environmental regions? The depolitization of water management. *Procedia: Social and Behavioral Sciences* 14: 24–27.

Hassenforder, E., Clavreul, D., Akhmouch, A., et al. (2019). What's the middle ground? Institutionalized vs. emerging water-related stakeholder engagement processes. *International Journal of Water Resources Development* 35(3): 525–542.

Heldt, S., Budryte, P., Ingensiep, H.W., et al. (2016). Social pitfalls for river restoration: how public participation uncovers problems with public acceptance. *Environmental Earth Sciences*, 75(13). https://doi.org/10.1007/s12665-016-5787-y.

Henze, J., Schröter, B., and Albert, C. (2018). Knowing me, knowing you: capturing different knowledge systems for river landscape planning and governance. *Water* 10(7): 934.

Junker, B., Buchecker, M., and Müller-Böker, U. (2007). Objectives of public participation: which actors should be involved in the decision making for river restorations? *Water Resources Research* 43(10). https://doi.org/10.1029/2006WR005584.

Landström, C., Whatmore, S.J., Lane, S.N., et al. (2011). Coproducing flood risk knowledge: redistributing expertise in critical "participatory modelling." *Environment and Planning A* 43(7): 1617–1633.

Lane, S.N., Landström, C., and Whatmore, S.J. (2011). Imagining flood futures: risk assessment and management in practice. *Philosophical Transactions of the Royal Society A: Mathematical, Physical and Engineering Sciences* 369: 1784–1806.

Latour, B. (2004). *Politics of Nature*. Cambridge, MA: Harvard University Press.

Le Lay, Y.-F., Piégay, H., and Rivière-Honegger, A. (2013). Perception of braided river landscapes: implications for public participation and sustainable management. *Journal of Environmental Management* 119: 1–12.

Light, A. and Higgs, E.S. (1995). The politics of ecological restoration. *Environmental Ethics* 18(3): 227–247.

Luyet, V., Schlaepfer, R., Parlange, M.B., et al. (2012). A framework to implement: stakeholder participation in environmental projects. *Journal of Environmental Management* 111: 213–219.

Maynard, C.M. (2013). How public participation in river management improvements is affected by scale. *Area* 45(2): 230–238.

McDonald, A., Lane, S.N., Haycock, N.E., et al. (2004). Rivers of dreams: on the gulf between theoretical and practical aspects of an upland river restoration. *Transactions of the Institute of British Geographers* 29(3): 257–281.

Mees, H., Crabbé, A., Alexander, M., et al. (2016). Coproducing flood risk management through citizen involvement: insights from cross-country comparison in Europe. *Ecology and Society* 21(3): 7.

Niehuis, P.H. and Leuven, R.S.E. (2001). River restoration and flood protection: controversy or synergism. *Hydrobiologia* 444: 85–99.

Nijland, H.J. and Cals, M.J.R. (eds) (2000). *River Restoration in Europe Practical Approaches: Conference on River Restoration: Wageningen, 2000*. Lelystad, Netherlands: Institute for Inland Water Management and Waste Water Treatment/RIZA.

Norton, L., Elliott, J.A., Maberly, S.C., et al. (2012). Using models to bridge the gap between land use and algal blooms: an example from the Loweswater catchment, UK. *Environmental Modelling and Software* 36: 64–75.

OFEFP (Office fédéral de l'environnement, des forêts et du paysage) and OFEG (Office fédérale des eaux et de la géologie) (2003). *Idées directrices Cours d'eau suisses: Pour une politique de gestion durable de nos eaux*. Bern, Switzerland: Office fédéral de l'environnement, des forêts et du paysage, Office fédéral des eaux et de la géologie.

OFEV (Office fédéral de l'environnement) (2008). *MManuel RPT dans le domaine de l'environnement: Communication de l'OFEV en tant qu'autorité d'exécution*. Bern, Switzerland: Office fédéral de l'environnement.

OFEV (Office fédéral de l'environnement) (2018). *Manuel sur les conventions-programmes 2020–2024 dans le domaine de l'environnement*. Bern, Switzerland: Office fédéral de l'environnement.

OFEV (Office fédéral de l'environnement) (2019). *Manuel Processus participatif dans les projets d'aménagement de cours d'eau: D'acteurs concernés à acteurs impliqués: Connaissance de l'environnement No. 1915*. Berne, Switzerland: Office fédéral de l'environnement.

Pahl-Wostl, C. (2006). The importance of social learning in restoring the multifunctionality of rivers and floodplains. *Ecology and Society* 11(1): 10.

Pahl-Wostl, C., Jeffrey, P., Isendahl, N., et al. (2011). Maturing the new water management paradigm: progressing from aspiration to practice. *Water Resources Management* 25(3): 837–856.

Parés, M. (2011). River basin management planning with participation in Europe: from contested hydro-politics to governance-beyond-the-state. *European Planning Studies* 19(3): 457–478.

Pearce, B.J. and Ejderyan, O. (2020). Joint problem framing as reflexive practice: honing a transdisciplinary skill. *Sustainability Science* 15(3): 683–698.

Petts, J. (2007). Learning about learning: lessons from public engagement and deliberation on urban river restoration. *Geographical Journal* 173(4): 300–311.

Reichert, P., Borsuk, M., and Hostmann, M. (2007). Concepts of decision support for river rehabilitation. *Environmental Modelling & Software* 22(2): 188–201.

Rohde, S., Hostmann, M., Peter, A., and Ewald, K.C. (2006). Room for rivers: an integrative search strategy for floodplain restoration. *Landscape and Urban Planning* 78(1): 50–70.

Rowe, G. and Frewer, L.J. (2004). Evaluating public-participation exercises: a research agenda. *Science, Technology & Human Values* 29(4): 512–556.

Ruef, F. (2015). Participation in river restorations or how are environmental subjects formed? Master thesis. University of Fribourg.

Sivapalan, M., Savenije, H.H.G., and Blöschl, G. (2012). Socio-hydrology: a new science of people and water. *Hydrological Processes* 26(8): 1270–1276.

Smith, B., Clifford, N.J., and Mant, J. (2014). The changing nature of river restoration. *Wiley Interdisciplinary Reviews: Water* 1(3): 249–261.

Stirling, A. (2008). "Opening up" and "closing down": power, participation, and pluralism in the social appraisal of technology. *Science, Technology & Human Values* 33(2): 262–294.

Swyngedouw, E. (2011). Depoliticized environments: the end of nature, climate change and the post-political condition. *Royal Institute of Philosophy Supplement* 69: 253–274.

Thaler, T. and Levin-Keitel, M. (2016). Multi-level stakeholder engagement in flood risk management: a question of roles and power: lessons from England. *Environmental Science & Policy* 55: 292–301.

Tsouvalis, J. (2016). Latour's object-orientated politics for a post-political age. *Global Discourse* 6(1–2): 26–39.

Tsouvalis, J. and Waterton, C. (2012). Building "participation" upon critique: the Loweswater Care Project, Cumbria, UK. *Environmental Modelling & Software* 36: 111–121.

Tunstall, S.M., Tapsell, S.M., and Eden, S.E. (1999). How stable are public responses to changing local environments? A "before" and "after" case study of river restoration. *Journal of Environmental Planning and Management* 42(4): 527–547.

UN (United Nations) (1992). *Agenda* 21: Programme of Action for Sustainable Development. United Nations Conference on Environment and Development (UNCED). New York: United Nations Department of Public Information.

UNECE (United Nations Economic Commission for Europe) (1998). *Convention on Access to Information, Public Participation in Decision-making and Access to Justice in Environmental*

Matters: Results of the Fourth Ministerial Conference in Aarhus, Denmark on the 25th of June 1998. Aarhus, Denmark: UNECE.

Utz, S., Buletti, N., Ejderyan, O., et al. (2017). Processus participatifs pour la mise en œuvre des projets d'aménagement de cours d'eau en Suisse. *Bulletin de l'ARPEA: Journal Romand de l'Environnement* 271: 41–50.

Waterton, C., Maberly, S.C., Tsouvalis, J., et al. (2015). Committing to place: the potential of open collaborations for trusted environmental governance. *PLOS Biology* 13(3): e1002081.

Whatmore, S.J. (2009). Mapping knowledge controversies: science, democracy and the redistribution of expertise. *Progress in Human Geography* 33(5): 587–598.

Wohl, E., Lane, S.N., and Wilcox, A.C. (2015). The science and practice of river restoration. *Water Resources Research* 51(8): 5974–5997.

World Bank (2006). *An Introduction to Integrated River Basin Management Integrated*. Washington DC: World Bank.

Wynne, B. (2008). Public participation in science and technology: performing and obscuring a political–conceptual category mistake. *East Asian Science, Technology and Society: An International Journal* 1: 99–110.

Zaugg, M., Ejderyan, O., and Geiser, U. (2004). Normen, Kontext und konkrete Praxis des kantonalen Wasserbaus: Resultate einer Umfrage zu den Rahmenbedingungen der kantonalen Ämter oder Fachstellen für Wasserbau bei der Umsetzung der eidgenössischen Wasserbaugesetzgebung. *Schriftenreihe Humangeographie* 19. Zürich: Geographisches Institut, Abteilung Humangeographie, Universität Zürich. DOI: 10.5167/uzh-90186.

Zingraff-Hamed, A., Greulich, S., Wantzen, K., et al. (2017). Societal drivers of European water governance: a comparison of urban river restoration practices in France and Germany. *Water* 9(3): 206.

Part IV

Evaluation of Socioeconomic Effects

9

What is the Total Economic Value of River Restoration and Why is it Important?

John C. Bergstrom[1] and John B. Loomis[2]

[1] University of Georgia, Athens, GA, USA
[2] Colorado State University, Fort Collins, CO, USA

9.1 Introduction

Total economic value (TEV) is the theoretically correct measure of benefit change associated with river restoration policy changes and projects. The components of TEV arise from the interactions between people and river-based ecosystem goods and services. Examples of ecosystem goods and services associated with river restoration include restoring self-sustaining fish and wildlife populations including threatened or endangered species, provision of clean water for drinking and recreation, and flood control.

In Section 9.2, we provide a comprehensive discussion of the interactions between people and rivers that give rise to economic values. We then define TEV and its components and describe methods for measuring these values. In Section 9.3, we make the case that the theoretically preferable method for valuing the TEV of river restoration is the single-study, holistic approach rather than adding up separate values from different studies. We then provide a summary of existing studies that valued river restoration TEV using the preferred single-study, holistic method, and discuss application of such values to river restoration policy and management decisions using a cost–benefit analysis (CBA) case study.

CBA is one of the most widely applied economic decision-making tools used to help inform policy and management decisions. For example, CBA may be used to determine if a river restoration project is economically justifiable (e.g. benefits are at least as much as the costs). Thus, reliable estimates of river restoration TEV are needed to help ensure the application of CBA to river restoration projects results in decisions that are consistent with improving social well-being. Another important use of the economic valuation is in an environmental impact statement (EIS) of a program or project where the goal is to facilitate consideration of economic values as one of the many relevant metrics a decision maker might weigh when choosing restoration alternatives. Some concluding comments are provided in Section 9.4.

River Restoration: Political, Social, and Economic Perspectives, First Edition. Edited by Bertrand Morandi, Marylise Cottet, and Hervé Piégay.
© 2022 John Wiley & Sons Ltd. Published 2022 by John Wiley & Sons Ltd.

9.2 Defining and valuing the total economic value of river restoration

9.2.1 Interactions between people and rivers

Kummu et al. estimate that, respectively, about 90 and 50% of the world's population lives within 10 km and 3 km of a surface freshwater body including lakes and rivers. Furthermore, they estimate that for about 66, 15, and 6% of the world's population the closest freshwater body was a small, medium, and large river, respectively. Thus, much of the world's population has likely had some sort of interaction with rivers at times and perhaps throughout their lives. Such interactions, which ultimately generate economic values, include fishing, swimming, birdwatching, boating, and obtaining drinking water for themselves, other people, and animals (e.g. pets and livestock).

All over the world, interactions between people and rivers affect the health and well-being of individuals and groups in communities every day. In some communities, people walk, ride, or drive to a nearby river and draw water directly using in buckets and other containers to meet daily household drinking, cooking, and washing needs. In other communities, water for meeting daily household needs is piped in from rivers and accessed by turning a kitchen or bathroom tap. In many communities, people walk, ride, or drive to rivers for various recreational activities, such as swimming, bathing, boating, wildlife observation, and simply enjoying the scenery. In addition to providing wildlife habitat, vegetation along rivers and sometimes hydrologically connected adjacent wetlands can also benefit communities by providing a natural buffer to capture and disperse rising water levels, thereby protecting communities from damaging floods.

Now let's suppose that in one community such as that just described several residents meet at a local restaurant on a Sunday night and start talking about their weekend activities over dinner as they sip glasses of ice water just delivered to the table by their server. Taylor mentioned first that on Saturday afternoon she and her kids went for a swim at a nearby river "swimming hole." Morgan then shared that on Saturday morning he went hiking on a nature trail along the same river with an environmental organization group to do some birdwatching. Then Adrian said, "Hey, you must have been the group I saw on the riverbank with binoculars when I was canoeing down the river!" In a moment of reflection, a fourth friend, Parker, who spent Sunday afternoon lounging at a riverside park enjoying the sights and sounds of the river, pronounced, "Isn't it great that we live near such a nice river where we can do all these fun things for free!" Lifting his water glass, Morgan replied chuckling, "I'll drink to that with this free glass of great-tasting water, which maybe you all don't know, also comes to us from the same river!"

If the four friends in the story are like many people, they may overlook or at least take for granted that all of the benefits they receive from the river near their community are dependent on an ecologically healthy and functioning river ecosystem. Ecologists and economists use the technical term "ecosystem goods and services" to describe things that ecosystems provide, often for free, that are useful and valuable to people in an economic sense. But, if some ecosystem good or service is "free," that is it has no market price, how is it possible to estimate an economic value on that good or service? We discuss the overall answer to this question later in this chapter. However, before we discuss *how* to measure the economic values of nonmarketed goods and services (e.g. goods and services without a market price), we first need to understand *what* values we want or need to measure.

Table 9.1, based on previous related research (e.g. Bergstrom and Loomis 1999, 2017; Brown et al. 2007; Loomis 2006), represents a classification of ecosystem goods and

Table 9.1 Classification of ecosystem goods and services supported by rivers.

Type of Interaction		Location of Interaction	
		On-site (in situ)	**Off-site**
Rival	Direct	A	B
		Fishing trips/days under congestion	Eating commercially harvested wild fish or game at a restaurant
		Hunting trips/days under congestion	Enjoying commercially harvested wild fish or game dining experiences at a restaurant
		Wild fish or game harvest in situ	Drinking river piped in from a river at home or away from home
		Boating trips/days under congestion	
		Drinking river water in situ	
	Indirect	C	D
		Satisfaction gained from rival, on-site river recreational trips by other individuals in the present or future	Satisfaction gained from other individuals eating commercially harvested wild fish or game at a restaurant and enjoying commercially harvested wild fish or game dining experiences at a restaurant, in the present or future
		Satisfaction gained from other people drinking river water in situ in the present or future	Satisfaction gained from other people drinking water piped in from a river at home or away from home, in the present or future
Non rival	Direct	E	F
		On-site fish and wildlife observation	At-home viewing photos of fish and wildlife taken at a riverine ecosystem by a commercial photographer
		On-site fish and wildlife photography	Nonriparian real property flood protection
		Viewing aesthetically pleasing river scenery	
		Riparian real property flood protection	
	Indirect	**G**	**H**
		Satisfaction gained from nonrival, on-site river recreational trips by other individuals in the present or future	Thinking about natural resource supported by rivers
			Satisfaction gained from participation in nonrival, off-site activities by other individuals in the present or future
			Community flood control provided by vegetation in healthy, riverine ecosystems

services involving interactions between individuals and rivers that generate TEV. Examples of ecosystem goods supported by rivers are species of fish and wildlife. Species of fish and wildlife can be tangibly measured in terms of biomass (e.g. numbers and weight of game fish and game waterfowl). A distinguishing feature of ecosystem goods is that these goods are typically something that a person could hold in their hand, such as a single rainbow trout. Fish and wildlife are also the good people seek to obtain on a fishing trip (e.g. catching a fish to take home) or on a birdwatching trip (e.g. taking a photo of a rare bird).

An example of ecosystem services supported by rivers are recreational fishing and hunting experiences. These ecosystem services can be measured in terms of recreation trips or visitor days. A distinguishing feature of ecosystem services is, although measurable, these services are not tangible in the sense of being an object a person can physically handle (Bergstrom and Loomis 2017; Brown et al. 2007).

As can be seen in the example, the analyst must be careful to avoid double counting the same benefit twice. For example, the economic value of a trout fishing trip at a river would include (economist would technically say, "capitalize") the value of trout caught on the trip. Thus, if for some reason one wanted to know the economic value of the trout caught on the trip, special techniques would need to be employed to separate out the value of the trout caught from the trout fishing trip experience. For example, there can be a fishing experience even without bringing home a fish or with catch and release fishing.

In Table 9.1, ecosystem goods and services supported by rivers are classified according to whether the interactions between individuals and rivers are rival or nonrival in consumption, whether these interactions are direct or indirect, and whether the location of these interactions between individuals and rivers are on-site (in situ) or off-site. These categorizations are consistent with well-established definitions of private and public goods and their characteristics in the economic literature (see Bergstrom and Randall 2016). "Rival use" means that the use of ecosystem goods and services supported by rivers by one individual reduces the quantity or quality of the good or service available to other individuals. For example, when one angler catches a fish and does not return it to the river alive (e.g. not practicing "catch and release"), there is one fewer of that ecosystem good (the caught fish) available to any other angler to catch. Rival use is most easily understood in the case of tangible, ecosystem goods. However, use of ecosystem services can also be rival. For example, because of congestion, there may only be a limited number of fishing and hunting spots along a particular river reach, or the hunting permits are rationed in some way (e.g. a lottery).

Rival and nonrival interactions (Table 9.1) are further divided into activities involving direct or indirect interactions between individuals and river-based resources. Direct interaction, for example, would involve some sort of physical contact with river resources such as swimming in a river or boating on a river. Indirect interaction, for example, would involve a person viewing photos or videos of river resources at her home that may be their own or on television.

Examples of direct, on-site, rival ecosystem goods and services supported by rivers are recreational fishing and hunting experiences (ecosystem services) constrained by congestion, in situ recreationally harvested fish and wildlife species (ecosystem good), and drinking river water in situ (Table 9.1, cell A). Examples of direct, off-site, rival ecosystem goods and services are eating commercially harvested fish or waterfowl (ecosystem good) at a restaurant, enjoying commercially harvested fish or wild game dining experiences at a restaurant, and drinking river water piped in from a river at home or away from home (Table 9.1, cell B). A fish or wild game dining experience is in the nature of a service produced by a household following a household production framework laid out in Bockstael and McConnell (1983).[1]

Rival ecosystem goods and services supported by rivers involving indirect on-site interactions include the satisfaction an individual gains from other people participating in hunting and fishing in the present or future, and the satisfaction gained from other people drinking river water in situ in the present or future (Table 9.1, cell C). Rival ecosystem goods and services supported by rivers involving indirect off-site interaction include the satisfaction an individual gains from other people eating commercially harvested fish or wild game at a restaurant in the present or future, and the satisfaction gained from other people drinking water piped in from a river in the present or future (Table 9.1, cell D). Although the ecosystem goods and services supported by rivers falling in cells C and D do not result in direct rival use of natural resources, we classify them as rival, consumptive goods and services because rival use indirectly occurs on the part of other people in the present or future. In economic terms, this linkage between individual utilities represents spatial or temporal externality relationships between individuals.

Nonrival ecosystem goods and services supported by rivers involve active and passive uses where use (consumption) by one individual does not reduce the quantity or quality of the good or service available to any other individual. Nonrival ecosystem goods and services supported by rivers involving direct on-site interaction with river-based resources, including fish and wildlife observation and photography (Table 9.1, cell E). A specific example of this type of ecosystem good or service supported by rivers is birdwatching, an activity enjoyed by millions of people (Cordell and Herbert 2002). An example of a nonrival ecosystem good or service supported by rivers involving direct off-site interaction with river-based resources is viewing a collection of photographs at home that were taken at a riverine ecosystem by a commercial photographer (cell F), for example viewing photographs in *National Geographic* magazine of tropical fish and wildlife taken on-site in the Amazon River basin.

Nonrival ecosystem goods and services supported by rivers involving indirect, on-site interaction with river-based resources include the satisfaction an individual receives from participation by other individuals in fish and wildlife observation and photography in the present or future (cell G). Because an individual receives satisfaction through participation by other individuals, cell G's activities represent interdependent relationships between people. For example, people may gain satisfaction from knowing and hearing about their family members' experiences viewing and photographing fish and wildlife at a riverine ecosystem.

Nonrival ecosystem goods and services supported by rivers involving indirect off-site interaction with river-based resources includes satisfaction gained from simply thinking about a natural resource continuing to live in a riverine ecosystem (cell H). Cell H's interactions also include the vicarious satisfaction an individual gains from participation in nonrival, off-site activities by other individuals in the present or future, which represents another type of interdependent relationship between people. For example, people may also gain satisfaction from knowing and hearing about their family members enjoying thinking about exotic natural resources that exist in the Amazon River basin. Finally, cell H's interactions also include community flood control services provided by vegetation in healthy, riverine ecosystems.

9.2.2 Economic values and valuations

The term "total economic value" was coined by Randall and Stoll (1983) and refers generally to the complete set of active and passive use values that a consumer places on ecosystem goods and services. In the context of river restoration, active use value (AUV) is the economic value of the utility gained from on-site activity using river-dependent ecosystem goods and services. Passive use value is the economic value of the utility gained off-site from passively using river-dependent ecosystem goods and services. Because use is passive, including the mere act of enjoying positive thoughts about a natural resource, economists may also refer to passive use values as "nonuse values" (Bergstrom and Randall 2016).

Theoretically, changes in river-dependent ecosystem goods or services are measured in terms of what economists call a "person's willingness to pay (WTP)" for an increment in ecosystem goods or services provided by a river restoration project, or their minimum willingness to accept (WTA) compensation in money to forgo the increment in ecosystem goods and services provided by a river restoration project (Bergstrom and Randall 2016; Brookshire et al. 1980). Given most river restoration projects are focused on improving the environmental condition of rivers through restoration, the difficulty economists have had empirically measuring WTA, and the fact that WTP is more commonly used as an economic value (benefit) measure by government agencies worldwide, in this chapter we focus our attention on WTP.[2]

Observations or statements of WTP for changes in river-based, ecosystem goods and services, however, may not always reflect economic motives and tradeoffs. For example, an individual may contribute money to the World Wildlife Fund to help save endangered river fish species because of an emotional commitment to a particular environmental ethic, the value of which is not measurable in terms of a standard economic WTP measure. Philosophers and theologians also suggest that an individual may be motivated to preserve a resource, not out of a social obligation but because they believe resources have a right to exist in and of themselves, independent of individual or social values (Bergstrom and Reiling 1997). Although we recognize the importance of noneconomic motives for supporting river restoration, this chapter is concerned with economic motives associated with river restoration reflected in TEV.

The components of TEV, which involve specific types of active and passive use values, are generated by the interactions between people and rivers shown in Table 9.1 and described here. In order to estimate dollar values for the components of TEV, or TEV as a whole, economists employ specialized valuation techniques. If an ecosystem good or service has a market price, economists can use standard market valuation techniques to estimate WTP for a change in the ecosystem good or service. For example, data on the market price and quantity of commercial whitewater rafting trips taken on a river where a dam was removed to restore a free-flowing river could be used to estimate demand functions and WTP for those rafting trips.

If an ecosystem good or service does not have a market price, economists employ what are known as "nonmarket valuation techniques" to estimate WTP. Nonmarket valuation techniques are divided into two categories: revealed preference methods and stated preference methods. The two most common revealed preference methods are the travel cost method (TCM) and the hedonic price method (HPM).

The TCM only captures use value, for example WTP for a recreational fishing trip to a restored river and/or WTP for an improvement in recreation fish catch resulting from a river restoration project. The TCM measures WTP by using variations in different visitor travel costs and number of trips to trace out the overall demand function. From the demand function, WTP is calculated as the area under the demand function and in excess of the travel costs incurred (the travel cost themselves are a cost to the visitor and not a benefit). Theoretically, a travel cost demand function captures the value of planning a recreation trip, the on-site recreation experience, and follow-up enjoyment of the trip at-home including reminiscing about the trip, and possibly eating a meal of fish caught on the trip.

The HPM also only captures use value, for example WTP for the aesthetic value added to an individual's home and property located along a restored river. Using real-estate price and home/property attribute data, the HPM employs statistical techniques to estimate how much consumers pay for increments in home/property attributes including environmental attributes. Thus, the HPM could be used to estimate WTP for improvements in environmental attributes of private property resulting from a public river restoration project (Bergstrom and Randall 2016; Bergstrom and Loomis 2017; Brouwer and Sheremet 2017).

Revealed preference techniques leave a behavioral trail that economists can follow in order to estimate economic values for nonmarket commodities and are thus most commonly used to measure active use values. In the case of the TCM, for example, the types of interaction between an individual and river-based resources which leads to behavioral trails are shown in Table 9.1, cells A, B, E, and F. These interactions, for example, include "put and take" recreational fishing involving activities in cell A, residential water consumption activities in cell B, and wildlife observation (e.g. birdwatching) involving activities in cells E and F. Consumer actions associated with activities in cells A, B, E, and F, such as actual trip-making behavior and expenditures, can be used to estimate recreation demand functions and economic values such as WTP for recreation trips.

In the case of interactions between individuals and river-based resources in cells in Table 9.1 other than cells A, B, E, and F, it is difficult for economists to observe a behavioral trail from which to estimate consumer demand functions and economic values. In fact, in the case of passive use or nonuse values, behavioral trails do not exist. Thus, in the case of quantifying passive use values, stated preference valuation techniques that do not rely on actual, observed consumer behavior (e.g. expenditures on travel, real-estate) must be employed.

Stated preference techniques, including the contingent valuation method (CVM) and the choice experiment method (CEM), are commonly used to measure both active and passive use values, and comprehensive TEV composed of both active and passive use values.[3] In the context of river restoration, the CVM, for example, would ask consumers to state WTP for changes in ecosystem goods and services resulting from river restoration projects directly using open-ended questions, or indirectly using close-ended (e.g. dichotomous choice) questions. The CEM, for example, would ask consumers to choose between alternative river restoration projects with different physical attributes (e.g. kilometers of restored riverbank vegetation) and costs to consumers. The resulting choice data can then be analyzed statistically to estimate WTP for changes in ecosystem goods and services resulting from river restoration projects (Bergstrom and Randall 2016; Bergstrom and Loomis 2017; Brouwer and Sheremet 2017).

We are now in a position in this chapter to discuss specific types of active and passive use values and methods for estimating these values. In situ, direct, and rival or nonrival interactions between individuals and river-based resources (Table 9.1, cells A and E) lead to on-site active use values such as WTP for changes in the quantity or quality of river-dependent recreational trips resulting from a river restoration project. On-site active use values can be measured using the TCM, HPM, CVM, or CEM.

Off-site, direct, and rival or nonrival interactions between individuals and river-based resources (cells B and F) lead to off-site active use values such as WTP for eating commercially harvested wild fish at a restaurant. Another example is viewing photos or videos featuring river-based resources at-home, such as WTP to watch a television documentary about wild Pacific and Atlantic salmon runs in North American rivers. If we are able to observe actual market prices for commercially harvested fish or wild game, restaurant dining experiences, and commercial photographs/videos, it may be feasible to estimate WTP for these goods and services using market valuation techniques. If such data are not available, WTP for these goods or services could be estimated using the CVM or CEM. Since WTP for wild fish or game restaurant dining experiences and photographs/videos capture the value of the featured fish or game, as well as the costs to cook and serve the fish or game along with the ambience of the restaurant, one must be careful about double-counting or inflating the value attributable to the fish or game itself.

In situ, indirect, and rival or nonrival interaction with river-based resources (Table 9.1, cells C and G) lead to on-site use values such as WTP to provide other people from current or future generations with opportunities to participate in on-site recreational activities at a restored river. On-site benefits can be rival and indirect (cell C) such as benefits received by an individual from providing others in current or future generations with opportunities to engage in consumptive activities (e.g. on-site hunting or fishing). On-site benefits can also be nonrival and indirect (cell G) such as benefits received by an individual from providing others in current or future generations with opportunities to engage in nonconsumptive recreation (e.g. on-site observation of wildlife without the intent of harvest). Off-site benefits can be rival and indirect (cell D), such as the satisfaction gained from others of the current or future generations consuming wild game at a restaurant. Likewise, off-site benefits can be nonrival and indirect (cell H) such as the benefits gained by thinking about natural resources supported by rivers, whether by the individual themselves, or other individuals of the current or future generations. External benefits occur because the utility of one individual (in the present) is positively affected by the use of river-based resources by other individuals in the present (which economists term "*intragenerational* bequest value") or in the future (which economists term "*intergenerational* bequest value") (Bergstrom and Reiling 1997; Bergstrom and Loomis 2017; McConnell 1983; Randall and Stoll 1983). Bequest values fall into the category of passive use values (since use is passive, not active). Because they are passive use values with no behavioral trail for economists to follow (as there is, for example, with the TCM), bequest values must be measured using stated preference methods such as the CVM and CEM.

Off-site, indirect, and nonrival interactions between river resources and people (Table 9.1, cell H) lead to another major type of passive use value termed "existence value." Existence value, for example, is reflected by a person's WTP for the knowledge that endangered fish and wildlife species continue to exist in restored riverine ecosystems. Krutilla (1967) is usually credited with introducing the concept of existence value in the economics literature. In this classic article, entitled "Conservation Reconsidered," Krutilla suggests that people may economically value natural resources simply because they exist. This existence value is different from typical economic values because it is not dependent on the current or future use of the resource (Krutilla 1967; Krutilla and Fisher 1975). In the 50 years since Krutilla's article, existence value has become well established conceptually and empirically, including successful measurement studies using nonmarket valuation techniques such as the CVM (Amirnejad et al. 2006; Louriero and Loomis 2013; Madariaga and McConnell 1987; Randall and Stoll 1983; Smith 1987; Turpie 2003).

9.3 Estimation and application of river restoration total economic value

9.3.1 Separate versus single study approaches

When conducting a CBA of river restoration projects, the conceptually correct measure of the benefits is the TEV from the restoration (Loomis 2006). One approach to measuring this TEV is to conduct separate studies to measure each of the components of TEV affected by the restoration project, and then add up these values to estimate the TEV of the project. Another approach is to measure all of the components of TEV affected by the restoration project at once using an "all encompassing" or holistic approach. A number of previous studies reviewed by Bergstrom and Loomis (2017) and Brouwer and Sheremet (2017) reported economic value estimates for separate components of river restoration TEV and TEV as a whole.

There are several pitfalls associated with the approach of estimating river restoration TEV by adding up component value estimates from separate studies. In general, there is a key concept in economic valuation which parallels a key concept in ecology: the sum of the parts does not necessarily equal the whole. In ecology, this phrase means that the complex physical and biological interrelationships between the biotic and abiotic components of an ecosystem when viewed as a whole are not accurately reflected by the mere "sum total" of the roles these components play in an ecosystem at a smaller scale such as an individual organism or community. In economics, this phrase refers to the fact that the sum of values of the components of TEV measured in separate will not generally equal the TEV as measured in a single, holistic study (Hoehn 1991; Hoehn and Randall 1987).

River restoration projects involve reclaiming or restoring riverine ecosystem functions such as ecologically healthy chemical cycling, water flow, and water temperature regulation. Restoring ecosystem functions, in turn, simultaneously supports many ecosystem goods and services. For example, a river restoration project designed to restore an endangered fish species in a river will likely simultaneously restore other ecosystem goods and services such as other fish species harvested by recreational anglers, recreational swimming opportunities, and perhaps even making the river water safe for human consumption. In this case, it would be very difficult to design separate valuation studies that would isolate the value of restoring endangered species from all other restoration values. It would be even more difficult to design studies to isolate, say, all four values (endangered species, game fish, swimming, and drinking water) from each other.

As an example, suppose we were commissioned by a government agency to estimate the separate values of a restoration project including endangered species, game fish, swimming, and drinking water using stated preference techniques. The CEM would involve setting up different valuation scenarios which vary in the levels of the endangered species, game fish, swimming, and drinking water attributes – in particular, some go up and some go down across the attributes. However, there is an ecological disconnect with this type of set-up since all four of these ecosystem goods and services would likely simultaneously move in the same direction as a result of restoration of ecological functions in the river necessary for restoring any single one of them. In other words, we have a collinearity problem which complicates estimating the separate values of each of these components of TEV using the CEM.

A similar problem would exist with using the CVM to estimate the economic values of the endangered species, game fish, swimming, and drinking water benefits of a restoration project. Suppose, for example, four separate CVM surveys (or one survey divided into four subgroups) were conducted to measure the economic value of each of these four ecosystem goods and services. In order to avoid double counting, respondents in each of the four survey groups would need to be told to value a change in just one of four ecosystem goods or services, and to keep the level of the other three constant at the level specified by the researcher. But, what levels of the other three would make sense to the perceptive respondent? For example, holding the levels of the other three at zero would not make sense, since the perceptive respondent would know that restoration of one of the four ecosystem goods or services would likely improve the other three. Thus, for example, when stating WTP for restoring the endangered species, the perceptive respondent may also include their WTP for game fish.

Even supposing the independent restoration of different ecosystem goods and services makes ecological sense, there are still significant challenges to obtaining valid estimations of river restoration TEV using the approach of adding up estimates of separate TEV components. In particular, Hoehn and Randall (1987) show that in order to obtain a valid estimate of TEV from separate estimates of the components of TEV the separate components must be valued sequentially where, for example, WTP for one particular component is based on the order in which it is valued in a valuation sequence due to substitute and complement effects. In practice, however, such coordinated sequential valuation exercises are very rare. Rather, it is more typical that previous valuation studies of separate components of river restoration TEV have been conducted by different researchers and different times in an uncoordinated manner. In these cases, adding up estimates of TEV components would yield an incorrect estimate of TEV, for example double counting and therefore overestimation when substitution effects are not taken into account (Hoehn 1991).

Because of the problems and challenges involved with the separate study, adding up approach, we argue that the preferred approach for measuring river restoration TEV is the single study, holistic approach. The need for holistic or "all encompassing" valuation points to another disadvantage of revealed preference methods, such as the TCM. These methods are limited to valuation of only the use value portion of TEV. An advantage of stated preference methods such as contingent valuation and choice experiments is the ability to directly capture both the use and nonuse portions of TEV as a whole in a single study.

9.3.2 River restoration TEV estimates and applications

Table 9.2 shows estimates of river restoration TEV using the conceptually valid single study, holistic approach. These studies measured all of the components of TEV derived from the interactions between people and river resources illustrated in Table 9.1 and discussed in Section 9.2.2 using stated preference techniques. Estimates of the TEV of river restoration range from $4 to $350[4] per household per year with an average of about $87 per household per year. Annual household TEV for river restoration averages out to about $24 per river km.

TEV estimates such as those presented in Table 9.2 are important for capturing the full benefits of river restoration projects to better inform restoration policy and management decisions. River restoration projects are relatively expensive endeavors which are typically paid for using public funds (e.g. financed by taxpayers). Thus, a reasonable question decision-makers in charge of public funds may ask is: "Are the public benefits of an expensive river restoration project worth the public costs?" One of the most common means for answering this question is CBA. CBA is a test of whether a river restoration policy change or project would be deemed an "improvement" if the benefits (TEV) of gainers of the policy change or project are greater than not only the monetary cost but any lost benefits to those made worse off by the restoration (e.g. in the case of removing a dam, those who enjoy the dam created reservoir recreation).

One of the empirical challenges in measuring the benefits of river restoration is determining the geographic distribution of the benefits (costs) received by gainers (losers). For example, in Table 9.2, the shortest river restoration section valued was about two hundred meters in Switzerland (Logara et al. 2019). Thus, this restoration project might just have local benefits to a few thousand people.

In Table 9.2, the longest river restoration section valued was about 4000 km of the Colorado and Green rivers in Colorado and Utah. Because of the relatively large regional scale of this restoration, and the fact that restoration involved increasing the populations of federally listed threatened and endangered species that provide nonuse values, restoration of these rivers could have national benefits and hence conceivably have benefits to millions of people. In addition to the kilometers of restored river, part of the determination of the geographic distribution of benefits and costs of a river restoration project may have to do with the source of the funding, for example county, state, or federal government sources (Loomis et al. 2000).

The river restoration CBA conducted by Becker et al. (2018) provides an example of applying TEV estimates to a relatively local river restoration project. The study area was an approximately 25 km stretch of the Kishon River in the Haifa District of Israel. In 2013, the Kishon Water Authority was allocated a budget of $60 million for a three-year restoration project which included the following four parts: (i) riverbed restoration (water purification and resolving bad odors), (ii) ecological restoration (restoring flora and fauna to the river),

Table 9.2 Estimates of the total economic value (TEV) of river restoration (USD, 2018).

Author	Study area	Valuation technique*	Total economic value (TEV)	Length of River Restored (km)	TEV per km	Ecosystem goods and services valued
Alam (2008)	Buriganga River, Bangladesh	CVM	$18.02	Not specified	n/a	Recreation, water quality, ecological functions
Becker et al. (2018)	Kishon River, Israel	CVM	$15.90	25	$0.64	Recreation, water quality, ecological functions
Brouwer et al. (2016)	Danube River, Austria	CEM	$134.62	35	$3.85	Boating, fishing, flood control
Brouwer et al. (2016)	Danube River, Hungary	CEM	$39.22	50	$0.78	Boating, fishing, flood control
Brouwer et al. (2016)	Danube River, Romania	CEM	$58.30	195	$0.30	Boating, fishing, flood control
Collins et al. (2005)	Deckers Creek, West Virginia	CEM	$201.40	38	$5.00	Fish, fishing, swimming, aesthetics
Getzner, M. (2012)	Mur River, Austria	CVM	$21.20	96	$0.22	Recreation, rare species, biodiversity, ecological functions
Getzner, M. (2012)	Mur River, Austria	CVM	$26.50	191	$0.14	Recreation, rare species, biodiversity, ecological functions
Giraud et al. (2001)	Green and Colorado rivers	CVM	$258.50	3953	$0.07	Threatened and endangered desert fish species, fish habitat
Holmes et al. (2004)	Little Tennessee River, North Carolina	CVM	$36.04	10	$3.60	Boating, fishing, swimming, biodiversity
Lee (2012)	Youngsan River, Korea	CVM	$2.12	20	$0.11	Recreation, water quality, ecological functions
Lehtoranta et al. (2017)	Forest streams, Finland	CVM	$43.46	240	$0.18	Trout fishery, natural flood cycle, water quality, endangered species, biodiversity
Logara et al. (2019)	Thur River, Switzerland	CVM	$57.24	0.5	S114.48	Walking, swimming, picnicking, biodiversity

Study	Location	Method	Value	Number	Value	Ecosystem services
Logara et al. (2019)	Töss, Switzerland	CVM	$65.72	0.2	$328.60	Walking, swimming, picnicking, biodiversity
Loomis (1996)	Elwha River, Washington	CVM	$115.54	112	$1.03	Salmon
Loomis et al. (2000)	South Platte River, Colorado	CVM	$30.74	72	$1.62	Soil erosion control, water quality, fish and wildlife, recreation
Martínez-Paz et al. (2014)	Segura River, Spain	CVM	$19.08	2	$9.54	Water quality, recreation, ecological flows
Ojeda et al. (2007)	Yaqui River, Mexico	CVM	$23.32	Not specified		Water quality, fish and wildlife, biodiversity, recreation
Perni et al. (2011)	Segura River, Spain	CR	$93.28	34	$2.74	Fishing, swimming, ecological flow
Sanders et al. (1990)	Colorado River	CVM	$356.09	895	$0.40	Maintain free-flowing river from damming and other development for nonuse values, rafting, kayaking, fishing
Polizzi et al. (2015)	River Pajakkajoki, Finland	CVM	$203.52	12	$16.96	Recreation, culture, ecological functions
Vásquez and Rezende (2018)	Paraíba do Sul River, Brazil	CVM	$116.60	Not specified	n/a	Recreation, water quality
Weber and Stewart (2008)	Middle Rio Grande, New Mexico	CVM	$58.30	27	$2.16	Fish and wildlife, native trees, biodiversity
Weber and Stewart (2008)	Middle Rio Grande, New Mexico	CEM	$183.38	27	$6.79	Fish and wildlife, native trees, biodiversity
Zhongmin et al. (2003)	Hei River, China	CVM	$4.24	Not specified	n/a	Soil erosion control, water quality, fish and wildlife, reducing sandstorms, reducing land salinization
Average			$87.29		$23.77	

*CVM: contingent valuation method; CEM: choice experiments; CR: contingent ranking (variation of CVM).

(iii) connectivity and accessibility (creating paths connecting different areas of Kishon park for both cyclists and family hiking), and (iv) aquatic sports facilities (e.g. rowing, paddle boating). The primary beneficiaries of the project were assumed to be residents of the Haifa District.

This study also demonstrates the theoretically appropriate method for estimating TEV holistically in a single study. Respondents were asked in the CVM survey to state their total WTP (or TEV) for the restoration project. For the approximately 160 000 households in the Haifa District, total annual WTP (TEV) was estimated at about $2.5 million. Respondents were also asked to divide their total annual WTP (TEV) into annual WTP for the four parts of the restoration project. The resulting WTP estimates for the four parts were: (i) $0.8 million for riverbed restoration, (ii) $0.7 million for ecological restoration, (iii) $0.4 million for connectivity and accessibility, and (iv) $0.6 million for aquatic sports facilities.

These estimates of total WTP (TEV) and WTP for the four restoration project parts provides insight into the division of TEV into its use and nonuse (passive use) components. The riverbed restoration and ecological restoration parts of the restoration projects (parts i and ii) provide both use and nonuse values which total $1.5 million annually. The connectivity, accessibility, and aquatic sports facilities parts (parts iii and iv) provide primarily use values which total $1 million annually. These numbers suggest that about 60% of TEV for the Kishon River restoration project is composed of nonuse values and 40% is composed of use values.

Using the TEV estimates of the river restoration project generated by the CVM, the present value of the total benefits of the Kishon River restoration project over a 50-year time horizon were estimated at $50 million. The present value of total restoration costs over the 50-year time horizon including all four parts of the restoration project were estimated at $63 million. A 50-year time horizon was selected for calculating present values of costs and benefits as the environmental engineers working for the restoration company who won the contract for completing the restoration work indicated that the restoration measures would last at least 50 years. The total benefits are less than the total costs by about $13 million. In addition, the $63 million price tag exceeds the original Kishon Water Authority restoration budget of $60 million. The negative net benefits of the complete restoration project would make it more difficult for the Kishon Water Authority to request and obtain additional restoration funds to cover the $63 million cost of complete restoration.

Because the Kishon Water Authority's restoration budget of $60 million does not cover the costs of complete restoration, the Authority is faced with the decision of how best to allocate the limited budget. This decision can be aided by the estimates of benefits (WTP) for the four parts of the restoration project. The estimated present value of the costs of the four parts of the restoration project are: (i) $26 million for riverbed restoration, (ii) $1 million for ecosystem restoration, (iii) $7 million for connectivity and accessibility, and (iv) $29 million for aquatic sports facilities. Comparing the benefits and costs estimates for each part or combinations of the different parts of the restoration project can help the Authority to determine which partial restoration strategy provides the highest net benefits given the total restoration budget constraint. For comparison purposes, the present value of the benefits of the four parts of the restoration project are: (i) $16 million for riverbed restoration, (ii) $14 million for ecosystem restoration, (iii) $8 million for connectivity and accessibility, and (iv) $12 million for aquatic sports facilities.

When evaluating the benefits and costs of river restoration project, estimates of TEV such as those shown in Table 9.2 and described in the Kishon River case study should be used to measure the total benefits. However, in certain situations, CBA of river restoration policy changes and projects may focus only on a portion of the TEV of river restoration. For example, a CBA of a policy change or project designed to enhance recreational fishing may only focus on use values which can be measured by revealed preference techniques such as the TCM or stated preference techniques such as contingent valuation. Examples of these types of revealed and stated preference technique valuation studies of river restoration are reviewed by Bergstrom and Loomis (2017) and Brouwer and Sheremet (2017).

In cases where the restoration occurs on a small section of a stream of primarily local interest to just a few small towns, the extent of the economic analysis may be correspondingly scaled down. While the components of TEV need to be qualitatively acknowledged in the analysis, it may be that nonuse values are relatively small in the aggregate relative to the cost of a survey to measure them. The primary benefits may largely be use values. Likewise, a carefully thought out evaluation of the categories and magnitude of change in the benefits with and without river restoration might suggest the change in several of the benefits are likely too small to warrant a full TEV analysis and the TEV analysis should be focused on benefit categories that change the most.

For example, the split of TEV into use and nonuse values varies by the relative potential for recreation (e.g. water quality at accessible rivers versus remote wild and scenic rivers in the US or endangered species that have a "no take provision" restricting commercial and often recreational fishing). In the case of protecting potential wild and scenic rivers in Colorado from development, Sanders et al. (1990) found that only 20% of a respondent's TEV was related to recreation use. In another study, Giraud et al. (2001) found that the TEV of protecting four threatened and endangered desert fishes in 3952 km of Colorado and Green rivers had a TEV of $167 at the time of the study and $258 in current dollars. This TEV is purely nonuse value, such as option, existence, bequest values, since the US Fish and Wildlife Service does not allow any harvesting of these species unless the user has a valid research permit issued by the agency.

In contrast, maintaining river and lake water quality in Utah, about 90% of the total benefits was on-site, active use value (Nelson et al. 2015). This higher percentage of use values was due to two factors: (i) the higher WTP of users than nonusers for preventing degradation of water quality in order to maintain the current level water quality and (ii) the fact that 74% of Utah's households reported they were visitors to Utah water resources either through direct water contact or participating in recreation immediately adjacent to water resources, where the aesthetics of water resources contributed to their recreation value. Because on-site, active use values require travel to a river site in Utah, the benefits will be spread over a relatively smaller population as compared to the case where people experience both use and nonuse values associated with river restoration.

9.4 Conclusions

The primary message of this chapter is that the most complete measure of the benefits of river restoration is TEV. In most instances, TEV is the appropriate measure of the total benefits of river restoration which should be used to evaluate river restoration policy and management decisions from an economic perspective. TEV is composed of both use and nonuse (passive use) values. Failure to consider both use and nonuse values (e.g. considering use values only) can lead to faulty CBA and subsequent policy and management decisions that do not improve social well-being. For example, ignoring nonuse values may often leave out a relatively large portion of TEV resulting in an underestimate of the total benefits of a restoration project – as a result, the restoration project may be incorrectly rejected based on a comparison of the total benefits and costs of the project. Some restoration projects, however, may affect primary use values. In such cases, measurement of use values only in a scaled-down valuation study may be adequate for river restoration project policy and management decisions.

In conclusion, the library of existing studies of the TEV of river restoration including estimates of its component use and nonuse values has expanded considerably in recent years. These studies provide a guide for conducting primary data valuation studies of the benefits of river restoration. In most cases, it will likely be adequate for informing policy and management decisions to measure TEV in an "all encompassing" or holistic manner, which is also theoretically preferable since the full benefits or restoration are captured.

Notes

1 Fish or wild game harvested by a person during a recreational trip can also be eaten at home (ecosystem good) and contribute to at-home recreationally harvested fish and wild game dining experiences (ecosystem service). However, the value of recreationally harvested fish and wild game is included in the value of a recreational hunting or fishing trip and is therefore not included in Table 9.1, to avoid double-counting.
2 The technical reason WTP is used as a benefit measure is that theoretically it measures consumer surplus, or the net value or benefits of a good service above and beyond what a consumer pays for the good or service (Bergstrom and Randall 2016).
3 Contingent ranking (CR) is a less commonly used stated preference technique for measuring TEV. With this method, survey respondents are presented, for example, with a set of alternative public good "packages" (e.g. public attributes provided and price) and asked to rank the packages in order of preference from "most preferred" to "least preferred." For an example, see Bateman et al. (2006).
4 All currency values are for USD in 2018. The conversion of USD to ILS (Israeli new shekel) is 1:0.29 as of 11 January 2020.

References

Alam, K. (2008). Cost benefit analysis of restoring Buriganga River, Bangladesh. *International Journal of Water Resources Development* 24(4): 593–607.

Amirnejad, S., Khalilian, S., Assareh, M., et al. (2006). Estimating the existence value of North Forests of Iran using a contingent valuation method. *Ecological Economics* 58: 665–675.

Bateman, I.J., Cole, M.A., Georgiou, S., et al. (2006). Comparing contingent valuation and contingent ranking: a case study considering the benefits of urban river water quality improvements. *Journal of Environmental Management* 79: 221–231.

Becker, N., Greenfeld, A., and Shamir, S. (2018). Cost–benefit analysis of full and partial river restoration: the Kishon River in Israel. *International Journal of Water Resources Development* 35(5): 871–890.

Bergstrom, J.C. and Loomis, J. (1999). Economic dimensions of ecosystem management. In: *Integrating Social Sciences with Ecosystem Management* (eds H.K. Cordell and J.C. Bergstrom), 181–193. Champaign, IL: Sagamore Publishing.

Bergstrom, J.C. and Loomis, J. (2017). Economic valuation of river restoration: an analysis of the valuation literature and its uses in decision-making. *Water Resources and Economics* 17: 9–19.

Bergstrom, J.C. and Randall, A. (2016). *Resource Economics: An Economic Approach to Natural Resource and Environmental Policy*. Cheltenham: Edward Elgar Publishing.

Bergstrom, J.C. and Reiling, S. (1997). Does existence value exist? In: *Multiple Objective Decision Making for Land, Water, and Environmental Management* (eds S.A. El-Swaify and D.S. Yakowitz), 481–492. St. Lucie, FL: St. Lucie Press.

Bockstael, N.E. and McConnell, K.E. (1983). Welfare measurement in the household production framework. *The American Economic Review* 73(4): 806–814.

Brookshire, D.S., Randall, A., and Stoll, J. (1980). Valuing increments and decrements in natural resource service flows. *American Journal of Agricultural Economics* 62(3): 478–488.

Brouwer, R., Bliem, M., Getzner, M., et al. (2016). Valuation and transferability of the non-market benefits of river restoration in the international Danube River Basin using a choice experiment. *Ecological Engineering* 87: 20–29.

Brouwer, R. and Sheremet, O. (2017). The economic value of river restoration. *Water Resources and Economics* 17: 1–8.

Brown, T., Bergstrom, J., and Loomis, J. (2007). Defining, valuing, and providing ecosystem goods and service. *Natural Resources Journal* 47(2): 331–376.

Collins, A., Rosenberger, R., and Fletcher, J. (2005). The economic value of stream restoration. *Water Resources Research* 41(2): W02017.

Cordell, H.K. and Herbert, N.G. (2002). The popularity of birding is still growing. *Birding* February: 54–61.

Getzner, M. (2012). The regional context of infrastructure policy and environmental valuation: the importance of stakeholders' opinions. *Journal of Environmental Economics and Policy* 1(3): 255–275.

Giraud, K., Loomis, J., and Cooper, J. (2001). A comparison of willingness to pay estimation techniques from referendum questions: application to endangered fish. *Environmental and Resource Economics* 20: 331–346.

Hoehn, J.P. (1991). Valuing the multidimensional impacts of environmental policy: theory and methods. *American Journal of Agricultural Economics* 73: 1255–1263.

Hoehn, J.P. and Randall, A. (1987). A satisfactory benefit cost indicator from contingent valuation. *Journal of Environmental Economics and Management* 73: 289–299.

Holmes, T., Bergstrom, J., Huszar, E., et al. (2004). Contingent valuation, net marginal benefits and the scale of riparian ecosystem restoration. *Ecological Economics* 49: 19–30.

Krutilla, J.V. (1967). Conservation reconsidered. *The American Economic Review* 57: 777–786.

Krutilla, J.V. and Fisher, A.C. (1975). *The Economics of Natural Environments: Studies in the Valuation of Commodity and Amenity Resources*. Baltimore, MD: Johns Hopkins University Press for Resources for the Future.

Kummu, M., de Moel, H., Ward, P., et al. (2011). How close do we live to water? A global analysis of population distance to freshwater bodies. *PLOS ONE* 6(6): e20578.

Lee, J.-S. (2012). Measuring the economic benefits of the Youngsan River restoration project in Kwangju, Korea, using contingent valuation. *Water International* 37(7): 859–870.

Lehtoranta, V., Sarvilinna, A., Väisänen, S., et al. (2017). Public values and preference certainty for stream restoration in forested watersheds in Finland. *Water Resources and Research* 17: 56–66.

Logara, I., Brouwera, R., and Paillex, A. (2019). Do the societal benefits of river restoration outweigh their costs? A cost benefit analysis. *Journal of Environmental Management* 232: 1075–1085.

Loomis, J. (1996). Measuring the economic benefits of removing dams and restoring the Elwha River: results of a contingent valuation survey. *Water Resources Research* 32(2): 441–447.

Loomis, J. (2006). Importance of including use and passive use values of river and lake restoration. *Journal of Contemporary Water Research & Education* 134: 4–8.

Loomis, J., Kent, P., Strange, L., et al. (2000). Measuring the total economic value of restoring ecosystem services in an impaired river basin: results from a contingent valuation survey. *Ecological Economics* 33: 103–117.

Louriero, M. and Loomis, J. (2013). International public preferences and provision of public goods: assessment of the passive use values in large oil spills. *Environmental and Resource Economics* 56: 521–534.

Madariaga, B. and McConnell, K. (1987). Exploring existence value. *Water Resources Research.* 23(5): 936–942.

Martínez-Paz, J., Pellicer-Martínez, F., and Colino, J. (2014). A probabilistic approach for the socioeconomic assessment of urban river rehabilitation projects. *Land Use Policy* 36: 468–477.

McConnell, K. (1983). Existence and bequest value. In: *Managing Air Quality and Scenic Resources at National Parks and Wilderness Areas* (eds R.D. Rowe and L.G. Chestnut). Boulder, CO: Westview Press.

Nelson, N., Loomis, J., Jakus, P., et al. (2015). Linking ecological data and economics to estimate the total economic value of improving water quality by reducing nutrients. *Ecological Economics* 118: 1–9.

Ojeda, M., Mayer, A., and Solomon, B. (2007). Economic valuation of environmental services sustained by water flows in the Yaqui River Delta. *Ecological Economics* 65(1): 155–166.

Perni, A., Martínez-Paz, J., and Martínez-Carrasco, F. (2011). Social preferences and economic valuation for water quality and river restoration: the Segura River, Spain. *Water and Environment Journal* 26: 274–284.

Polizzi, C., Simonetto, M., Barausse, A., et al. (2015). Is ecosystem restoration worth the effort? The rehabilitation of a Finnish river affects recreational ecosystem services. *Ecosystem Services* 14: 158–169.

Randall, A. and Stoll, J. (1983). Existence value in a total valuation framework. In: *Managing Air Quality and Scenic Resources at National Parks and Wilderness Areas* (eds R.D. Rowe and L.G. Chestnut). Boulder, CO: Westview Press.

Sanders, L., Walsh, R., and Loomis, J. (1990). Toward empirical estimation of total value of protecting rivers. *Water Resources Research* 25(7): 1345–1357.

Smith, V.K. (1987). Nonuse values in benefit cost analysis. *Southern Economic Journal* 51: 19–26.

Turpie, J. (2003). The existence value of biodiversity in South Africa: how interest, experience, knowledge, income and perceived level of threat influence local willingness to pay. *Ecological Economics* 46: 199–216.

Vásquez, W.F. and de Rezende, C.E. (2018). Willingness to pay for the restoration of the Paraíba do Sul River: a contingent valuation study from Brazil. *Ecohydrology & Hydrobiology* 19(4): 610–619.

Weber, M. and Stewart, S. (2008). Public values for river restoration option on the Middle Rio Grande. *Restoration Ecology* 17(6): 762–771.

Zhongmin, X., Guodong, C., Zhiqiang, Z., et al. (2003). Applying contingent valuation in China to measure the total economic value of restoring ecosystem services in Ejina Region. *Ecological Economics* 44: 345–358.

10

Valuation of Ecosystem Services to Assess River Restoration Projects

Xavier Garcia[1], Stefanie Müller[2], and Matthias Buchecker[2]

[1] *Barcelona, Institute of Regional and Metropolitan Studies, Cerdanyola del Vallès, Spain*
[2] *Swiss Federal Research Institute WSL, Unit Economics and Social Sciences, Birmensdorf, Switzerland*

10.1   Introduction

The fact that decisions about the restoration of ecosystems are often conflictive makes it necessary to understand the motivations and values that nourish these conflicts. In this sense, there is growing agreement on the need to assess the anthropocentric values of ecosystems restoration in order to support decision-making and enhance stakeholders' awareness and involvement (de Groot et al. 2013; Tallis and Lubchenco 2014). In the river restoration academic and policy arena, assessment of anthropocentric values has gained ground in large part due to the development of the ecosystem services (ES) concept (Palmer and Filoso 2009). This fact has been reflected in the increasing number of studies that have applied ES valuation approaches to assess river restoration projects (see Garcia et al. 2016a for a review).

The main objective of this chapter is advancing our understanding on how ES can serve researchers and practitioners to assess river restoration projects. Concretely, we provide insight into how valuation of ES can be reinforced as a major criterion for assessment. To do that, we explore some of the main theoretical implications of valuing ES. We therefore introduce an ES conceptual framework and link it to river restoration issues. Some important considerations while assessing river restoration projects based on ES are reviewed. These are: (i) the selection of ES to be valuated and the assessment of conflicts, that is taking into account both ES that become enhanced or depleted; (ii) the consideration of temporal and spatial heterogeneities in ES valuation; and (iii) the selection of proper ES valuation methods to assess river restoration projects. Based on these, an analytical framework to assess river restoration projects based on ES valuation is proposed. Three case studies of river restoration in urban contexts are presented and compared based on this framework. Finally, a conclusion section summarizes the main findings and operational implications.

10.1.1 Ecosystem services and river restoration

Millennium Ecosystem Assessment (MEA) describes ES plainly as the benefits people obtain from ecosystems, classifying them into supporting, provisioning, regulating, and cultural (MEA 2005). Supporting services underpin the supply of all the ES. Provisioning services encompass basically all the services which are tradable in markets such as the abstraction of water for agriculture (Garcia et al. 2016b). Regulating services represent the benefits that society obtains from regulation of ecosystem processes, such as the biological control of mosquitoes (Garcia and Pargament 2015) or the pollution purification (Acuña et al. 2013). Finally, cultural services are all those whose contribution provides human enjoyment and a positive influence on human health, such as the aesthetic appreciation toward restored riparian landscapes (Junker and Buchecker 2008).

A recent ES conceptual framework that is focused on a more systemic description of complex relationships between people and nature has been developed by the Intergovernmental Platform on Biodiversity and Ecosystem Services (IPBES) (Díaz et al. 2015). According to IPBES, ES are the set of benefits that humanity obtains from nature, and consider provisioning, regulating and cultural services as the principal categories of benefits people receive from ecosystems (Figure 10.1). Among other features,

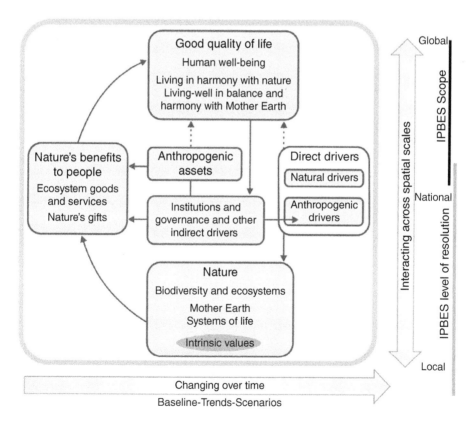

Figure 10.1 The IPBES Conceptual Framework. *Source:* Díaz, S., Demissew, S., Carabias, J., et al. (2015). The IPBES Conceptual Framework: connecting nature and people. *Current Opinion in Environmental Sustainability* 14: 1–16.

IPBES emphasizes the multiple interdependences of the human-derived capital to realize ES. As such, ES are regarded as being co-produced by nature and people properties and dynamics (Jones et al. 2016).

Based on this framework, river restoration may be deemed an anthropogenic driver decided or performed by institutions and governance systems (such as water resources management agencies or river basin authorities). The main purpose of river restoration is improving the ecological integrity of river ecosystems, as required by principles and regulations such as the European Union Water Framework Directive or the US Clean Water Act. Complementarily, and with the aim of fulfilling the needs and expectations of a majority of citizens and stakeholders (Nassauer et al. 2001), river restoration should contribute to the development or maintenance of anthropogenic assets that are vital for the realization of provisioning, regulating, and cultural services, either in urban or rural contexts (Findlay and Taylor 2006; Palmer et al. 2014). Examples of such anthropogenic assets are river water pumping systems, managed aquifer recharge systems, cycle/pedestrian green paths, river docks, and birdwatching towers, among others.

10.1.2   River restoration and ecosystem services conflicts

One of the most challenging aspects of river management actions aiming at increasing the supply of specific ES is that they may result in the depletion of others. For instance, the channelization of a river reach may increase the floodplain land that can be used for different purposes. However, it can intensify bank erosion downstream (Emerson 1971), and impact severely the cultural ES (e.g. aesthetic appreciation) of the modified riparian landscape (Junker and Buchecker 2008). Similarly, river restoration actions imply conflicts among ES. For instance, restoration actions aimed at developing riparian buffers in agricultural environments may contribute to protect water quality from diffuse pollution, as well as enhancing recreational opportunities. However, it will decrease the amount of land that could be farmed or built upon (Kenwick et al. 2009). Therefore, although usually river restoration enhances benefits of society as a whole, effective policy-making in river restoration requires the assessment of conflicts among ES. This can support the prioritization process of potential planning and management alternatives (Sanon et al. 2012), and to prevent conflicts among stakeholders with different priorities and interests (Fliervoet et al. 2013).

10.1.3   Temporal and spatial heterogeneities of ecosystem services

Structures and dynamics of social and ecological systems interact at various spatial and temporal scales (Cash et al. 2006). This is reflected in the geographical and temporal scale of the ecosystem functions (i.e. the potential that ecosystems have to deliver a service, such as the maintenance of a viable fish population) (Fisher et al. 2009), the geographical scale of influence of policy decisions (i.e. national, regional, or local) of institutions and governance systems, or the trade-offs among different stakeholders derived from a decision affecting the ecosystems (Hein et al. 2006). These temporal and spatial heterogeneities, which characterize the provision of ES, may generate many uncertainties in their valuation, which can determine the effectiveness of the decisions to be made (Fryirs and Brierley 2009).

For instance, there is quite large uncertainty associated to the time lapse and geographical extent of the impacts derived from overfertilization of soils, and how this will impact the well-being of communities at different geographical locations (e.g. groundwater pollution) (Basso et al. 2016). Also in river restoration, the provision of ES benefits is highly dependent on the spatial and temporal structures and dynamics of river ecosystems and society. A river restoration action in a specific section may take a significant time until improvement of the ecological integrity is measurable (Fuchs and Statzner 1990), and a bit longer until stakeholders perceive the benefits of this restoration (Åberg and Tapsell 2013). Besides, provision of ES resulting from local scale river restoration actions can also often exceed that geographical limit (Gilvear et al. 2013). This implies that assessment of river restoration projects requires adopting a multiscale perspective in order to include all possible ES benefits and stakeholders involved, instead of assuming a spatially and temporally predetermined approach (Potschin and Haines-Young 2011).

10.1.4 Values and valuation of ecosystem services of river restoration

How stakeholders perceive river restoration is related to their personal and interpersonal value systems (Kolkman et al. 2007). For instance, historical weirs and other in-channel structures may be perceived by some groups as impacting elements that alter the biological and hydromorphological processes, and by other groups as anthropogenic assets that structure the riverscapes and provide multiple cultural services, such as aesthetic experiences or place attachment (Vallerani 2016). According to the IPBES conceptual framework, nature holds basically intrinsic and anthropocentric values. Intrinsic values are "inherent to nature, independent of human judgment, and therefore beyond the scope of anthropocentric valuation approaches" (Díaz et al. 2015, p. 14). Anthropocentric values can be broken down into instrumental and relational. The instrumental values of ecosystems are "the direct and indirect contributions of nature's benefits to the achievement of a good quality of life" (Díaz et al. 2015, p. 14), and they embrace all the direct and indirect use values, and nonuse values, as conceptualized by the total economic value framework (Pearce and Moran 1994).

Relational values are the "preferences, principles, and virtues associated with relationships, both interpersonal and as articulated by policies and social norms" (Chan et al. 2016, p. 1462), and in this context especially respond to those relationships among people and between people and nature. A common example is the contribution of environmental volunteering to social cohesion or the desire to care for a particular place (Measham and Barnett 2008). In addition, relational values conceptualize the processes through which cultural ES are inherently related with the realization of the other types of ES (Fish et al. 2016), such as the social relations and identities embedded in fishing activities (Worster and Abrams 2005). Unlike nonuse values, relational values are not economic values and, therefore, cannot be expressed in monetary terms (Small et al. 2017).

In river restoration, considering instrumental and relational values is of major importance when assessing the public support for these projects. This can be illustrated with the "Room for the River" program, a Dutch flood mitigation initiative that focuses on increasing the depth of rivers, storing water, relocating dikes, creating high water channels or lowering floodplains, among other actions. This program seeks to re-establish the ecological integrity of a river ecosystem and improve the water flow regulation to reduce flood risk. However, the public may oppose this type of project because some of the actions involved (e.g. removal of floodplain trees) may go against their perception of naturalness of such landscapes and thus their desire to protect them (de Groot and de Groot 2009).

How to assess the contribution of river restoration in terms of provision (or diminishment) of ES is still an active research field. Seminal studies of river restoration assessment through the lens of ES have focused on the economic valuation of the instrumental values (e.g. Loomis et al. 2000; Holmes et al. 2004). Apart from the estimation of market values provided by provisioning or regulating ES (e.g. cost of water treatment or sediment removal in reservoirs (Acuña et al. 2013; Honey-Rosés et al. 2013), most of these studies implemented nonmarket valuation methods (basically stated and revealed preference methods) to estimate the benefits of river restoration in terms of cultural ES (Loomis et al. 2000; Becker and Friedler 2013). In fact, estimated benefits in terms of cultural ES have often been decisive when justifying the economic costs of river restoration projects, especially in urban contexts (Garcia et al. 2016b). This has had some relevant implications in river restoration ES valuation studies. Principally, the use of nonmarket valuation methods has tended to favor the assessment of cultural ES, which are the easiest to quantify, such as recreational opportunities or aesthetic appreciation, marginalizing other relevant cultural ES, such as spiritual benefits or place identity, which are not primarily instrumental but relational (Milcu et al. 2013). This may have influenced negatively on river restoration policy-making by relying on valuation methods that emphasized the need to enhance the recreational or aesthetic services of the project over other cultural services that are difficult to quantify, but are of enormous importance to the local population (Ryan 1998). This failure of nonmarket valuation methods in recognizing the importance of some cultural ES has propelled the use (sometimes in a complementary way) of other noneconomic based methods (Kumar and Kumar 2008). In this line, research aimed at exploring a community's experiences, interests, preferences, or expectations toward river restoration has applied noneconomic qualitative- (Davenport et al. 2010; Vollmer et al. 2015) and quantitative-based (Ryan 1998; Junker and Buchecker 2008) methods. Recently, the mapping of ES using a public participation geographic information system (PPGIS) has also been implemented in river areas. PPGIS refers to the use of geographic information system (GIS) methods and technologies to engage the public in spatial decision-making (Sieber 2006). In this context, this method has been applied to better understand how the structures and processes of river areas contribute to human well-being, and on this basis to contribute to the identification of management priority areas (Zhu et al. 2010; Garcia et al. 2017, 2018). As a general conclusion, the ES valuation methods to be applied to assess river restoration projects have to meet not just the expectations of the decision-makers but also those of the other involved stakeholders.

10.2 Analytical framework of ecosystem services valuation

The main objective of this chapter is to provide useful insight for researchers and practitioners into the valuation of ES for assessing river restoration projects. For this purpose, we examine the previously mentioned implications of valuing ES in river restoration based on three case studies. We used an analytical framework to compare each river restoration

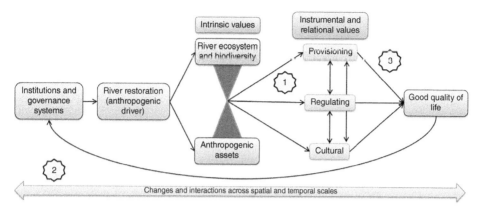

Figure 10.2 Analytical framework to compare the assessment of river restoration projects based on ES valuation.

assessment based on ES valuation (Figure 10.2). This framework has been elaborated based on the IPBES conceptual framework (Díaz et al. 2015) and the challenges identified in the literature. This analytical framework reflects the responsibility of institutions and governance systems of implementing and/or managing river restoration projects. Restoration supports the improvement of the river ecosystem integrity and biodiversity (strengthening of the intrinsic values), and optionally, the development or improvement of anthropogenic assets that support the provision of ES in accordance with complementary project objectives. As a result, an array of provisioning, regulating, and cultural ES are generated or depleted through the interaction of the restored functions of the river ecosystem and biodiversity, and the anthropogenic assets. These ES are interdependent, contribute to a good quality of life, and hold instrumental and relational values.

Three main challenges in valuing ES of river restoration are identified within the framework (indicated with numbers in the figure): (1) the selection (by those who conduct the assessment) of ES to be taken into account and the assessment of possible conflicts; (2) the adoption of a temporal and spatial multiscale perspective in the valuation of ES; and (3) the suitability of the ES valuation methods used.

10.3 Case studies of ecosystem services valuations in river restoration project assessments

The three case studies selected for the analysis are the Yarqon River in Israel, the Caldes stream in Spain, and the Wigger river in Switzerland (Figure 10.3). The Yarqon River is situated in the center of Israel and flows through the most densely populated area of the country, the Tel Aviv Metropolitan Area. The river is approximately 28 km long, has an average annual flow of 0.28 m^3/s, and the area of its basin is about 1800 km^2, having almost two-thirds of the river basin located in the West Bank (Palestine) (Acuña and Garcia 2019). The Caldes stream is an intermittent stream located in the Metropolitan Region of Barcelona, at a distance of approximately 20 km to Barcelona City. It is a tributary of the

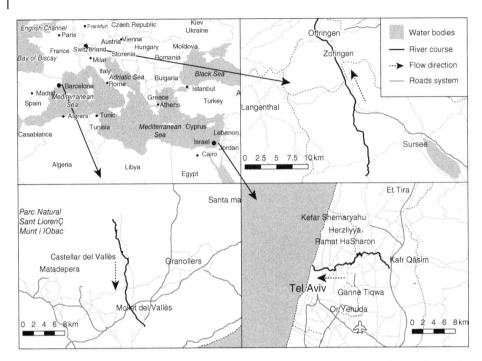

Figure 10.3 Location of the three river restoration case studies: Yarqon River (Israel), the Caldes stream (Spain), and the Wigger river (Switzerland).

Besòs River, with a basin of 111 km², stream length of 22.6 km, and an average annual flow of 0.25 m³/s (Benages-Albert and Vall-Casas 2014). Finally, the Wigger river is located in the Swiss cantons of Lucerne and Aargau, and is a tributary of the river Aare. The Wigger is around 41 km long, has an average annual flow of 0.8 m³/s, and the area of its basin is about 400 km² (Müller et al. 2017).

The main reason for a comparison of these restoration projects is that all these case studies are located in urban or peri-urban contexts. In addition, the valuation of ES is assessed at different scale levels, that is at the local level and beyond. In contrast, the ES valuation methods that were used to assess each river restoration project differ substantially, in accordance with the particularities of the assessment purposes of each project. This last feature has allowed enriching the comparative analyses of these case studies. Table 10.1 synthesizes each case study for comparison purposes.

10.3.1 Assessing the economic feasibility of river restoration: the Yarqon River

After the creation of the state of Israel in 1948, the increase in urban and agricultural water demand, as well as in the flow of poorly treated sewage discharging to the river, had a severe ecological impact on the Yarqon. This was aggravated by the usage of poorly treated wastewater for irrigation, becoming an important nonpoint source of water pollution for the river. At the end of the 1980s, the public concern about the health issues of the polluted

Table 10.1 Characteristics of restoration projects of the Yarqon, Caldes and Wigger rivers.

Case studies	Yarqon River	Caldes Stream	Wigger River
Country	Israel	Spain	Switzerland
Time between restoration and assessment	10 years	Restoration actions are implemented at different points in time	Depending on section: one year/five years after the project
Territorial scale	Tel Aviv metropolitan area	Lower stream corridor (four municipalities)	The lower part of the watershed (six municipalities)
Assessment techniques applied	Cost–benefit analysis	PPGIS	Standardized survey combined with PPGIS and illustrated river scenarios
Ecosystem services considered	Provisioning, regulating, and cultural	Provisioning, regulating, and cultural	Regulating and cultural
Stakeholders responsible for evaluation	Public administration/ academics	Academics	Academics
Main purpose for assessment of river restoration	Economic feasibility	Support a public participation process	Support a public participation process

Yarqon, and the emerging demand for recreational areas for central Israel city-dwellers, increased the political pressure for restoring this river. In 1988, the Yarqon River Authority (YRA) was created, becoming the first river authority in Israel dedicated to drainage works, restoring the ecological functions of the river, and adapting it for leisure and recreational purposes. In order to improve its ecological integrity and amenity value, the Yarqon River Rehabilitation Project (YRRP) was approved in 2003. The YRRP included the following components: (i) increasing water quantity and quality in the river by reallocating clean and recycled water, (ii) cleaning the river channel, (iii) restoring its biodiversity and recover threatened species such as Yarqon bleak fish (*Acanthobrama telavivensis*), and (iv) improving the recreational and aesthetic values of the river to become an essential asset for the Yarqon Park and other recreational areas. In 2012, within the framework of the Sustainable and Integrated Urban Water System Management (SANITAS) project, financed by the European Union, an individual research project was initiated with the purpose of assessing the economic feasibility of the YRRP, considering costs and benefits related with the provision of ES.

For this purpose, a cost–benefit analysis integrating market and nonmarket ES valuation methods was undertaken (Garcia et al. 2016b). Both costs and benefits were estimated as net present value (NPV)[1] over a period of 30 years. In this case, the costs included the capital costs of implementing restoration measures, concretely the upgrading of two wastewater treatment plants, the construction of a wetland, and the restoration of the riparian zone. Furthermore, the opportunity costs[2] of water reallocation were estimated based on the foregone net revenues from taking water out of agricultural production. The benefits derived from the cultural ES at local scale were estimated from the increase in housing rent prices caused by the water quality improvement due to the river restoration, comparing it with the previous water quality situation before the restoration. This was conducted using the hedonic pricing method.[3] The change in cultural ES provision at the regional level in monetary value was quantified here using a benefit transfer method, particularly a value function transfer variant (Lovett et al. 1997).[4] Finally, the gene-pool protection benefits were also estimated based on the replacement cost of the breeding center created to recover the endangered endemic Yarqon bleak. The investments in the breeding facilities assured the existence of a healthy population and the viability of the subsequent reintroduction actions.

The NPV of the costs of the YRRP for the period 2003–2033 amounted to approximately $191 million (Figure 10.4). The highest cost component of the YRRP was the foregone benefits to farmers, with a NPV of approximately $108 million (56.3% of the total). The implementation of the restoration actions was the second highest cost, at $36 million (18.8%). The annual YRA expenses and the additional water treatment operational and maintenance costs associated with the upgrades to the wastewater treatment plants (WWTP) contributed similarly to the total costs, with a NPV of $26 million (13.6%) and $21 million (11.2%), respectively.

For the NPV of the benefits of the YRRP for the period 2003–2033 amounted to approximately $329 million. For the same period, regional cultural ES produced the greatest benefits, with a NPV of $183 million, or 55.51% of the total benefits. The second largest contributor to total benefits is local cultural ES ($145 million, 44.42%). Finally, the total present value of gene-pool protection was $242,726 (0.074%). Consequently, cost–benefit

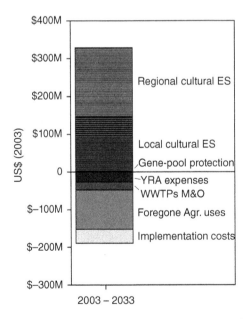

Figure 10.4 Sum of the present values of the ES benefits and costs elements over the 2003–2033 period. "WWTPs M&O" stands for the additional water treatment operational and maintenance costs of the upgraded wastewater treatment plants.

analysis results showed that the YRRP had a marginal value of approximately $139 million for the 2003–2033 period.

10.3.2 Mapping the ES benefits and disbenefits in restored river landscapes: the Caldes stream

The Caldes stream is situated in the highly urbanized province of Barcelona in Spain. The stream stretch considered in the case study flows across municipalities with a population of approximately 70,000 inhabitants. From the 1960s to the 1990s, the suburban sprawl phenomenon that encompassed the entire metropolitan region caused the progressive development of urban and industrial areas. For many years, the impacts derived from this urban development process have severally degraded the historical rural and natural landscapes, and distorted the human–environment interaction and resource management practices of these communities strongly associated with the traditional Mediterranean agricultural activities. Motivated by this and other environmental-based concerns, from the start of the 1980s, public administrations and citizen associations started promoting the protection, environmental improvement, and social use of the open spaces within the stream corridor area. This was mainly driven by a hardening of the wastewater treatment regulation in Europe (e.g. Directive 91/271/EEC concerning urban wastewater treatment), and consequently in Catalonia, that fostered the development of the required infrastructure to restore the water quality and ecological status of the Caldes stream. At the same time, the Catalan government promoted a number of measures to decrease the flood risk in this highly vulnerable area. These, and the accumulation of public land in the riparian areas provided by the urban industrial developments, initiated the creation of a system of open spaces along the Caldes stream that nowadays is highly appreciated and visited by many city-dwellers (Figure 10.5).

Despite that, the prevalent human pressure effects on these landscapes, and the derived controversial perceptions, represent serious management issues for local decision-makers. Some of these controversial perceptions are flood risk management versus ecological restoration of habitats, or demand for recreation versus visitor use management. In the framework of a national research project for the development of a "Decision-making framework for the participatory rehabilitation of urban river corridors," this study aimed at describing and spatially identifying the most relevant positive and negative landscape values based on ES in the Caldes stream corridor.

This study is based on a PPGIS approach in which data collection was qualitative (Garcia et al. 2017). In total, 53 stakeholders were interviewed. The study used semistructured interviews grid with a set of open-ended questions which incurred participants to identify, sketch, and describe places with landscape values. Predefined categories of positive and negative landscape values were used to guide participants' responses, even though they could identify, sketch, and describe similar or different values. These categories were based on MEA (2005) cultural ES categories on the one hand and factors negatively influencing social perceptions (e.g. disservices, nuisances, or threats) of river landscapes and other urban open spaces on the other. All interviews were transcribed for interpretative text analysis to classify all types of landscape values mentioned by the interviewees. During the interviews, printed aerial photograph maps (orthophotomaps) were used as a cartographic

Figure 10.5 Images of common landscapes that can be found in the Caldes stream corridor. *Source:* Courtesy of Dr. Marta Benagues-Albert.

base to assist participants to identify and sketch places with positive and negative landscape values. At the end of the interviews, all the places sketched by the participants were digitized by the interviewers. In total, 1052 places with associated values were mapped to analyze the spatial distribution of these positive and negative landscape values. Frequency of mapped landscape values was also analyzed in this study as a proxy measure for the perceived importance of landscape values (Beverly et al. 2008).

In total, 14 categories of positive landscape values, and 13 of negative landscape values, were identified and described. Table 10.2 presents the frequency and total area of distribution according to the intensity raster maps for the corresponding landscape value categories (Figure 10.6). Regarding positive landscape values, the three most frequently mapped categories, which are thus presumably more relevant in this context, were the "recreational/tourism" (n = 269), "cultural heritage" (n = 267) and "aesthetic/scenic" (n = 233). The "recreational/tourism" category defines places used for recreational and ecotourism activities. These places are mainly located in the forest and agricultural parks of the corridor, as well as in the green paths developed along the stream's middle reach. "Cultural heritage" refers to places to relevant local history and culture, which are mostly located within the old towns and all these areas with agricultural activity such as urban orchard-/gardens. "Aesthetic/scenic" places are those particularly appreciated by their beauty. These are also frequently attached to open and urban areas and the middle and upper reach of the stream.

Table 10.2 Frequency and total area of distribution of positive and negative landscape values in the Caldes stream corridor.

Type of landscape values	Features (n)	Total area (Ha)
Positive values		
Aesthetic/scenic	233	7326.20
Cultural heritage	267	3131.59
Connection	35	2522.37
Ecological	41	4376.30
Economic	26	948.76
Educational	63	5508.61
Lookout	24	1573.24
Naturalness	110	7388.93
Provisioning	18	5517.56
Air pollution regulation	11	1406.38
Recreational/tourism	269	8766.79
Social interaction/relations	132	3189.44
Special places	130	6411.34
Spiritual/religious	30	719.92
Negative values		
Aesthetic unpleasantness	88	1799.39
Negative behaviors	29	396.89
Barriers	33	305.83
Barriers for wildlife	9	119.08
Flood risk	36	413.03
Insects or other animals	10	41.28
Invasive species	6	64.98
Other risks	11	1954.43
Pollution	47	2866.41
Sense of insecurity	10	136.84
Smell	44	389.57
Uncleanliness	94	3832.31
Unsustainability	12	295.26

In the case of negative landscape values "uncleanliness" (n = 94), "aesthetical unpleasantness" (n = 88), and "pollution" (n = 47) were the categories most often mapped. "Uncleanliness" describes areas that are negatively perceived as unclean or neglected. These are frequently located in the mountains and forested northern area of the corridor and along the stream's middle reach. "Aesthetical unpleasantness" is assigned to areas that are visually unattractive. Industrial areas and power lines are negatively perceived in this

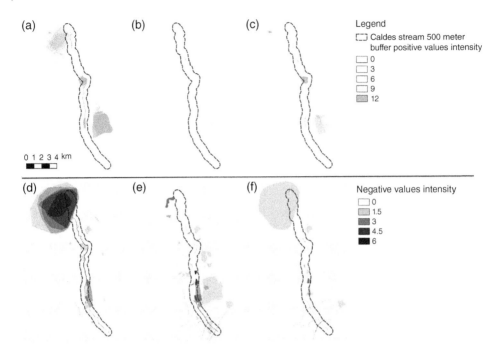

Figure 10.6 Intensity raster maps of positive – "recreational/tourism" (a), "cultural heritage" (b), "aesthetic/scenic" (c) – and negative – "uncleanliness" (d), "aesthetical unpleasantness" (e), and "pollution" (f) – landscape values along the Caldes stream corridor.

sense. Finally, areas with "pollution" problems of any type are focused also in the industrial areas and road infrastructures of the study area.

According to Figure 10.6, there appeared to be particular areas where these positive and negative landscape values co-existed. Consequently, we could identify potential environmental impacts or risks, thereby communicating management priorities. For instance, the "recreational/tourism" landscape value spatially overlapped with the "uncleanliness" perception, especially in the northern hilly area and for the most part of the stream channel. The "aesthetic unpleasantness" and "pollution" negative values spatially co-existed with "recreational/tourism" and "aesthetic/scenic" in the middle-lower section of the stream channel and corridor area, where industrial and urban impacts were most noticeable. Results demonstrate that the consolidation of the open spaces system along the Caldes stream and the improvement of the ecological status of the stream have enhanced the provision of valuable ES that people have increasingly demanded. Still, it also indicated that controversial perceptions co-exist in some places, where management actions should be prioritized.

10.3.3 Recording residents' recreation use an their assessments of river scenarios benefits: the Wigger river

The Wigger is a middle-sized river that originates in the Napf ridges in central Switzerland (canton of Luzerne) and flows after 40 km into the Aare river. Its lower part belongs to the canton of Aargau and has served since early industrialization for energy production and economic use. As a consequence, it was channelized and so has lost its natural riverbed. In the last decades, the lower Wigger valley also faced a massive urbanization process with a sharp increase of the population and new motorways crossing the area. River management has therefore to deal with three challenges: an increased flood risk due to the higher potential damage, an increased need for outdoor recreation options, and a low ecological quality of the river space. The latter has become more of a public issue in Switzerland since the New Federal Water Protection Act was revised in 2011, stipulating now that (i) flood management has to be combined with ecological enhancement and (ii) rivers need to be given a minimum space (Zimmermann et al. 2017). To meet these challenges, two sections of the Wigger have been restored in the last five years. For two larger sections in the center of the lower valley area, technical plans for restoration have also been elaborated. However, the affected three municipalities have shown little interest in implementing these plans due to a high conflict potential. In order to push the process forward and stimulate a participatory decision-making process, the Cantonal Department for Landscape and Waters requested the Swiss Federal Research Institute for Forest, Snow and Landscape Research (WSL) to conduct a survey about local population's attitudes toward different river management alternatives.

This survey included (i) a wide range of questions around people's river-related attitudes (meanings, beliefs, preferences) and their outdoor recreation behavior, (ii) graphically illustrated river management scenarios to be assessed according to ES-related criteria, and (iii) a PPGIS exercise using a topographical map of the region, on which the respondents indicated their recreation areas, preferred recreation routes, and favorite places (Müller et al. 2017). This survey was sent to a random sample (n = 2500; response rate 21%) of the regional population (overall 50 000 inhabitants) based on a spatial sampling strategy which ensured that all six affected municipalities were evenly represented and that 50% of the sampled people lived within a distance of 700 m from the river.

The aggregated data on respondents' perceived meaning of the closest river section revealed that the Wigger river was, first, considered as ecologically valuable area and, second, as a nearby recreation area. In spite of the high noise level caused by the motorway running along the Wigger river, "place of silence and contemplation" follows very closely as the third rank. The identity-related aspects appeared to be of similar relevance, in particular the aspect "part of home." Functional meanings of river such as "technical building," "drainage line," or "source of risk," however, ranked at the very bottom and their relevance was rated as negative. Interestingly, only the ratings of two meanings appeared to differ significantly between the samples of the restored and the nonrestored river sections: those of "economic use" and "living space." When asked about their relationship to the Wigger in general, respondents appeared to be more attracted by the river (mean value 3.84; scale 1/5) than convinced of its ecological intactness (mean value 3.42; scale 1/5) or connected to the river (mean value 3.3; scale 1/5). Significant differences between the municipality samples only appeared for reported outdoor recreation use of the Wigger river that seems to be most sensitive to river enhancement measures.

This is confirmed by the results of the PPGIS exercise. The map indicating the most often used routes for outdoor recreation highlights that the restored sections of the Wigger form hotspots for outdoor recreation use (see Figure 10.7). Interestingly, however, also

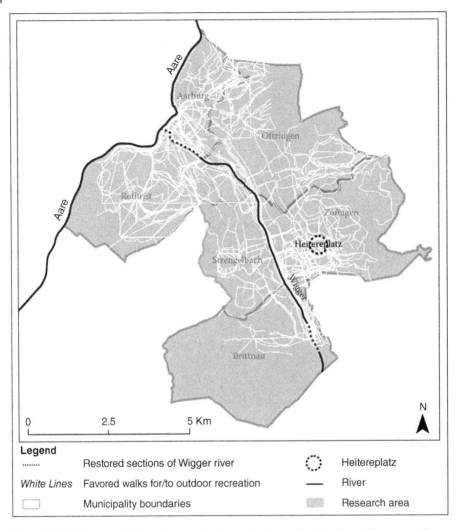

Figure 10.7 Aggregated data of respondents reported use of routes for outdoor recreation (n = 225) along the Wigger river.

nonrestored river sections such as the one within the municipality of Zofingen are intensively used for outdoor recreation, in spite of the high noise level in this area. They seem to serve for city residents as corridors to more attractive recreation areas.

The analysis of respondents' favorite places revealed that after a very popular mystical place at the forest edge, called Heidenplatz, the two restored sections of the Wigger were the areas most often indicated as favorite places. The strong spatial overlap of favorite places, expressing high attachment, and recreation use frequency suggests that these cultural ES are strongly interrelated.

Corresponding to those findings, the analysis of respondents' assessments of river scenarios revealed that respondents expected stronger benefits of river restoration regarding ideal river notions (attractiveness) and naturalness rather than outdoor

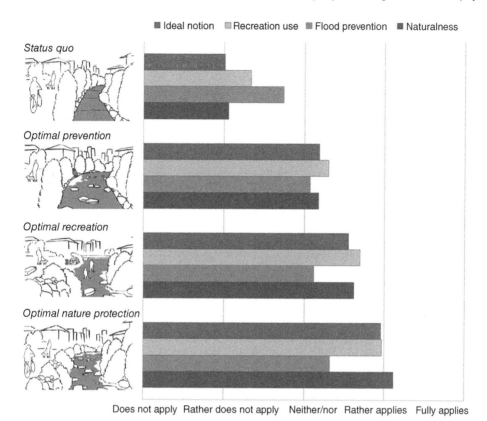

Figure 10.8 Respondents' assessment of their satisfaction with river scenarios of the Wigger river inside the settlement area (n = 225) according to four ES-related criteria: conforms with my ideal notion of rivers; would I use for recreation use; protects me from floods; looks natural to me.

recreation and in particular flood prevention, which is rated as the most positive aspect in the status quo scenario (Figure 10.8). This is also confirmed by data on verbal items referring to the importance of river restoration measures, where a more natural river design and a higher variety of plants was clearly preferred to improvements of outdoor recreation qualities.

10.4   Case studies analysis: valuation of ES for assessing river restoration projects

In this section, based on the reported case studies, we analyze how valuation of ES can serve to assess river restoration projects according to the three main challenges highlighted in our analytical framework (Figure 10.2). These ES valuation challenges are: (i) the selection of ES and assessment of conflicts, (ii) temporal and spatial scale adopted, and (iii) the ES valuation methods.

10.4.1 Addressing the selection of ES and the assessment of conflicts in river restoration project assessments

Even though different ES were selected for valuation, such as flood regulation (e.g. Wigger), gene-pool protection (e.g. Yarqon) or air pollution regulation (e.g. Caldes), the river restoration assessments conducted in each particular case study focused primarily on the valuation of cultural ES. One of the principal reasons for this is that these three case studies are located in urban contexts. Despite the potential for ecological restoration of urban rivers is limited, the benefits from restoration in terms of cultural ES provision are numerous, owing to their higher demand in densely populated areas (Wohl et al. 2005; Vermaat et al. 2016). A second reason is that the restoration projects assessed involved the complementary development of anthropogenic assets to support the provision of cultural ES. Many of the functions required to support cultural ES are not so strongly dependent on the recovery of ecological integrity but on the development and maintenance of anthropogenic assets of riparian landscapes (Junker and Buchecker 2008). For instance, recreational opportunities in riparian areas strongly depend on the development of facilities such as urban parks, parking areas, or bike and footpaths, and less on their ecological recovery, as has been shown in the Wigger river case study. Despite this, rivers that are ecologically healthier tend to be aesthetically more appreciated (Junker and Buchecker 2008; McCormick et al. 2015), and scenic beauty of a riparian area is also strongly related to the recreational values, as seen in the Caldes stream example. Hence, recovering the ecological integrity of urban rivers enhances the recreational benefits and those of other interrelated cultural ES. Another reason is that recreation represents a good opportunity or nexus for managing the interaction of the public and ecosystems (Daniel et al. 2012). Therefore, augmenting the recreational functions of a river can contribute, for instance, to improving the public's acceptance of a restoration project that often includes controversial ecologically oriented goals.

In all the three river restoration case studies, ecosystem service conflicts were revealed, that is not only the beneficial outcomes of restoring these ecosystems to some stakeholders was recognized but the disbenefits or negative consequences for some other stakeholders was also recognized. For instance, water reallocation to improve the Yarqon River flow may compromise water availability for agricultural production. In the Caldes stream, the expansion of the riparian and in-channel vegetation was perceived negatively (e.g. uncleanliness) by some people, because this did not meet their aesthetic expectations (Gobster et al. 2007) or was often linked to the causation of flooding. Conflicts were also assessed for the different Wigger river management scenarios (e.g. between nature protection and recreation use, in particular within the settlement area). However, the study revealed that in the optimal nature protection restoration scenario perceived benefits of the ES assessed would all be enhanced, in comparison with the status quo, including flood protection services.

10.4.2 Approaching the temporal and spatial scale of river restoration project assessments

As previously mentioned, all three case studies assumed that restoration actions' effects on ES should be assessed at the territorial level. Nevertheless, only Caldes and Wigger assessed these effects considering spatial heterogeneity through the application of a PPGIS method. In the Wigger river study, this was also accomplished through a selective sampling of riparian dwellers in the restored and the nonrestored river sections. These analyses allowed a better understanding of how the differences related to the biophysical characteristics of the river landscapes, some of these positively modified by the restoration intervention, influenced the valuation of ES. This demonstrated, for instance, that the restored sections of the same rivers provided more recreational opportunities and aesthetic benefits to people than those that were not environmentally improved. In addition, PPGIS analyses in the Caldes case have shown the co-existence ES in different places of contradictory values that reflect actual conflicts to be considered in the management of this river area. Temporal influence of restoration actions on ES was only investigated in the Yarqon River case, indicating that, after a 30-year period, the restoration project become beneficial in economic terms. This has shown that there is a significant time lapse between the implementation of the river restoration projects and offsetting the required economic investments.

10.4.3 Selecting an adequate ecosystem services valuation method for river restoration project assessments

Regarding the selected ES valuation method, the Yarqon River case study was the only one that used economic market and nonmarket approaches to estimate the instrumental values of ES of the restoration project. Therefore, other more relationally or intrinsically based values related to ES were not properly recognized. For instance, the protection of the endangered fish species, the Yarqon bleak (*Acanthobrama telavivensis*), was only considered with a downward monetary estimate that does not reflect its relational and intrinsic values. Similarly, these market and nonmarket ES valuation approaches could not take the identity values of those urban farmers into account that might see their livelihood compromised by a reduction in land and water availability. Instead, PPGIS valuation of the Caldes stream corridor, prompted the identification of cultural heritage values as one of the most significant in the territory. In the Wigger river, the assessment of naturalness and place attachment values toward the river in its different reaches allowed important insights into the citizens' values and beliefs toward the restoration of this river to be obtained, but these did not include provision services. In summary, the use of market and nonmarket valuation methods for the assessment of river restoration, considering the monetary costs and benefits (willingness to pay) estimates, can provide a straightforward mean to communicate to decision-makers on the success (or failure) of these kinds of projects. However, river restoration assessment studies that include noneconomic-based methods have the advantage of comprising a wider range of ES resistant to monetary valuation, which notably affects people's quality of life. This is a key issue for ensuring public support toward these projects and hence their social feasibility (Buijs 2009).

10.5 Conclusions

In this chapter, we analyzed three case studies that tackled the assessment of river restoration projects by means of valuing ES. The first analysis involved the case of the Yarqon River, characterized by an improvement and increase of the river's flow quality and quantity, the restoration of its biodiversity, and the enhancement of the amenities of the river and surrounding green areas. The second case focused on the Caldes stream and its socioenvironmental transformation to address the negative impacts of urban and industrial development. Finally, the Wigger river case study illustrated the partial restoration of its ecological and amenity values including recreation use and flood protection requirements. Focusing on three relevant challenges identified in the presented analytical framework, that is (i) the selection of ES and assessment of conflicts, (ii) temporal and spatial scale adopted, and (iii) ES valuation methods applied, we have provided useful insight into the valuation of ES for assessing river restoration projects. In summary, the selection of ES to be evaluated is closely related to the restoration objectives, in these case studies, strongly shaped by the interests of managing institutions in providing cultural ES for urban dwellers. Improving the ecological integrity of river ecosystems (the principal objective of river restoration projects) enhances the provision of some ES while reducing others. This demonstrates that stakeholders' values and beliefs toward restoration projects cannot be taken for granted and, therefore, that there is need for, as far as possible, scrutinizing a wide range of ES and other potential disbenefits in these assessments, by involving different stakeholders in the processes. Adopting a territorial perspective in the valuation of ES allowed assessing spatially explicit differences in the provision of several ES that might reflect the influence of site-specific restoration actions. In fact, approaches that assimilate this territorial perspective can also serve to detect conflicting ES values or risks that spatially co-exist and that can indicate management priority areas. Adopting a temporal perspective in the ES valuation provides a better understanding of the dynamic relationship between the achievement of detectable improvements in the integrity of the river ecosystem and a noticeable provision of ES. Finally, the application of economic market and nonmarket approaches to estimate the instrumental values of ES of the restoration project can be used to communicate in a comprehensible manner the profitability of river restoration investments. However, many values of river restoration will go unnoticed if the exclusive use of these approaches are taken for granted in the assessment of river restoration projects. For this reason, future research on ES valuation for river restoration assessment should benefit from the combination of economic and social science approaches.

Acknowledgments

We are deeply grateful to Dr. Marta Benagues-Albert for participating in the interviewing process of the Caldes stream study, and for permitting us to use her pictures to compose Figure 10.5 of this manuscript.

Notes

1 NPV is the present value of the benefits and costs estimated over a time period and including a discount rate or return that could be earned in alternative investments. To estimate it in this case, a discount rate of 4% was applied, and the results were expressed in 2003 US$.

2 Are the benefits given up when choosing one alternative over another.

3 This method is based on the value of a surrogate good or service to measure the implicit price of a nonmarket good. Commonly, the hedonic pricing method uses housing prices to estimate the value of an environmental attribute, such as the water quality of a water body.

4 This method adapts the monetary values of ecosystem services estimated for the location where the original study was conducted (study site) to a new location (policy site).

References

Åberg, E.U. and Tapsell, S. (2013). Revisiting the River Skerne: the long-term social benefits of river rehabilitation. *Landscape and Urban Planning* 113: 94–103.

Acuña, V., Díez, J.R., Flores, L., et al. (2013). Does it make economic sense to restore rivers for their ecosystem services? *Journal of Applied Ecology* 50(4): 988–997.

Acuña, V. and Garcia, X. (2019). Managing ecosystem services under multiple stresses. In: *Multiple Stressors in River Ecosystems* (eds S. Sabater, A. Elosegi, and R. Ludwig), 303–313. Amsterdam: Elsevier. https://doi.org/10.1016/B978-0-12-811713-2.00017-0.

Basso, B., Dumont, B., Cammarano, D., et al. (2016). Environmental and economic benefits of variable rate nitrogen fertilization in a nitrate vulnerable zone. *Science of the Total Environment* 545: 227–235.

Becker, N. and Friedler, E. (2013). Integrated hydro-economic assessment of restoration of the Alexander-Zeimar River (Israel–Palestinian Authority). *Regional Environmental Change* 13(1): 103–114.

Benages-Albert, M. and Vall-Casas, P. (2014). Vers la recuperació dels corredors fluvials metropolitans: El cas de la conca del Besòs a la regió metropolitana de Barcelona. *Documents d'Anàlisi Geogràfica* 60(1): 5–30.

Beverly, J.L., Uto, K., Wilkes, J., et al. (2008). Assessing spatial attributes of forest landscape values: an Internet-based participatory mapping approach. *Canadian Journal of Forest Research* 38(2): 289–303.

Buijs, A.E. (2009). Public support for river restoration: a mixed-method study into local residents' support for and framing of river management and ecological restoration in the Dutch floodplains. *Journal of Environmental Management* 90(8): 2680–2689.

Cash, D., Adger, W.N., Berkes, F., et al. (2006). Scale and cross-scale dynamics: governance and information in a multilevel world. *Ecology and Society* 11(2): 8.

Chan, K.M., Balvanera, P., Benessaiah, K., et al. (2016). Opinion: why protect nature? Rethinking values and the environment. *Proceedings of the National Academy of Sciences* 113(6): 1462–1465.

Daniel, T.C., Muhar, A., Arnberger, A., et al. (2012). Contributions of cultural services to the ecosystem services agenda. *Proceedings of the National Academy of Sciences* 109(23): 8812–8819.

Davenport, M.A., Bridges, C.A., Mangun, J.C., et al. (2010). Building local community commitment to wetlands restoration: a case study of the Cache River wetlands in southern Illinois, USA. *Environmental Management* 45(4): 711–722.

Díaz, S., Demissew, S., Carabias, J., et al. (2015). The IPBES Conceptual Framework: connecting nature and people. *Current Opinion in Environmental Sustainability* 14: 1–16.

de Groot, R.S., Blignaut, J., van der Ploeg, S., et al. (2013). Benefits of investing in ecosystem restoration. *Conservation Biology* 27(6): 1286–1293.

de Groot, M. and de Groot, W.T. (2009). "Room for river" measures and public visions in the Netherlands: a survey on river perceptions among riverside residents. *Water Resources Research* 45(7). https://doi.org/10.1029/2008WR007339.

Emerson, J.W. (1971). Channelization: a case study. *Science* 173(3994): 325–326.

Findlay, S.J. and Taylor, M.P. (2006). Why rehabilitate urban river systems? *Area* 38(3): 312–325.

Fish, R., Church, A., and Winter, M. (2016). Conceptualising cultural ecosystem services: a novel framework for research and critical engagement. *Ecosystem Services* 21: 208–217.

Fisher, B., Turner, R.K., and Morling, P. (2009). Defining and classifying ecosystem services for decision making. *Ecological Economics* 68(3): 643–653.

Fliervoet, J.M., van den Born, R.J.G., Smits, A.J.M., et al. (2013). Combining safety and nature: a multi-stakeholder perspective on integrated floodplain management. *Journal of Environmental Management* 128: 1033–1042.

Fryirs, K. and Brierley, G.J. (2009). Naturalness and place in river rehabilitation. *Ecology and Society* 14(1): 20.

Fuchs, U. and Statzner, B. (1990). Time scales for the recovery potential of river communities after restoration: lessons to be learned from smaller streams. *Regulated Rivers: Research & Management* 5(1): 77–87.

Garcia, X., Barceló, D., Comas, J., et al. (2016a). Placing ecosystem services at the heart of urban water systems management. *Science of the Total Environment* 563: 1078–1085.

Garcia, X., Benages-Albert, M., Pavón, D., et al. (2017). Public participation GIS for assessing landscape values and improvement preferences in urban stream corridors. *Applied Geography* 87: 184–196.

Garcia, X., Benages-Albert, M., and Vall-Casas, P. (2018). Landscape conflict assessment based on a mixed methods analysis of qualitative PPGIS data. *Ecosystem Services* 32: 112–124.

Garcia, X., Corominas, L., Pargament, D., et al. (2016b). Is river rehabilitation economically viable in water-scarce basins? *Environmental Science & Policy* 61: 154–164.

Garcia, X. and Pargament, D. (2015). Rehabilitating rivers and enhancing ecosystem services in a water-scarcity context: the Yarqon River. *International Journal of Water Resources Development* 31(1): 73–87.

Gilvear, D.J., Spray, C.J., and Casas-Mulet, R. (2013). River rehabilitation for the delivery of multiple ecosystem services at the river network scale. *Journal of Environmental Management* 126: 30–43.

Gobster, P.H., Nassauer, J.I., Daniel, T.C., et al. (2007). The shared landscape: what does aesthetics have to do with ecology? *Landscape Ecology* 22(7): 959–972.

Hein, L., van Koppen, K., de Groot, R.S., et al. (2006). Spatial scales, stakeholders and the valuation of ecosystem services. *Ecological Economics* 57(2): 209–228.

Holmes, T.P., Bergstrom, J.C., Huszar, E., et al. (2004). Contingent valuation, net marginal benefits, and the scale of riparian ecosystem restoration. *Ecological Economics* 49(1): 19–30.

Honey-Rosés, J., Acuña, V., Bardina, M., et al. (2013). Examining the demand for ecosystem services: the value of stream restoration for drinking water treatment managers in the Llobregat River, Spain. *Ecological Economics* 90: 196–205.

Jones, L., Norton, L., Austin, Z., et al. (2016). Stocks and flows of natural and human-derived capital in ecosystem services. *Land Use Policy* 52: 151–162.

Junker, B. and Buchecker, M. (2008). Aesthetic preferences versus ecological objectives in river restorations. *Landscape and Urban Planning* 85(3): 141–154.

Kenwick, R.A., Shammin, M.R., and Sullivan, W.C. (2009). Preferences for riparian buffers. *Landscape and Urban Planning* 91(2): 88–96.

Kolkman, M.J., van der Veen, A., and Geurts, P. (2007). Controversies in water management: frames and mental models. *Environmental Impact Assessment Review* 27(7): 685–706.

Kumar, M. and Kumar, P. (2008). Valuation of the ecosystem services: a psycho-cultural perspective. *Ecological Economics* 64(4): 808–819.

Loomis, J., Kent, P., Strange, L., et al. (2000). Measuring the total economic value of restoring ecosystem services in an impaired river basin: results from a contingent valuation survey. *Ecological Economics* 33(1): 103–117.

Lovett, A.A., Brainard, J.S., and Bateman, I.J. (1997). Improving benefit transfer demand functions: a GIS approach. *Journal of Environmental Management* 51(4): 373–389.

McCormick, A., Fisher, K., and Brierley, G. (2015). Quantitative assessment of the relationships among ecological, morphological and aesthetic values in a river rehabilitation initiative. *Journal of Environmental Management* 153: 60–67.

MEA (Millennium Ecosystem Assessment) (2005). *Ecosystems and Human Well-Being*. Washington DC: Island Press.

Measham, T.G. and Barnett, G.B. (2008). Environmental volunteering: motivations, modes and outcomes. *Australian Geographer* 39(4): 537–552.

Milcu, A., Hanspach, J., Abson, D., et al. (2013). Cultural ecosystem services: a literature review and prospects for future research. *Ecology and Society* 18(3): 44.

Müller, S., Buchecker, M., Gaus, R., et al. (2017). Wie soll die Wigger in der Region Zofingen in Zukunft gestaltet werden? *Wasser Energie Luft* 109(3): 181–189.

Nassauer, J.I., Kosek, S.E., and Corry, R.C. (2001). Meeting public expectations with ecological innovation in riparian landscapes. *JAWRA Journal of the American Water Resources Association* 37(6): 1439–1443.

Palmer, M.A. and Filoso, S. (2009). Restoration of ecosystem services for environmental markets. *Science* 325(5940): 575–576.

Palmer, M.A., Filoso, S., and Fanelli, R.M. (2014). From ecosystems to ecosystem services: stream restoration as ecological engineering. *Ecological Engineering* 65: 62–70.

Pearce, D. and Moran, D. (1994). *The Economic Value of Biodiversity*. London: Earthscan.

Potschin, M.B. and Haines-Young, R.H. (2011). Ecosystem services: exploring a geographical perspective. *Progress in Physical Geography* 35(5): 575–594.

Ryan, R.L. (1998). Local perceptions and values for a midwestern river corridor. *Landscape and Urban Planning* 42(2): 225–237.

Sanon, S., Hein, T., Douven, W., et al. (2012). Quantifying ecosystem service trade-offs: the case of an urban floodplain in Vienna, Austria. *Journal of Environmental Management* 111: 159–172.

Sieber, R. (2006). Public participation geographic information systems: a literature review and framework. *Annals of the Association of American Geographers* 96(3): 491–507.

Small, N., Munday, M., and Durance, I. (2017). The challenge of valuing ecosystem services that have no material benefits. *Global Environmental Change* 44: 57–67.

Tallis, H. and Lubchenco, J. (2014). Working together: a call for inclusive conservation. *Nature* 515(7525): 27–28.

Vallerani, F. (2016). Modernismo e engenharia hidráulica como estratégias de desenvolvimento: o caso das vias navegáveis do Vêneto em um novo estado italiano (1866–1966). *Revista Movimentos Sociais e Dinâmicas Espaciais* 5(1): 184–204.

Vermaat, J.E., Wagtendonk, A.J., Brouwer, R., et al. (2016). Assessing the societal benefits of river restoration using the ecosystem services approach. *Hydrobiologia* 769(1): 121–135.

Vollmer, D., Prescott, M.F., Padawangi, R., et al. (2015). Understanding the value of urban riparian corridors: considerations in planning for cultural services along an Indonesian river. *Landscape and Urban Planning* 138: 144–154.

Wohl, E., Angermeier, P.L., Bledsoe, B., et al. (2005). River restoration. *Water Resources Research* 41(10). https://doi.org/10.1029/2005WR003985.

Worster, A.M. and Abrams, E. (2005). Sense of place among New England commercial fishermen and organic farmers: implications for socially constructed environmental education. *Environmental Education Research* 11(5): 525–535.

Zhu, X., Pfueller, S., Whitelaw, P., et al. (2010). Spatial differentiation of landscape values in the Murray River region of Victoria, Australia. *Environmental Management* 45(5): 896–911.

Zimmermann, V., Zemp, H., Kräuchi, N., et al. (2017). Gewässerraum als politischer Zankapfel. *Wasser Energie Luft* 109(3): 173–179.

11

Public Perspectives of River Restoration Projects

Riyan van den Born[1], Bernadette van Heel[1], Kerstin Böck[2], Arjen Buijs[3], and Matthias Buchecker[4]

[1] *Institute for Science in Society, Radboud Universiteit, Nijmegen, The Netherlands*
[2] *Universität für Bodenkultur Wien, Institut für Hydrobiologie und Gewässermanagement (IHG), Vienna, Austria*
[3] *Forest and Nature Conservation Policy Group, Wageningen University, Wageningen, The Netherlands*
[4] *Swiss Federal Research Institute WSL, Unit Economics and Social Sciences, Birmensdorf, Switzerland*

11.1   Introduction

River restoration increasingly aims at fulfilling multiple objectives, such as flood safety, ecological improvement, and recreation (Verbrugge et al. 2017). A growing focus on the relevance of river restoration to local communities has initiated a tendency for experimenting with more community-based forms of restoration, with explicit efforts to include stakeholders and especially residents in the process of decision-making, implementation, and evaluation (Smith et al. 2014). Since the 2000s, the importance of social sciences in river management has become more broadly acknowledged (Eden and Tunstall 2006; Junker et al. 2007; Warner et al. 2013; Smith et al. 2014; Bennett et al. 2017; Poppe et al. 2018).

Recent experiences with river restoration projects signal the need for inclusive ways of working. Typically, restoration projects give ample attention to the hydrological and ecological aspects, while residents' perspectives of the changing physical and social environment are often forgotten, considered too superficially or involved too late (Eden and Tunstall 2006; Bennett et al. 2017; Muhar et al. 2018). Consequently, many of these projects face implementation problems (Kondolf and Yang 2008; Wohl et al. 2015; Fox et al. 2016; Heldt et al. 2016). Indeed, the lack of attention for residents' relationship to their river environments may be an important reason for residents' concerns about river restoration projects and the delays of implementations these implicated (Eden and Tunstall 2006; Buijs et al. 2011; Wohl et al. 2015; Muhar et al. 2018). The restoration of the Jamison Creek in California (USA) is an example of a project where residents' perspectives were ignored. A university team conducted an analysis for restoring the river based on hydrological criteria but did not include local knowledge and perceptions into the design of the project. A local group of residents opposed this design and developed their own plan with a smaller channel, as this better suited their understanding of the river's function (e.g. as a fish habitat)

River Restoration: Political, Social, and Economic Perspectives, First Edition. Edited by Bertrand Morandi, Marylise Cottet, and Hervé Piégay.

and their knowledge about fish stocks. The responsible authority ended up choosing the locals' plan over the solution designed purely on the grounds of hydrological considerations. As a result, the university team dropped out of the project (Kondolf and Yang 2008).

Wohl et al. (2015) argue that involving residents in river restoration is important because of two overarching reasons. First of all, to ensure that residents' perspectives are considered in the definition of objectives, design, and evaluation of the project, because their life space and life quality is directly affected by river restorations (Junker et al. 2007). Second, in practice a successful implementation of river restoration depends on public support and acceptance, which is challenged by excluding residents from planning (Wohl et al. 2015). Including residents' perspectives of their river environment can be done through participatory processes (Villaseñor et al. 2016; Verbrugge et al. 2017) and through social science studies (and any combination of the two). In this chapter we focus on the latter and illustrate how empirical research on residents' perspectives of river environments can help to improve project planning, design, implementation, and evaluation.

In particular, we investigate residents' perspectives on their river environments through the lens of their relationships to local river environments or, more generally, their place relationships as a basis for river restoration planning. In literature, place relationships have often been expressed by the construct of sense of place defined as the meanings people attach to a setting or environment (Jorgensen and Stedman 2001). This approach highlights that residents' perspectives on environments or issues are strongly shaped by the links they perceive to their life, and not just by their knowledge of functional aspects, as the perception approach normally suggests (Maidl and Buchecker 2019).

Because of the tendency to focus on functional, mainly hydrological, and ecological issues, integrating residents' place relationships in restoration projects is not an easy task.

In cases where residents' perspectives are considered in restoration projects, the operationalization is often too limited, focusing only on specific functions of river environments such as recreation and accessibility (Eden and Tunstall 2006; Palmer et al. 2007). Moreover, issues related to the symbolic meanings of river environments representing residents' place relationships are rarely considered or only included superficially (Fox et al. 2016). Similarly, when utilizing the concept of ecosystem services, cultural services comprising recreation and all forms of people–place relationships tend to be neglected or included only as an afterthought or operationalized as predominantly functional (Vermaat et al. 2016).

Based on experiences from two research studies (Müller et al. 2017; Verbrugge et al. 2017), this chapter aims to describe different approaches to study residents' perceptions of and relationships to their river environment. Through case studies we also highlight and discuss the relevance and added value of including these approaches in restoration project planning, implementation, and evaluation.

11.2   Theoretical foundations of public perspectives

In this section we present our conceptual framework and review some of the most important concepts in research of residents' perspectives in the context of river restoration. Thereby we have to consider the double nature of humans, on the one hand as existential beings that need resources to survive and, on the other, as social beings that need relationships with others to achieve a meaningful life. Accordingly, Habermas (1981) distinguishes in his theory of communicative action two ways of experiencing the environment: as a "system" in which functions, resource use, and scientific concepts such as biodiversity and ecosystem services are in focus, and as a "lifeworld," where people live their daily lives and interact with each other and their natural environment and experience it in all kinds of social practices. In a corresponding sense, we distinguish between two modes of residents' perspectives: residents' perception of river functions and their place relationships. These modes are complementary, but there are also transition zones where the two modes interact, in particular when river functions refer to generalized forms of relationship, such as spirituality (e.g. belonging to the universe or nature), or when place relationships or place meanings refer to existential needs fulfilled by personal practices, such as outdoor recreation options in the local area.

11.2.1 Public perception of river functions

If laypeople have to assess social functions of environmental systems such as flood protection, energy production, food production, outdoor recreation, or aesthetics, they relate these functions to their existential interests, values, or vulnerabilities (Jurt Vicuña Muñoz 2009). In this situation, purpose-oriented values determine their assessments or perceptions (Buchecker et al. 2007).

Rivers and floodplains may contribute to a diverse array of functions for local people, including recreation, nature conservation, beauty, and safety. Often, a distinction is made between instrumental and noninstrumental, or intrinsic, values. Instrumental values relate to valuing natural elements such as rivers for the purposes they serve for humans. These functional values relate to the extraction of goods and services from the river, including recreational values and safety.

When we look at beauty as a functional value, we can record the aesthetic values residents attach to landscapes. Although aesthetic value is subjective, empirical studies suggest assessing it from several characteristics, such as seasonal variety, presence of water, and difference and rareness of species living in the area (De Vries et al. 2007).

A number of studies have shown that the attractiveness of a river relates to how natural the river is perceived to be (Cockerill 2016; Junker and Buchecker 2008; Gregory and Davis 1993). Perceptions of naturalness also influence aesthetic values. People who prefer well-maintained areas possibly prefer channelized rivers while people who prefer more pristine nature may prefer landscapes where rivers can run their natural course (van den Born 2007). In a study by Tunstall et al. (2000) comparing three river restoration case studies in the United Kingdom (UK), the authors conclude that aesthetic value is among the main benefits residents perceive.

Recreation is often one of the secondary goals in a river restoration project; including residents' perceptions of recreational opportunities of the river is a very practical way of checking whether the recreational aims are achieved. Functions of the river for recreation can, among others, be measured with residents' perceptions of accessibility of an area and the type and frequency of their recreational behavior in the area (Buijs et al. 2004). Aside from studying how an area is used for recreation, the accessibility and appreciation of the area for these recreational purposes should also be studied (Verbrugge and van den

Born 2018). For many residential groups, in particular families, recreation is considered the main argument for supporting river restoration (Verbrugge et al. 2017). We refer to these functions as "recreational values."

The third functional value we consider is safety. Perceptions of safety are often investigated under the theoretical term "flood risk perception." "Flood risk perception" refers to the perceived flood probability and the perceived consequences of a flood (Bubeck et al. 2012). Residents' perception of flood risk is influenced by four factors: the perceived likelihood or frequency of flood events; the types and trustworthiness of information sources; personal factors like gender, profession, and education; feelings about previously experienced floods; and finally contextual factors like economic issues, family and community composition, and closeness to the waterfront (Wachinger et al. 2013). However, "risk perception" is an ambiguous term and more emotional variables (e.g. "fear" and "worry") should also be included (Bubeck et al. 2012).

Next to these functional values of nature, we also need to mention a value that is usually not considered a function: the intrinsic value of nature. Intrinsic value is often regarded as nature having a value "independently of the existence of any being who evaluates" (Benson 2000, p. 5). Nature can, however, as well be considered to have a spiritual function (e.g. Osbaldiston 2011) symbolizing an all integrating, intact world. Research on traditional and restored floodplains along the Rhine river (the Netherlands) shows that its intrinsic value is important to most residents. As the intrinsic value people attach to a floodplain influences the perceived overall quality of a landscape (Buijs 2009), it is important to include this value in social science research for river restoration.

11.2.2 People–place relationships

If laypeople assess places of their residential environment or life space, they refer to self-related meanings of this place. There are many different ways of naming, defining, conceptualizing, and operationalizing the meaning a place has for residents (Manzo 2003; Williams and Vaske 2003; Hernandez et al. 2014; Ganzevoort and van den Born 2019). In this chapter we use sense of place as an overarching construct that comprehends all relevant representations humans associate with a place or environment, including individual, social, and cultural meanings; identities; and values (Jorgensen and Stedman 2001; Kianicka et al. 2006). In environmental psychology, sense of place has been defined as a "multidimensional construct representing beliefs, emotions and behavioural commitments concerning a particular geographic setting" (Jorgensen and Stedman 2006, p. 316). Thereby, place identity, place dependence, and place attachment are considered subdimensions. "Place identity" indicates how attributes of an environment contribute to someone's identity (Proshansky et al. 1983; Jorgensen and Stedman 2006; Devine-Wright 2009). "Place dependence" describes how well a place allows one to fulfil personal, mainly self-related, needs such as social contacts or hobbies in comparison to other places (Jorgensen and Stedman 2001). Finally, "place attachment" can be seen as a positive, emotional bond between someone and their environment (Hidalgo and Hernandez 2001; Jorgensen and Stedman 2001). There are several ways of bonding with places; "social bonding" describes bonding to place through meaningful social relationships and shared experiences in that place, "nature bonding" describes connection to the natural environment, and "narrative bonding" describes bonding through landmarks that reflect people's relation with the place through its history, culture, mythology, and stories (Verbrugge and van den Born 2018).

In human geography and anthropology, sense of place goes beyond individual place relationships and includes social and cultural meanings (Hay 1999; Kianicka et al. 2006). In this sense, also collective experiences, social identities, and cultural values are represented and can be considered as further subdimensions of sense of place.

The importance of including sense of place in river restoration planning is illustrated by Fox et al. (2016) research about dam removal in New England (USA). According to these authors, residents hold a strong sense of place for the river associating it as a common good, which results in strong opposition for reasons much more complex than mere functionality. Residents expressed that "dams are a central part of the community . . . they are a visual cue to the past" (Fox et al. 2016, p. 98), which seems to indicate narrative bonding. Also, residents had a strong attachment to and identification with the river environment, illustrated by a resident's citation used in the title of Fox et al.'s (2016) publication as: "You kill the dam, you are killing part of me."

11.3 Two empirical examples of how to include residents' perspectives

This section describes two single case studies that illustrate different ways of including residents' perspectives in river restoration research. The first case study from Switzerland focuses on residents' perceptions of restored river sections in a before survey (an after survey is envisioned after the planned revitalization project), but also includes assessments of completed revitalization projects. The second case study from the Netherlands addresses sense of place in the context of river restoration through a before and after survey approach.

11.3.1 Residents' perceptions of the peri-urban river Wigger in Switzerland

This case study considers the lower catchment of the middle-sized river Wigger in the Swiss Midlands. This catchment is characterized by a highly diverse landscape structure including six municipalities with urbanization levels ranging from rural to highly industrialized. Within this catchment area, two sections of the river Wigger were restored ("revitalized," according to Swiss legislation) in the last years, meaning that in these sections flood protection measures were combined with ecological enhancement measures (Figure 11.1). The uppermost section within the municipality of Brittnau was restored in 2010 in the context of a flood protection project. In this project, the riverbed was deepened, considerably enlarged, nature-like designed (dispersed blocks), and fixed with dispersed ramps at certain locations in order to stabilize it: a river design that allows only for very limited natural dynamics. The section in the lowest part of the Wigger, located in the municipality of Aarburg, was restored in 2015 as a compensation measure for the extension of a national motorway running along the river. Here, the riverbed was shifted by 40 meters, designed in a highly structured way using fascine work and groynes, and fixed with two ramps. Aside from these two restored sections, additional plans are under development for two sections between these two restored sections (see Figure 11.1), which are currently completely

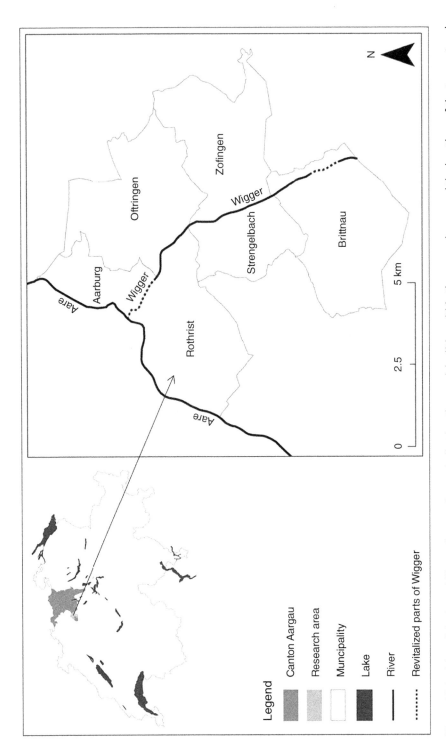

Figure 11.1 Map of the lower Wigger river catchment indicating the municipalities within the research area and the locations of the two restored sections. *Source*: Vector200, swisstopo(DV033594), FSO GEOSTAT, Federal Office of Topography, 2018.

canalized and feature high eco-morphological impairments (Kulli et al. 2013). Initially, the municipalities showed little interest in implementing the project, because they anticipated resistance from the landowners. Furthermore, the same municipalities have delayed a project for realizing an agglomeration park, "Wiggerpark," to enhance the outdoor recreation options in this highly urbanized region. To establish a baseline for evaluating the river restoration and to push planning processes forward, the responsible authorities of the canton Aargau mandated the Swiss Federal Research Institute WSL to conduct a population survey on the development of the Wigger river and outdoor recreation. One of the objectives of this survey was to measure residents' perspectives of the Wigger river and especially of restored river sections.

11.3.1.1 Measurement of residents' perspectives of the Wigger river

Based on a qualitative pre-study, a standardized questionnaire was designed. It included questions about statements on place and river attachments. These two items were measured using Likert scales.

The questionnaire was sent to a random sample of 2500 residents living in the six municipalities of the lower Wigger valley, including urban settlements such as Zofingen and Oftringen and rural settlements such as Brittnau and Strengelbach (Figure 11.1). A spatial sampling approach was used to ensure that (i) the six municipalities were represented according to their population size and (ii) that 50% of the sampled residents lived within 700 m of the Wigger river, so that the segment of the population most affected by the restoration measures would be represented.

Overall, the survey resulted in 507 valid responses, a response rate of 21%. The characteristics of the sample are in considerable agreement with the ones of the regional population (Table 11.1), with male residents being slightly overrepresented and an overrepresentation of elderly residents and respondents with the Swiss nationality.

11.3.1.2 Residents' relationships to the Wigger river

The residents of the Wigger valley reported stronger attachments to their municipality than to the Wigger river. However, the difference between residents' attachment to their municipality and to the Wigger river varied considerably between the six municipalities (Table 11.2). Residents living in the middle section felt a significantly stronger attachment to their municipality, in particular in Zofingen and Strengelbach. The highest attachment to the Wigger

Table 11.1 Characteristics of the regional population and the survey sample.

Characteristics	Population (BFS 2013)	Sample (%)
Female	50%	46
Age >64	16%	19
Service sector	66%	71
>5 years of residence	—	84
Swiss nationality	65%	90

Table 11.2 Respondents' place and river relationships (mean values; scales: 1 = low; 5 = high) by municipality subsamples. Analysis of variance between municipal samples: ***p <0.001; ** p <0.005.

	Attachment to municipality	Attachment to Wigger	Attracted to Wigger	Wigger intact nature	Know Wigger	Recreation use Wigger
Aarburg	3.67	3.15	3,82	3.42	3.32	3.38
Rothrist	3.88	3.08	3.93	3.39	3.49	2.87
Oftringen	3.84	3.15	3.77	3.49	3.18	2.78
Strengelbach	4.00**	3.47	4.03	3.61	3.28	3.12
Zofingen	4.20***	3.40	3.71	3.29	3.63**	3.42***
Brittnau	3.88	3.62	4.10	3.59	3.60	3.86**

Table 11.3 Perceived relevance of values to residents by municipality subsamples. Question: "Which meaning has the closest section of the Wigger river for you personally?" (mean values; scales: 1 = does not apply; 5 = fully applies). Analysis of variance between municipal samples: ** p <0.005.

	Aarburg	Rothrist	Oftringen	Strengelbach	Zofingen	Brittnau
Intact nature	3.42	3.39	3.49	3.61	3.29	3.59
Outdoor recreation	4.13	3.88	3.64	3.91	3.93	3.94
Nature experience	3.70	3.64	3.64	3.31	3.43	3.69
Source of risk	2.27	2.43	2.37	2.59	2.58	2.43
Economic use	3.02**	2.47	2.88	2.54	2.27	2.37
Silence and contemplation	3.97	3.78	3.74	3.97	3.75	3.94
Home place	3.51	3.72	3.76	4.17	3.84	4.08
Drainage function	2.78	2.89	3.06	3.00	2.93	2.67
Living space	3.69	3.19**	3.29	3.51	3.58	3.67
Technics/building	2.58	2.34	2.70	2.54	2.44	2.33
Childhood memory	2.43	2.58	2.59	2.94	2.73	2.40
Attractive element	3.98	3.74	3.69	3.74	3.55	3.88

river was, however, found in the sample of the restored section of Brittnau, whereas the difference between municipality and Wigger attachment was particularly high in the Oftringen sample, where the Wigger is in a bad ecological state and least accessible by trails. Compared to these fact-compatible findings, the differences between the municipal samples in terms of perceived attractiveness and perceived intactness of the Wigger appeared to be low and insignificant, with the lowest values appearing in the Zofingen sample.

In contrast to respondents' place–relationship data, the differences between the subsamples in respondents' rating of relevant meanings of the Wigger river appeared to be rather marginal (Table 11.3). Only two significant differences were found. First, the meaning "economic use" received significantly higher ratings in the Aarburg subsample, a municipality where hydropower is produced. Second, among inhabitants of Rothrist, who have much better access to the Aare river than to the Wigger, the meaning of "living space" was rated significantly lower than in the other subsamples. The restored sections of Brittnau and Aarburg scored slightly higher than the other sections only on the perceived relevance of the meanings "intact nature" and "outdoor recreation." Interestingly, the meanings of "home place" and "childhood memory," which refer to place attachment, appeared to be highest in the municipality of Strengelbach, in the territory where the Wigger remained unchanged. Generally, river meanings seem to be quite symbolic and thus rather resistant to change. Accordingly, in rural settings the most relevant meanings of the Wigger appeared to be "home place" and "silence and contemplation," whereas in more urban settings the meaning "outdoor recreation" appeared to be more important.

Respondents' ex-post assessment of the benefits provided by the restoration of the lowest and the uppermost section of the Wigger confirms that river restoration had relevant social effects, yet also highlights that these effects have a very limited spatial scope (Figure 11.2). The greater the distance from the home place to the restored river section, the lower the benefit was assessed by the respective municipal subsample (significant decrease for all items). In the municipalities most distant from the Brittnau river section, no positive change was reported, even though the respondents indicated having at least some knowledge of the project. The assessment of benefits related to the restoration of the section of Aarburg showed a very similar pattern. Interestingly, the reported benefits

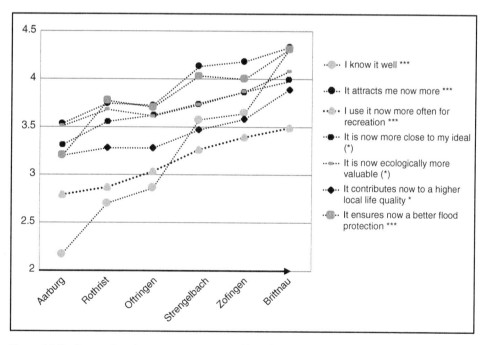

Figure 11.2 Respondents' ex-post assessment of benefits from the 2010 restoration of the Wigger section in Brittnau (mean values; scales: 1 = fully disagree, 3 = neither nor, 5 = fully agree) by municipal subsamples, arranged according to their geographical distance to the restored river section. Analysis of variance between municipal samples: ***$p < 0.001$; * $p < 0.05$; (*) $p < 0.1$.

of this restoration project achieved a higher level than those in Brittnau. However, it is not clear if this more positive assessment of benefits is due to the more fundamental restoration design in Aarburg, or to the shorter time period since the realization of the project. The case study of the Wigger river shows that river restoration brings about positive social effects for the residents, but that these effects are limited in their size and spatial reach. As the relationship between recreation use and perceived attractivity suggest, the perception of river functions and place relationship seem to be strongly interlinked.

Overall, the study highlights that rivers have high instrumental and symbolic values even under poor ecological situations (as this is the case in those sections of the Wigger not yet restored), and that most of these values can be substantially enhanced through river restorations, as the residents' assessments of the restoration effects suggest. In other words, no river is in such a bad state that it could not be restored; also, from a sociopsychological perspective, it maintains its value as a highly symbolic natural element.

11.3.2 Public perceptions of the construction of longitudinal training dams in the river Waal in the Netherlands

This case study concerns the construction of longitudinal training dams in the river Waal in the Netherlands, in the context of the "Room for the River" restoration project. Along a 10 km stretch between the villages of Ophemert and Wamel in the province of Gelderland, the groynes along the inner bed of the Waal have been replaced by longitudinal training dams placed parallel to the river flow (Figure 11.3). In the new situation, this part of the river is split into a main and secondary channel. The goal of this project is adaptation to both high and low water levels and to enable natural processes to develop. Because of the longitudinal training dams, the waterway becomes narrower and thus mitigates low water levels. Inversely, in case of high water levels, the longitudinal dams increase discharge capacity. They provide a more natural flow and more interaction between water and land, which in turn creates a more natural riverbank with more vegetation and nurseries for fish than the previous groynes (Collas et al. 2018). This way, sustainable and integrated river management is promoted by combining improvements for flood safety and ecological quality. Rijkswaterstaat (RWS), the Dutch Directorate for Public Works and Water, designed and initiated the construction of the longitudinal training dams. RWS is now responsible for management and maintenance of the Dutch waterways. The construction works were completed in December 2015.

Because of the dams' major impact on the river landscape, RWS intended to involve many different stakeholders, such as professional shippers, recreational boaters, anglers, and residents. Here we focus on residents of the nearby areas (the town of Tiel and the villages of Dreumel, Wamel, and Ophemert).

11.3.2.1 Evaluation of the river dams' impact on residents' perspectives of the Waal river

Surveys of local residents were held both before the planned construction works to collect baseline data (baseline December 2013–February 2014) and after the construction works were (October 2016–January 2017).

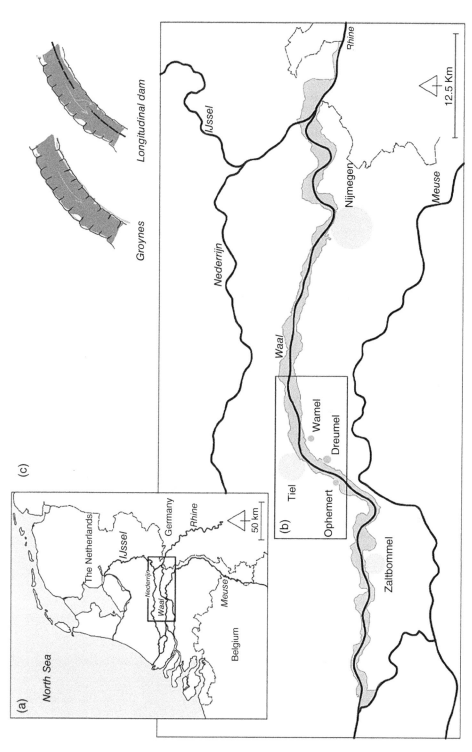

Figure 11.3 Location of the study area in the Netherlands (a), the location of the city of Tiel and the three villages (b) and illustrations of groynes and longitudinal training dams (c). *Source:* Adapted from Verbrugge, L.N., Ganzevoort, W., Fliervoet, J.M., et al. (2017). Implementing participatory monitoring in river management: The role of stakeholders' perspectives and incentives. *Journal of Environmental Management* 195: 62–69.

For both surveys, respondents could choose to fill in a hard copy or an online version. The postal questionnaire was sent to a random sample of residents of the town of Tiel (n = 2000) and to all residents of the villages of Wamel (n = 1043), Dreumel (n = 1472), and Ophemert (n = 678). Men, people aged 45 and older, and people who completed higher education were overrepresented among the respondents, compared to the population of the residential areas in 2015 (Table 11.4) (Verbrugge and van den Born 2018).

The baseline and final survey concerned the effects of the dams on functions of the Waal river, such as recreation, safety, and accessibility as well as scenic beauty, and people–place relationships, such as attachment to place (e.g. "In my opinion, the longitudinal dams make the Waal more beautiful to see"). In addition, trust in RWS was assessed. These variables were measured on a scale from –2 (negative/disagree) to 2 (positive/agree). In the baseline survey, that is before construction of the longitudinal training dams, these questions were introduced by explaining that it was about their expectations regarding the impact of the dam on naturalness, scenic beauty, accessibility, and flood safety. Regarding sense of place, four dimensions were measured: place identity (e.g. "I feel like this area is a part of myself"), narrative bonding (e.g. "I like to tell stories about this area"), place dependence (e.g. "The things I like to do, I can do best in this area"), and social bonding (e.g. "Belonging to the community in this area is important to me"). The four dimensions were measured with level of agreement to 17 statements on a scale of 1 (low) to 5 (high).

Table 11.4 Response rate and demographic composition of baseline and final survey.

	Baseline survey	Final survey
Response		
Number of respondents	1102	877
Response rate	21%	17%
Gender		
Male	59%	55%
Female	41%	40%
Education		
Lower education	31%	27%
Average education	35%	35%
Higher education	34%	35%
Age		
Average age	57 years	58 years
44 years old or younger	23%	17%
45–65 years old	46%	41%
65 years old or older	31%	35%
Living in the area for over 20 years	76%	72%
Born in the area	52%	46%

Source: Based on Verbrugge, L. and van den Born, R.J.G. (2018). The role of place attachment in public perceptions of a re-landscaping intervention in the river Waal (The Netherlands). *Landscape and Urban Planning* 177: 241–250.

11.3.2.2   Changes of residents' perceptions of the Waal river

In the baseline study, residents had neutral expectations regarding the effects of the longitudinal training dams on the "naturalness" of the river landscape; this decreased to a negative view in the final survey (Figure 11.4). The expectations of the effects on "scenic beauty" were neutral in the baseline study and negative in the final survey. The view on "accessibility" was neutral and did not significantly change in the final survey. With regard to "flood safety," people had positive expectations of the dam, both in the baseline and final survey. This is possibly related to the very high level of trust residents have in RWS to protect them against floods: in the baseline survey, over 80% of respondents agreed or strongly agreed with the statement "I have trust that RWS takes care of the protection against floods in the riverine area." The results of the final survey show that residents have high levels of trust when it concerns the expertise of RWS; the scores are, however, lower when it concerns their relationship with this organization. Recreation is important to the residents: in the baseline survey, 62% indicated that it is important or very important to have sufficient recreation possibilities in the riverine area. When asked how attractive they find the area for recreation, 64% in the baseline survey and 74% in the final survey indicated they find it (very) attractive.

11.3.2.3   Changes of residents' sense of place related to the Waal river

Figure 11.5 shows the results of the two surveys on the four dimensions of sense of place. Place identity slightly decreased in all residential areas; the longitudinal training dams are new attributes in the landscape residents cannot (yet) identify with after one year. Narrative bonding and

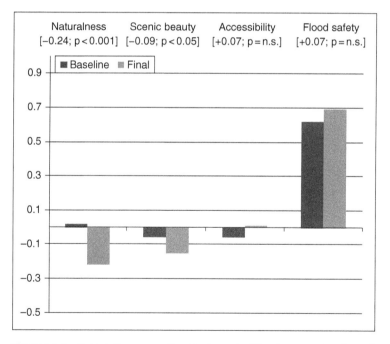

Figure 11.4   Expectations regarding the impact of the dam on naturalness, beauty, accessibility, and flood safety. The expected increase on each dimension was measured on a scale of −2 (strongly disagree) to 2 (strongly agree). The difference between the baseline and final survey, and the corresponding p value, is given in square brackets.

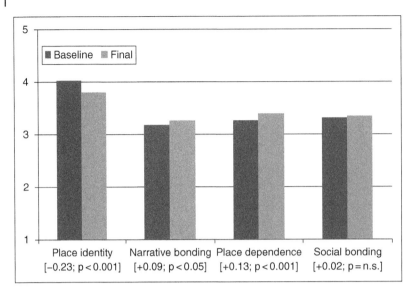

Figure 11.5 Average score in all four residential areas on four dimensions of sense of place. The four dimensions were measured on a scale of 1 (low) to 5 (high). The difference between the baseline and final survey, and the corresponding p value, is given in square brackets.

place dependence both slightly increased, yet not in all locations. Narrative bonding only significantly increased in Tiel, while it did not change in the villages. Possibly, the restoration project opened a dialogue in the community of Tiel, hence increasing the narrative bonding. This effect is not significant in the villages, perhaps because levels of narrative bonding were higher in the villages than in the city before construction of the longitudinal training dams. Place dependence only significantly increased in two of the places (Dreumel and Wamel). This result can only be accounted for if the number of facilities changed in the meantime, but this was not part of our study. There was no significant difference in social bonding between the baseline and final survey in any of the residential areas ($p \geq 0.506$) indicating that the dams were not a controversial topic in the community as they did not seem to reduce social cohesion.

11.4 Conclusions and Implications

Through these case studies, we aimed to describe how understanding residents' perceptions and their place relationships related to how rivers contribute to the planning, design, and implementation of river restoration projects. Based on results from the two case studies, this final section discusses the importance of including residents' perceptions of river functions and place relationships and the implications for scientists and practitioners in river restoration projects.

11.4.1 Importance of including residents' perspectives

Our case studies showcase how river restorations can affect residents' perspectives of rivers and how relevant the investigation of this aspect is. Moreover, they highlight that considering both residents' perceptions of river functions and their relationships to the affected riverine area is necessary in river management for three main reasons. First, including residents' perception of river functions is relevant in river restoration research because residents are affected by the change of river functions induced by river restoration works, and their perspectives should therefore be considered in the planning process. Second, including residents' place relationships to rivers is essential as these express residents' identification with and sense of responsibility for the residential river environment. Third, the relationship residents have with a riverine area changes when rivers are restored, and therefore these changes need to be investigated and reflected upon to increase public acceptance. Carefully designing river restorations based on residents' place meanings furthermore improves the benefits of the project's outcome to residents. We elaborate on these three reasons here.

11.4.1.1 Public perceptions of river functions

River restoration has an impact on the significance and quality of river functions. Empirical research has demonstrated that functions of the river for residents to be fostered by river restorations differ considerably from the ones of interest groups that are usually involved in river management planning (Junker et al. 2007). Our case studies also highlight that residents assess the relevance of recreation function as very high, unlike this has been found for involved interest groups. In contrast, residents tend to underestimate the significance of flood prevention, in particular if the last flood event dates back more than a decade like in both case studies. This is important information if public acceptance of a restoration project is an issue. Knowing the functions and values residents apply to the river is also valuable for the effective communication of restoration projects.

11.4.1.2 Residents' relationships toward rivers

River restoration causes changes in the landscape. This affects residents' experiences in and with the river environment. Our case studies revealed that river restorations affect the meanings residents attach to river environments – even though they express desired rather than perceived river qualities – and thus tend to be relatively stable, as the findings of the Wigger case study suggest. For instance, after restoration residents seem to attach more meanings and values to the river area that relate to possibilities for recreation and contemplation than to intactness of nature or attractiveness. However, the currently dominant economic view in river management limits the possibilities for including values that contribute to human quality of life (Wohl et al. 2005). Consequently, studies on public perceptions will make practitioners aware of the need to counteract this tendency and build on these noninstrumental meanings to do justice to these perspectives and increase public support.

Moreover, river restoration affects the intensity of residents' relationship with the river. As shown in our case studies, these changes can be both negative and positive. In our Waal case study, identification with the area and place attachment tend to decrease, owing to difficulties people may have with identifying themselves with the new landscape. This is in line with previous empirical research concluding that sense of place decreases significantly after river restoration (Buijs 2009). The history a place "breathes" is often not sufficiently taken into account, resulting in a loss of place identity (Drenthen 2009). In the longer run, place identity might, however, increase, owing to enhanced options for recreation and increased perceived naturalness. This assumption is supported by evidence from previous empiric research by Tunstall et al. (2000). Possibly, there is a strong interlinkage between the level of functional values such as ecological quality or recreation and place attachment, in particular in the sense of place dependence. The Waal case study furthermore shows that dialogue on and public participation in river restoration can increase narrative bonding. Systematic monitoring and evaluation of residents' sense of place will increase river managers' understanding of these interactions and enable them to optimize future projects.

11.4.1.3 The benefits of including residents' perspectives in river restoration projects

Studying residents' perceptions, meanings, and attachments, asking them for their input and listening attentively to their response, allows river managers and other practitioners to adapt the project to residents' values and sense of place. This includes adaptations regarding, for example, recreational use options, but also preserving or reconstructing certain landmarks, such as the piles of original stones that remind residents of the former groynes and thus maintain their place attachment or narrative bonding. Revealing residents' preferences of river functions and river-related meanings that express desired river qualities at an early planning stage also provide valuable information for designing river restorations or, as has been done in the Wigger case study, for motivating affected municipalities to involve local residents in river management planning, and for encouraging the residents to participate in this planning process. Measuring residents' attachment to and their recreation use of river sections is furthermore particularly relevant for monitoring and evaluating the life-quality-related success of river restoration projects. Such studies will allow researchers and practitioners to learn how to improve this increasingly relevant aspect of river restorations, in particular for highly urbanized regions, which was the main rationale of the Wigger study. Important for the detailed interpretations of such evaluations and the practical use of the findings is, as the limitations of the Waal study suggest, to record residents' river perspectives combining attitudinal and spatial data. River management planning has to deal with spatial decisions and thus need context specific data. Although evaluations of socio-spatial aspects of river restorations have been called for since many years (Woolsey et al. 2005), good examples are still very rare, also for methodological reasons. Today, however, tools are available to easily integrate mapping exercises in online surveys (Verbrugge et al. 2017).

11.4.2 Implications for scientists and practitioners

Despite a growing recognition of the importance of including and studying residents' perspectives in the context of river restoration, there are many barriers in science and practice to do so (Bennett et al. 2017; Muhar et al. 2018) such as the subjective nature of the data and the unclarity of how to use it for planning. In the following, we discuss some options for practitioners to reduce and overcome these barriers in the future. For example, practitioners might include social scientists in an early project phase as a basis to optimize the design of the project. This way, river managers gain insight in the meanings residents associate with the river and specific river elements and may consider them when planning the recreational infrastructure. Recent research in the field of environmental planning have shown that public acceptance has become a main barrier for successful implementation of projects. Through such studies, residents are also given a voice to inform river restoration planning, so that they are taken seriously. When considering perception study findings in river restoration, the project design can more effectively align with local perspectives, which in the long run could also save time and money, and contribute to public acceptance of restoration projects. If, in contrast, practitioners do not use residents' input, the trust gained by long-standing competent river management might be lost quickly and could be difficult to regain (De Vente et al. 2016).

If residents are involved in a perception study for a local project, they are furthermore stimulated to reflect their relationship with their river environment, its history, nature, and risks. Moreover, if they exchange their reflections in public talks and discussions, this can lead to sharing these narratives with each other, enriching residents' attachment to the place, but also increasing their risk awareness.

Finally, empirical studies are indispensable for informing public communication of the restoration project and the design of the public involvement. Reaching target groups with information requires an understanding of their concerns and expectations (Höppner et al. 2012). The limited reach of awareness about the completed restoration projects in the Wigger case study suggest the need for improved public communication: without target-group tailored communication, the high public investment will not be appreciated by the large majority of the affected population.

Science as well as practice can benefit from knowledge and theoretical insights on people–place relationships acquired through empirical research, especially in relation to a changing environment.

Acknowledgments

We would like to thank Stefanie Müller for her help in data collection for the first case study and for creating Figure 11.2. The work for the second case study was supported by the Dutch Technology Foundation STW and partly funded by the Ministry of Economic Affairs [grant number P12-14 (Perspective Programme)]. We thank Wessel Ganzevoort for his feedback on the manuscript, and both him and Laura Verbrugge for the data collection for the second case study. We would also like to thank our two anonymous reviewers for their valuable comments.

References

Bennett, N.J., Roth, R., Klain, S.C., et al. (2017). Mainstreaming the social sciences in conservation. *Conservation Biology* 31(1): 56–66.

Benson, J. (2000). *Environmental Ethics: An Introduction with Readings*. London: Routledge.

BFS (Bundesamt für Statistik). (2013). Gemeindeportraits. https://www.bfs.admin.ch/bfs/de/home/statistiken/regionalstatistik/regionale-portraets-kennzahlen/gemeinden/gemeindeportraets.html (accessed 5 December 2016).

Bubeck, P., Botzen, W.J., and Aerts, J.C. (2012). A review of risk perceptions and other factors that influence flood mitigation behavior. *Risk Analysis* 32(9): 1481–1495.

Buchecker, M., Kianicka, S., and Junker, B. (2007). Value systems: drivers of human–landscape interactions. In: *A Changing World: Challenges for Landscape Research* (eds F. Kienast, S. Ghosh, and O. Wildi), 7–27. Dordrecht, The Netherlands: Landscape Series, Springer.

Buijs, A.E. (2009). Public support for river restoration: a mixed-method study into local residents' support for and framing of river management and ecological restoration in the Dutch floodplains. *Journal of Environmental Management* 90(8): 2680–2689.

Buijs, A.E., Arts, B.J., Elands, B.H., et al. (2011). Beyond environmental frames: the social representation and cultural resonance of nature in conflicts over a Dutch woodland. *Geoforum* 42(3): 329–341.

Buijs, A.E., de Boer, T.A., Gerritsen, A.L., et al. (2004). *Gevoelsrendement van natuurontwikkeling langs de rivieren*. Wageningen, The Netherlands: Alterra.

Cockerill, K. (2016). Environmental reviews and case studies: public perception of a high-quality river: mixed messages. *Environmental Practice* 18(1): 44–52.

Collas, F.P.L., Buijse, A.D., van den Heuvel, L., et al. (2018). Longitudinal training dams mitigate effects of shipping on environmental conditions and fish density in the littoral zones of the river Rhine. *Science of the Total Environment* 619–620: 1183–1193.

De Vente, J., Reed, M.S., Stringer, L.C., et al. (2016). How does the context and design of participatory decision making processes affect their outcomes? Evidence from sustainable land management in global drylands. *Ecology and Society* 21(2): 24.

Devine-Wright, P. (2009). Rethinking NIMBYism: the role of place attachment and place identity in explaining place-protective action. *Journal of Community & Applied Social Psychology* 19(6): 426–441.

De Vries, S., Roos-Klein Lankhorst, J.R., and Buijs, A.E. (2007). Mapping the attractiveness of the Dutch countryside: a GIS-based landscape appreciation model. *Forest Snow and Landscape Research* 81: 43–58.

Drenthen, M. (2009). Ecological restoration and place attachment: emplacing non-places? *Environmental Values* 18(3): 285–312.

Eden, S. and Tunstall, S. (2006). Ecological versus social restoration? How urban river restoration challenges but also fails to challenge the science–policy nexus in the United Kingdom. *Environment and Planning C: Government and Policy* 24(5): 661–680.

Fox, C.A., Magilligan, F.J., and Sneddon, C.S. (2016). "You kill the dam, you are killing a part of me": dam removal and the environmental politics of river restoration. *Geoforum* 70: 93–104.

Ganzevoort, W. and van den Born, R.J.G. (2019). Exploring place attachment and visions of nature of water-based recreationists: the case of the longitudinal dams. *Landscape Research* 44(2): 149–161.

Gregory, K.J. and Davis, R.J. (1993). The perception of riverscape aesthetics: an example from two Hampshire rivers. *Journal of Environmental Management* 39(3): 171–185.

Habermas, J. (1981). *Theorie des kommunikativen Handelns*. Frankfurt, Germany: Suhrkamp.

Hay, R. (1998). Sense of place in a developmental context. *Journal of Environmental Psychology* 18(1): 5–29.

Heldt, S., Budryte, P., Ingensiep, H.W., et al. (2016). Social pitfalls for river restoration: how public participation uncovers problems with public acceptance. *Environmental Earth Sciences* 75(13): 1–16.

Hernandez, B., Hidalgo, M.C., and Ruiz, C. (2014). Theoretical and methodological aspects of research on place attachment. In: *Place Attachment: Advances in Theory, Methods and Applications* (eds L.C. Manzo and P. Devine-Wright), 125–137. New York: Routledge.

Hidalgo, M.C. and Hernandez, B. (2001). Place attachment: conceptual and empirical questions. *Journal of Environmental Psychology* 21(3): 273–281.

Höppner, C., Whittle, R., Brundl, M., et al. (2012): Linking social capacities and risk communication in Europe: a gap between theory and practice? *Natural Hazards* 64(2): 1753–1778.

Jorgensen, B.S. and Stedman, R.C. (2001). Sense of Place as an attitude: lakeshore owners attitudes toward their properties. *Journal of Environmental Psychology* 21(3): 233–248.

Jorgensen, B.S. and Stedman, R.C. (2006). A comparative analysis of predictors of Sense of Place dimensions: attachment to, dependence on, and identification with lakeshore properties. *Journal of Environmental Management* 79(3): 316–327.

Junker, B. and Buchecker, M. (2008). Aesthetic preferences versus ecological objectives in river restorations. *Landscape and Urban Planning* 85(3/4): 141–154.

Junker, B., Buchecker, M., and Müller-Böker, U. (2007). Objectives of public participation: which actors should be involved in the decision making for river restorations? *Water Resources Research* 43(10): W10438.

Jurt Vicuña Muñoz, C. (2009). Perceptions of natural hazards in the context of social, cultural, economic and political risks: a case study in South Tyrol. Doctoral dissertation, Verlag nicht ermittelbar, Birmensdorf.

Kianicka, S., Buchecker, M., Hunziker, M., et al. (2006). Locals' and tourists' sense of place. *Mountain Research and Development* 26(1): 55–63.

Kondolf, G.M. and Yang, C.N. (2008). Planning river restoration projects: social and cultural dimensions. In: *River Restoration: Managing the Uncertainty in Restoring Physical Habitat* (eds S. Darby and D. Sear), 43–60. Chichester: Wiley-Blackwell.

Kulli, M., Hackl, S., Mathys, D., et al. (2013). Leitbild der Wigger. Kanton Aargau, Departement Bau, Verkehr und Umwelt, Abteilung Landschaft und Gewässer, 28 S.

Maidl, E. and Buchecker, M. (2019). Social representations of natural hazard risk in Swiss mountain regions. *Geosciences* 9(2): 1–30.

Manzo, L.C. (2003). Beyond house and haven: toward a revisioning of emotional relationships with places. *Journal of Environmental Psychology* 23: 47–61.

Muhar, A., Raymond, C.M., van den Born, R.J.G., et al. (2018). A model integrating social-cultural concepts of nature into frameworks of interaction between social and natural systems. *Journal of Environmental Planning and Management* 61(5–6): 756–777.

Müller, S., Buchecker, M., Gaus, R., et al. (2017). Wie soll die Wigger in der Region Zofingen in der Zukunft gestaltet werden? Sozialräumliche Optimierung des planerischen Leitbilds durch eine Bevölkerungsbefragung. *Wasser Energie Luft* 109(3): 181–189.

Osbaldiston, N. (2011). The authentic place in the amenity migration discourse. *Space and Culture* 14(2): 214–226.

Palmer, M., Allan, J.D., Meyer, J., et al. (2007). River restoration in the twenty-first century: data and experiential knowledge to inform future efforts. *Restoration Ecology* 15(3): 472–481.

Poppe M., Weigelhofer G., and Winkler G. (2018). Public participation and environmental education. In: *Riverine Ecosystem Management* (eds S. Schmutz and J. Sendzimir), 435–458. Cham, Switzerland: Springer.

Proshansky, H.M., Fabian, A.K., and Kaminoff, R. (1983). Place-identity: physical world socialization of the self. *Journal of Environmental Psychology* 3(1): 57–83.

Smith, B., Clifford, N.J., and Mant, J. (2014). The changing nature of river restoration. *WIREs Water* 1: 249–261.

Tunstall, S.M., Penning-Rowsell, E.C., Tapsell, S.M., et al. (2000). River restoration: public attitudes and expectations. *Water and Environment Journal* 14(5): 363–370.

van den Born, R.J.G. (2007). *Thinking Nature: Everyday Philosophy of Nature in The Netherlands*. Nijmegen, The Netherlands: Radboud University Nijmegen.

Verbrugge, L. and van den Born, R.J.G. (2018). The role of place attachment in public perceptions of a re-landscaping intervention in the river Waal (The Netherlands). *Landscape and Urban Planning* 177: 241–250.

Verbrugge, L.N., Ganzevoort, W., Fliervoet, J.M., et al. (2017). Implementing participatory monitoring in river management: the role of stakeholders' perspectives and incentives. *Journal of Environmental Management* 195: 62–69.

Vermaat, J.E., Wagtendonk, A.J., Brouwer, R., et al. (2016). Assessing the societal benefits of river restoration using the ecosystem services approach. *Hydrobiologia* 769(1): 121–135.

Villaseñor, E., Porter-Bolland, L., Escobar, F., et al. (2016). Characteristics of participatory monitoring projects and their relationship to decision-making in biological resource management: a review. *Biodiversity and Conservation* 25(11): 2001–2019.

Wachinger, G., Renn, O., Begg, C. et al. (2013). The risk perception paradox: implications for governance and communication of natural hazards. *Risk Analysis* 33(6): 1049–1065.

Warner, J.F., van Buuren, A., and Edelenbos, J. (eds) (2013). *Making Space for the River: Governance Experiences with Multifunctional River Flood Management in the US and Europe*. London: IWA Publishing.

Williams, D.R. and Vaske, J.J. (2003). The measurement of place attachment: validity and generalizability of a psychometric approach. *Forest Science* 49: 830–840.

Wohl, E., Angermeier, P.L., Bledsoe, B., et al. (2005). River restoration. *Water Resources Research* 41(10). https://doi.org/10.1029/2005WR003985.

Wohl, E., Lane, S.N., and Wilcox, A.C. (2015). The science and practice of river restoration. *Water Resources Research* 51(8): 5974–5997.

Woolsey, S., Weber, C., Gonser, T., et al. (2005). *Handbuch für die Erfolgskontrolle bei Fliessgewässerrevitalisierungen*. Publikation des Rhone-Thur Projektes. Eawag, WSL, LCH-EPFL, VAW-ETHZ.

Part V

Diversity of Methods, Diversity of Knowledge

12

Social Surveys: Methods for Taking into Account Actors' Practices and Perceptions in River Restoration

Caroline Le Calvez[1], Silvia Flaminio[2], Marylise Cottet[3], and Bertrand Morandi[3]

[1] Université d'Orléans, EA, 1210, CEDETE, Orléans Cedex 2, France
[2] Institut de Géographie et Durabilité, Université de Lausanne, Géopolis, Lausanne, Suisse
[3] Université de Lyon, CNRS, ENS de Lyon, Environnement Ville Société, Lyon, France

12.1 Introduction

Rivers are not only biophysical environments but also lived, perceived and managed environments. The progressive recognition by public policies of the social dimensions of rivers (e.g. in Europe, the Water Framework Directive 2000) requires going beyond restoration approaches that focus only on ecological structures and processes (Westling et al. 2014). River restoration projects aim to transform river environments (and impact landscapes, human infrastructures, and activities), improve their ecological quality, and meet social needs (Dufour and Piégay 2009). The changes produced by river restoration are influenced by theories from environmental science regarding natural river structures and processes, as well as by social expectations of what a river should be (Baker et al. 2014). Such changes are guided, to the extent of their impact, by the relationship between people and rivers.

According to the scientific literature, taking into account the social dimensions of river restoration is an important issue for the implementation and success of restoration projects (Tunstall et al. 2000; Egan et al. 2011; Barraud et al. 2017; Johnson et al. 2018). The aim of this chapter is to highlight the interest in using social surveys to study the social dimensions of river restoration. Social surveys include questionnaires and interviews,[1] two methods which are based on the collection of oral and written information provided by individuals or groups during a formal interaction with one or several surveyors (e.g. Dunn 2000; Babbie 2010).[2]

The questions brought up in this chapter are both operational and methodological. Why use social surveys in relation to restoration projects? When and how can such methods be used? What are the specificities of the different methods (for data collection and analysis)? We will draw on a qualitative review of international literature to address these questions. First of all, we will provide some examples of scientific and operational contributions from social survey studies toward river restoration project planning, implementation and evaluation. Second, we will specifically look at interviews and questionnaires through a methodological lens.

12.2 Survey methods for studying the social dimensions of river restoration

Social survey studies in the field of river restoration use different approaches. They adopt a variety of spatial and temporal scales and consider a variety of actors according to the objectives and operational contexts of the studies.

1) Studies relying on social surveys have focused on different spatial scales. National and regional scales surveys have been carried out to address global trends in the relationship between people and rivers (Le Lay et al. 2008). On the other hand, local scale surveys have highlighted the specificities of the social context surrounding the restoration project (Le Calvez and Hellier 2017).

2) Social surveys can also be applied at different stages of restoration projects. First, in the preparatory phase of a project, they can contribute to the definition of its social context (Barraud et al. 2017). Second, they can be carried out before and during the project's implementation in order to understand and accompany social processes related to restoration (Flaminio et al. 2015). Finally, social surveys can be used after project completion to monitor the social effects of restoration and provide feedback (Tunstall et al. 2000; Buijs 2009). Therefore, social surveys can be considered part of a river restoration's "lifecycle."

3) The variety of surveys are also related to the different categories of people surveyed and their relationship to rivers. Some surveys focus on the public in general, whereas others focus on specific groups of actors, such as riverside residents, economic or political actors, river managers, or scientists. In this chapter, we define all of these people as "river actors." Based on different case studies, we highlight situations in which social surveys have been used and underline how they may help structure the primary idea of a "place-specific policy designed to give meaning to the site" (Cuaz et al. 1996, p. 361).

12.2.1 Understanding river practices to better anticipate the social effects of restoration projects

Social surveys can contribute to the understanding of the social context of river restoration and to the identification of the actors – even the less visible ones – concerned by the project and of their practices in relation to river environments. Indeed, restoring rivers requires anticipating, planning, and implementing changes in river practices.[3] For river managers and restoration promoters, the success of a project is partly based on their ability to anticipate and support the social changes – such as changes affecting river practices – that restoration induces for the different actors. Who will be the actors who, because of their practices, engage with the restored river? What practices will be affected by river restoration? Social surveys are a useful "preliminary activity" which may contribute to understanding the local context of restoration projects (Junker and Buchecker 2006, p. 248) in terms of practices.

These questions have guided the study led by Le Calvez (2017) concerning the restoration of ecological continuity along the Aulne and Seiche rivers (France). Le Calvez demonstrates that restoration projects concern many people who share the river areas through a range of different practices (e.g. recreation, economic production, contemplation) (Figure 12.1). She relies on social surveys to identify and characterize the diversity of river practices and to go beyond the mere description of river uses generally based on the most visible groups of actors, such as anglers or kayakers. Moreover, she underlines that, within these groups, not all actors practice their activities in the same way (e.g. a salmon angler does not seek the same environment as a carp angler).

Figure 12.1 Examples of recreational uses related to the Seiche river (a, d) and the Aulne river (b, c, e) (France). *Source:* Caroline Le Calvez 2013, 2015, 2016.

Based on 75 interviews, results of Le Calvez's social survey have also helped anticipate how river practices may be affected by river restoration. The interviews outlined a variety of practices which were going to be differently impacted by dam removal works conducted to restore river continuity. For example, dam removal may change recreational fishing practices, from static fishing to on-the-move fishing. This may affect not only anglers but also local angling societies. In the case of the Aulne and Seiche river restoration projects, some members of angling societies, who were concerned with membership decline even before the restoration works, feared that the dam removal would disrupt fishing habits and contribute even more to the angling societies' decline in membership. Among the kayaker group, those seeking the more sporty practice of kayaking in white waters viewed dam removal favorably. On the other hand, kayak instructors and kayak society members who had an economic interest in kayaking considered the disappearance of the dam reservoirs a problem, because beginner courses and activities for young kayakers might be threatened. The survey thus made it possible to specify the expectations of the different actors and highlighted their diversity.

Surveys help identify and define the different actors who should be integrated in the development of the projects, especially the least visible ones. In the previously mentioned survey, Le Calvez (2017) established a difference between actors affected by the restoration project and targeted actors, that is the actors identified a priori by river managers and often involved in restoration project decision-making committees (Larrue 2014). Some actors are affected by restoration projects but are not targeted during the restoration process – or at least not targeted enough – because they are not identified as affected actors. The a priori recognition of all affected actors is problematic because it often depends on restoration project managers' perceptions and river management practices. At the beginning of the Aulne river restoration project, the targeted actors invited to the steering committee meetings were institutional actors (e.g. organizations, such as the Departmental Fishing Federation or riverside residents' committees, water suppliers, and environmental organizations). Tourism actors, user and leisure organizations, and lock keeper-owners were not included in the steering committee. One challenge for project managers is to include these actors and take into account all the different river practices which can be impacted by the restoration project. Social surveys give more visibility to the "affected actors" and make it possible to understand their sociological profile and their relationship to the river, in particular through their practices.

12.2.2 Acknowledging people's expectations and opinions on restoration projects

Social surveys help us to understand people's expectations toward and opinions on projects by shedding light on their perceptions of the river. Questionnaires and interviews include questions that lead the respondents to formulate opinions and expectations regarding the restoration of rivers, which can then be taken into account when defining restoration guidelines. Social surveys make visible what the actors expect, both regarding the content of the project and the decision-making process (e.g. the nature and degree of actors' involvement).

The survey conducted by Junker et al. (2007) in Switzerland demonstrates that river restoration projects should rely not only on the participation of informed actors but also on a broader spectrum of people. The results contradict the commonly accepted idea that the public slows down the process and eventually abandons the original objectives of the restoration project (Junker et al. 2007). On the contrary, taking into account the public's opinions from the early stages of the project can limit the cost and implementation time of the restoration project for river managers (Woolsey et al. 2007).

To collect the opinions of a broad spectrum of riverside residents on restoration of the Yzeron urban river (France), Flaminio et al. (2015) designed a questionnaire which sought to identify the priorities that people would focus on if they were to conceive a project for the river they lived close to. The riverside residents were asked whether they would take measures to implement an ecological restoration of the riverbanks and bed and/or seek to renew the riverbank vegetation with native plant species (Figure 12.2). The results revealed that while most respondents would have sought to implement both measures they engaged the most with the restoration of the riverbanks and bed and were less concerned about the vegetation. The respondents were also asked questions in relation to an ongoing project involving the river and implemented by local authorities and river managers. They were asked about their knowledge and perception of the ongoing project with questions such as: "When was the first time you heard about the project to rework the banks and widen the riverbed?" or "How would you describe the project in one word?" At the same time, semi-structured interviews were conducted with some riverside residents to better understand the stakes and the residents' perception of river restoration.

While the set of questions on the project design showed that a large majority of riverside residents favored river restoration and that their expectations were close to the measures planned by the river managers, the interviews also helped understand the negative reactions to the ongoing project. Most of the negative reactions could indeed be linked to the long history of a project which took decades to be implemented and the "fatigue" felt by some respondents (Flaminio 2015). Such results show that opposition to restoration projects is not necessarily connected with specific measures but can be linked to the way projects are planned and implemented.

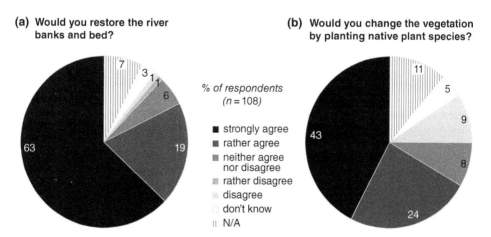

Figure 12.2 Answers from riverside residents to a questionnaire about their support for restoration actions on the Yzeron river (France).

Social surveys help us to understand the reasons for supporting change and to adapt the actions planned by river managers. They also contribute to the identification of what is negotiable and to the building of a project which is considered suitable by the different actors.

12.2.3 Monitoring and evaluating the impact of a river restoration project's implementation

Social surveys are also useful methods once the restoration works are completed. The success of a restoration project is not only based on ecological measures but also on the project's ability to offer a desirable local alternative to the current situation. While evaluations are regularly made to assess ecological evolutions using expert criteria, evaluations led by nonexpert actors are rare. To rely on a more integrative perspective, feedback on restoration projects should include evaluation from nonexpert actors. This evaluation can be produced through surveys.

In the scientific literature, some post-restoration surveys have demonstrated that actors evaluate positively the transformations undergone by restored rivers, primarily in terms of aesthetics and use (Tunstall et al. 1999, 2000; Åberg and Tapsell 2013; Westling et al. 2014). The studies carried out by Tunstall et al. (2000) on the Cole, Skerne, and Medway rivers (United Kingdom) led to such a conclusion. A questionnaire survey was conducted among the population living along the restored rivers. Interviews with residents, managers, and other actors involved in the projects made it possible to deepen the understanding of these actors' reactions to the projects. The sites were in general considered more attractive after restoration. However, the responses indicated a difference in appreciation between the restored urban and rural sections. The urban residents had higher expectations and finally appreciated more the outcome of the project; they were pleased with the greater visibility and accessibility of the sites. The rural residents were more reserved. Respondents pointed out that the effects of the restoration project were not clearly visible. Such a difference may be explained by the design of a restoration project with "no planting at the site and a slow natural regeneration process" (Tunstall et al. 2000, p. 368).

Before/after social surveys enable us to find out if there has been an evolution in the practices and perceptions of the actors as a result of the restoration project. Tunstall et al. (1999) conducted two surveys on the River Skerne based partly on the same sample of interviewed people: one survey before the restoration project in 1995 (n = 252 interviews) and the other 12 months after the project's implementation in 1997 (n = 260 interviews). Results showed that people – interviewed before and after – seemed to continue to visit the site as often as they usually would, and generally for the same purpose. However, one change regarding the purpose of the visit was highlighted: a higher proportion of the word "wildlife" in responses indicated that wildlife was more present in their experience of the river once the restoration project had been completed. Other scientific studies based on different time lapses but without before/after comparisons have also been carried out. For example, Westling et al. (2014) conducted a survey almost 15 years after the restoration works of the River Dearne (United Kingdom). The purpose of the study was to analyze the evolution of the landscape after restoration and the evolution of its perceptions in the long term.

These different case studies illustrate the diversity of the timeframes chosen by the investigators to evaluate the restoration projects. Using surveys to monitor or evaluate the social impact of projects raises the question of how much time should be left between a project's implementation and the follow-up survey, and the influence of this choice on the responses. Knowledge about the local context of the project and the subject of the survey (practices, perceptions of landscapes, etc.) are no doubt key criteria in defining the time needed to observe changes or constants. Nevertheless, repeating surveys at different stages could help us further to understand the restoration's temporalities and help characterize the evolution of practices, perceptions, and constants over time (Åberg and Tapsell 2013), that is the social response to restoration projects. Surveys may therefore benefit from being integrated into restoration monitoring just as ecological monitoring is.

12.3 Choosing between interviews and questionnaires for river restoration surveys

Interviews and questionnaires as survey methods follow similar steps, from defining the aims of the survey to presenting the results to the survey participants (Figure 12.3). Many handbooks offer developed insights into survey methods (see Bickman and Rog 2009; Babbie 2010; Gideon 2012; Fowler 2013). Despite a similar workflow, questionnaires and interviews each have specificities, described in the following sections.

12.3.1 Interviews and questionnaires: differences and complementarities

An interview is a verbal interchange between at least two people, but it can also involve more people (Dunn 2000). A focus group can be considered a particular type of interview involving several respondents (Bickman and Rog 2009). Interviews can be recorded and transcribed for in-depth analysis.[4] On the other hand, a questionnaire refers specifically to a series of standardized questions. The answers are filled in by the respondents themselves or by the surveyors.

Interview and questionnaire surveys therefore differ in their approach (Figure 12.4). Interviews can be considered a rather qualitative method, while questionnaires are most often quantitative. However, different types of interviews can be used based on different degrees of structuring (Dunn 2000).[5] Most of the time, interview questions are open-ended and let interviewees respond freely. Interview surveys often target a small number of respondents. Conversely, questionnaires are highly structured and standardized and target a large number of respondents.

The choice of one method over the other depends on the objectives and context of the study (Table 12.1). The interview, through the collection of verbatim reports, allows an in-depth understanding of complex social phenomena. Respondents communicate their perception of reality, world views, their value or belief system, the meanings they attribute to beings or behaviors, etc. Interviews "allow[s] interviewees to express the details and meanings of their experiences in their own terms and at their own pace" (Pratt 2009, p. 393). The questionnaire is best suited for quantifying, establishing patterns, making temporal or spatial comparisons, observing relationships between variables, and explaining social or

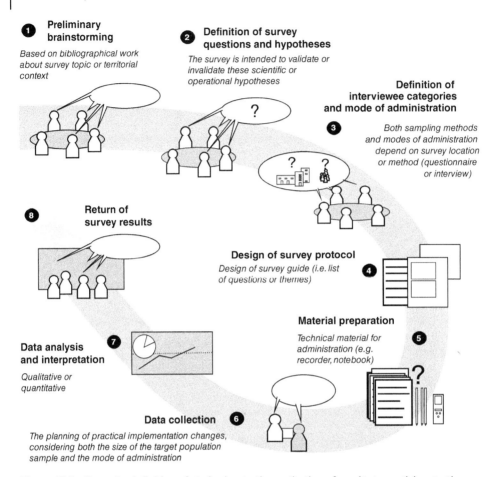

Figure 12.3 From the definition of study aims to the restitution of results to participants: the eight main steps of social surveys.

Interview		Questionnaire
Degree of structure		
Low		High
Standardization of questions		
Low		High
Type of questions		
Open-ended		Close-ended
Respondent's discourse		
Free		Question-focused
Target population		
Small		Large

Figure 12.4 Differences and complementarities of interviews and questionnaires: a graphical summary.

Table 12.1 Aims and arguments guiding the use of interviews and questionnaires: two examples based on French river restoration case studies.

Examples and chosen method	The Aulne and Seiche rivers: semistructured interviews (Le Calvez 2017)	The Rhône river and its tributary, the Ain river: questionnaires (Cottet et al. 2010)
Aims of the survey	Describe the users' practices and perceptions of the river Put these into perspective with the perceptions of institutional actors Characterize the opposition to the restoration of the Aulne and the Seiche rivers (dam removal) and deepen the understanding of the mechanisms and reasons for opposing removal projects Understand the type of conflicts regarding dam removal	Highlight the links between the biophysical characteristics of the rivers and their aesthetic appreciation Compare the perceptions of experts (scientists and managers) and nonexperts (students)
Arguments in favor of the type of survey chosen	The study did not seek to compare territories, but to highlight local arguments and context effects. Semi-structured interviewing is a proven method for this purpose Publications highlighted the French controversy about dam removal. It characterizes the arguments of opposing groups against dam removal at a national level	Preliminary interview surveys had been conducted in this area. The relationships (perceptions, practices, and values) between the inhabitants and the river were thus well known Previous scientific publications had highlighted close links between the appearance of the river and its aesthetic appreciation There was a desire to compare the perceptions of different populations, which implied the use of standardized questions

territorial characteristics. In general, when the survey seeks to explore a general topic or obtain in-depth information on a specific topic, interviews are used. On the other hand, when the survey aims to test hypotheses and quantify social phenomena, questionnaires are often preferred.

In river restoration studies, the two survey methods are frequently used together (Tunstall et al. 2000; Buijs 2009). Both methods provide relevant and reliable results when they are rigorously implemented. The choice of using qualitative and/or quantitative methods to collect and analyze data on perceptions and practices is also guided by the available resources in terms of expertise, time, and funding.

12.3.2 Interviewing and analyzing interview data

12.3.2.1 An adaptable method for collecting and producing a comprehensive set of data

When designing an interview survey, it is difficult to define a priori a precise number of interviews to be carried out. The number of interviews strongly depends on the aim of the survey and the local context. For practical reasons, it is difficult to carry out a broad set of interviews. As a result, surveys by interviews are not statistically representative of the target population. This is a criticism that is often erroneously addressed to this type of method. Nonetheless, the purpose of interviews is often less about measuring the regularity of a practice – as it often is for questionnaires – and more about adopting a comprehensive position which may allow us to better understand the variety of practices, perceptions, or opinions. The challenge therefore rests with the choice of respondents, which should be considered carefully: interviewees are often targeted so as to reflect the most diverse opinions possible within the target population studied or else to understand the opinions and expectations of a particular category of river actors. Different sampling methods can be used (e.g. reasoned choice method, referral sampling).

12.3.2.2 Qualitative content analysis

The challenge of the interview analysis is to make sense of the rich and complex content of the collected oral discourses. For this purpose, qualitative methods are mostly used, notably thematic content analysis (Berelson 1952).[6] This method relies on the systematic examination of the interviews' recordings or transcripts. It is based on the design of an analytical guide based on the main themes that appear in the interviewees' speech. These themes can be defined a priori (based on the hypotheses of the survey) or a posteriori (based on the observation of the empirical data). The combination of a priori and a posteriori approaches makes it possible to consider the themes that were expected a priori to be found but ultimately not observed, as well as the themes that were unexpected a priori but ultimately observed. The analytical guide is then used to code sections of interviews in a systematic way.

Recent publications on dam removal (Fox et al. 2016; Barraud et al. 2017) reveal the difficulty of legitimizing such restoration projects on a local scale and highlight that divergent perceptions of what a river should be and can be may coexist. Using a comprehensive approach with semistructured interviews, Le Calvez and Hellier (2017) used thematic content analysis to analyze their interview survey, presented in Section 12.2, on social opposition to the ecological continuity restoration project along the Aulne and Seiche rivers (France). The analysis focused on the motivations of respondents (n = 30) who contested dam removal projects. The content analysis guide was based on four types of attachment to the rivers (Figure 12.5). The analysis outlined links between respondents' attachment to the river, their sociological and practical profiles, and their perceptions of the restoration project.

The first attachment (or theme) concerns the will of certain actors (who traditionally played a role in river management, such as anglers or millowners) to determine the future of the river. In this group, opposing dam removal is a response to the threat of hydraulic infrastructure dismantling or to the transformation of the river. The second attachment expressed by opponents is an attachment to environments that are regarded as functional and of equal value to the expected environments after dam removal. People in this group of opponents feel they are not considered legitimate actors in the decision-making process regarding the transformation of the river environment. Thus, these opponents to restoration recognize the value of a river on which developments have been carried out. The third attachment mentions the role of the river in creating and maintaining a social bond to

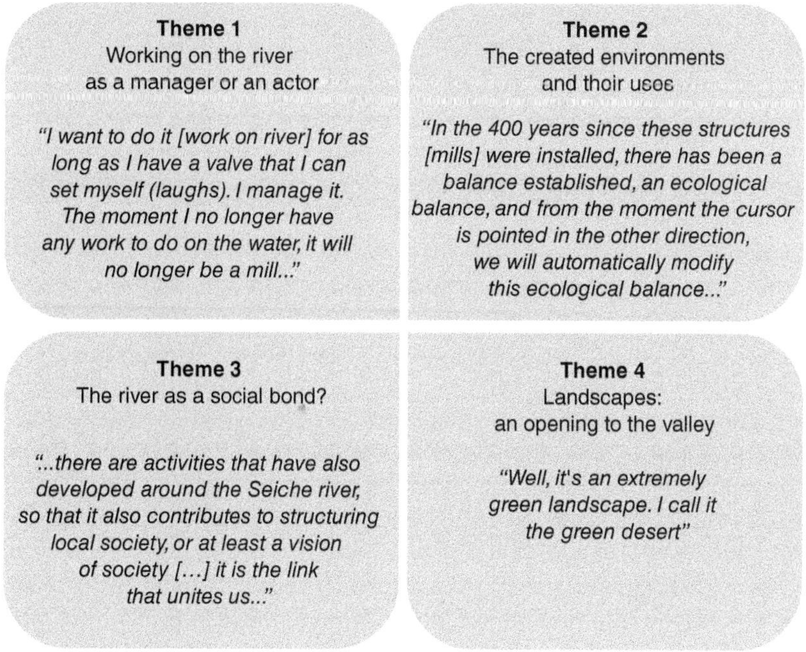

Theme 1 Working on the river as a manager or an actor *"I want to do it [work on river] for as long as I have a valve that I can set myself (laughs). I manage it. The moment I no longer have any work to do on the water, it will no longer be a mill..."*	Theme 2 The created environments and their uses *"In the 400 years since these structures [mills] were installed, there has been a balance established, an ecological balance, and from the moment the cursor is pointed in the other direction, we will automatically modify this ecological balance..."*
Theme 3 The river as a social bond? *"...there are activities that have also developed around the Seiche river, so that it also contributes to structuring local society, or at least a vision of society [...] it is the link that unites us..."*	Theme 4 Landscapes: an opening to the valley *"Well, it's an extremely green landscape. I call it the green desert"*

Figure 12.5 Types of river attachments identified through the thematic content analysis of 30 semistructured interviews with opponents to ecological continuity restoration projects on the Aulne and Seiche rivers (France).

protect the inhabitants against floods, for example, or for the inhabitants to coordinate its preservation. This fourth group evokes a sensory dimension and an aesthetic relationship to the river. Opponents want their river landscapes to be preserved.

12.3.3 Questionnaires: specificities of data collection and analysis

12.3.3.1 Standardized data collection
The challenge of a questionnaire survey is to collect a significant number of responses and to obtain a representative sample of people with respect to a "parent population"[7] so that the results of the survey can be generalized. Responses are gathered with regard to the specific criteria of the parent population (i.e. reproducing, at the sample level, the characteristics of the population). Different sampling methods exist to achieve representativeness and can be probabilistic (e.g. drawing at random) or empirical (e.g. quotas).

12.3.3.2 Quantitative data analysis
Owing to the mainly quantitative nature of the data collected (e.g. continuous data, categorical data), the analysis of the responses to a questionnaire is primarily statistical. There are exceptions for open-ended questions. In the latter case, responses can be considered textual data, which can be analyzed in a similar way to interviews (see Section 12.3.2.2). For close-ended responses, the challenge is to test the hypotheses of the survey by carrying out descriptive statistical analyses, question by question, but also by

cross-analyzing the responses so as to reveal dependency links between different varia-bles (e.g. between variables that qualify the profile of the respondent and the variables that qualify their opinions or expectations related to the river restoration project). The highly standardized nature of questionnaires also allows temporal or spatial analyses by comparing responses given to the same questions at different points in time or space. Spatial and temporal analyses are particularly relevant for highlighting the consequences of restoration on public practices or perceptions, or to measure the evolution of opinions related to the project.

In a study on two Dutch floodplain restoration projects, Buijs (2009) used questionnaires to understand why some projects encounter public opposition. He addressed a question-naire to residents (n = 562) living near restored and nonrestored floodplains, asking them to score different kind of meanings (e.g. "nature," "safety," "attractive recreational space," "cultural history") they attached to these environments. Using a clustering method, Buijs (2009, pp. 2685–2686) drew a distinction between respondents who framed restoration with regard to cultural heritage ("the attachment frame") and those who partially opposed restoration projects, respondents who focused on the natural character of the landscapes ("the attractive nature frame") and those who favored restoration, and respondents who mainly valued the agricultural function of the floodplain ("the rurality frame") and those who were the most opposed to restoration projects. The comparison between responses given in restored and nonrestored contexts also outlined a potential evolution of people's attitudes toward restoration. According to Buijs (2009, p. 2688): "people using an attach-ment frame are especially likely to reconsider negative opinions after restoration has been implemented."

12.3.4 A focus on the use of photography in social surveys

Using photographs in social surveys opens interesting opportunities in the field of river restoration. Photographs can be used as strategic material to accompany or even provoke the production of discourse in combination with the verbal questions.[8] They can constitute the very aim of the survey in the case of photo-elicitation interviews or photo-questionnaires. These methods can bring the topic of landscape to the heart of the survey. For example, in a survey conducted by Yamashita (2002), respondents were asked to take photographs of a Japanese river and comment on their shooting choices. The corpus of photographs pro-duced by the respondents made it possible, for instance, to "directly obtain and analyse what they [the respondents] see and what they assess in the landscape" (Yamashita 2002, p. 4) by connecting the content of the photograph and the discourse produced around it.

Generally, the analysis aims to understand why landscapes are valued. The use of photo-graphs in social surveys is particularly relevant for connecting the biophysical characteris-tics of the river and their perception. Many contributions highlighted the benefit of photo-questionnaires to address this link (Le Lay et al. 2008; Junker and Buchecker 2008; Cottet et al. 2013). In such surveys, photographs are sampled according to different criteria to test their impacts on perceptions (e.g. presence/absence of dead wood in Le Lay et al. 2008; restoration scenarios in Junker and Buchecker 2008). When photographs are taken or selected by the surveyors, it is necessary to be careful of the biases which can be introduced. For example, the framing or the use of color or black and white can influence respondents' answers. Weather conditions or seasons may also influence the respondents. Using many photographs of the same place, subject, landscape, etc. may be a solution to limit these biases.

On the Rhône river (France), Cottet et al. (2013) conducted a photo-questionnaire to study the perception of floodplain lake landscapes (a type of alluvial wetland) within the context of their restoration. The survey was addressed to people without (students) and with (managers, researchers, experts) ecological knowledge of floodplain lakes. The photo-questionnaire presented photographs of floodplain lakes which were chosen according to their nutrient level as an indicator of ecological quality.[9] Based on the photographs, interviewees were asked to score the aesthetic value and ecological health of different floodplain lakes (Figure 12.6). The results demonstrated that the aesthetic value and perceived

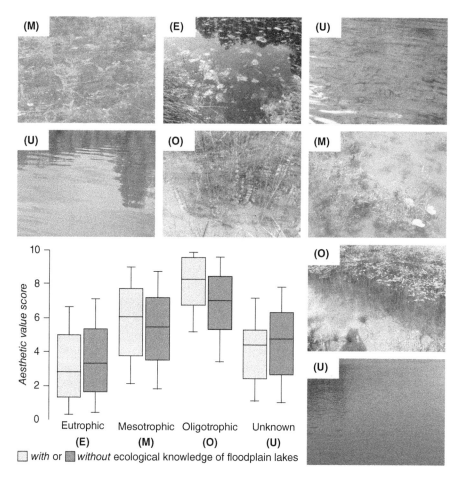

Figure 12.6 Photo-questionnaire survey to study the perceived aesthetic value and ecological health of floodplain lakes on the Rhône river (France).

ecological health of floodplain lakes were strongly linked and depended on the nutrient level of their water: the higher trophic the water, the less it was aesthetically appreciated. This perception went against the natural evolution of floodplain lakes. People without ecological knowledge wished to maintain floodplain lakes in a stable state, while restoration aimed to ensure ecological dynamics.

12.4   Conclusions

By "giving voice" to a wide range of river actors, social surveys act as first port of call for public participation. Social surveys offer river restoration managers an opportunity to "take the pulse" of the social context. In this respect, questionnaire and interview methods can only improve the success of restoration projects and make the projects more integrative. Questionnaires and interviews can bring out the social dimension of rivers, and more specifically they allow the analysis of the often very deep interactions between people and river environments. These methods also help identify opinions and expectations regarding river restoration projects and help us understand why different river actors support or contest these projects. For these reasons, social surveys are interesting operational tools both upstream of restoration projects – to understand the context of intervention and the expectations of actors – and downstream of the project – during its monitoring and to carry out a collective and integrated evaluation of the completed actions. These methods help achieve multiple objectives because they are greatly flexible. They are adaptable to the variety of scales, local contexts, actors, and questions raised. In order to highlight this diversity, our chapter has presented the scientific and operational interests of social surveys in the field of river restoration through a variety of case studies.

However, some questions remain once the surveys are completed. For example, how do the actors – river managers in particular – take ownership of the survey's results? Are these results truly taken into account within the projects? Are river restoration aims adjusted? Moreover, do the survey results change the role of the different river actors in the project, especially in the decision-making process? A survey among restoration project managers could shed light on these issues. The answers would most likely be affected by local contexts.

The political issues are clear when it comes to taking into account the social dimension in river restoration projects. Sometimes, contributions of social survey methods are viewed with suspicion by environmental professionals who are not familiar with them. A real change of perspective needs to be initiated and we hope that this chapter will contribute to this change. Training river managers and students in social survey methods is a compelling way to better integrate the social issues raised by river restoration (Figure 12.7). More generally, the challenge for researchers, river managers, and all river actors is to consider rivers not merely as ecosystems but rather as socioecological systems. Interviews and questionnaires methods, when they contribute to identifying obstacles and motivations in river restoration, greatly contribute to this evolution.

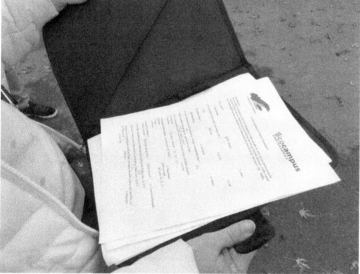

Figure 12.7 Social survey training is included in professional courses for river managers in France. During field trips, students practice questionnaire and interview methods. *Source:* Caroline le Calvez 2018.

Notes

1 These two methods meet different standards (see Section 12.3). Questionnaire and interview methods differ by the degree of standardization of the questions, by the number of people interviewed and by the type of data collected.

2 Observation is another method that can be used to understand the social dimensions of river restoration. However, it is less effective for collecting actors' opinions or expectations related to river restoration or analyzing the meanings actors attribute to their practices. In this chapter, we focus on questionnaire and interview methods.

3 "River practices" can be defined as individual or collective activities or uses related to rivers. In this chapter, we consider people who have one or several river practices "river actors."

4 Recording requires the consent of the interviewee(s). When authorized, recording is useful since it facilitates the analysis by allowing the interviewer to listen to the recordings or read the interview transcripts. Transcripts play a key role in qualitative content analysis or quantitative textual data analysis. However, transcribing interviews is time consuming (as an indication, one hour of oral speech usually takes three hours to transcribe).

5 There are different types of interviews: (i) the semi-structured interview is the one most commonly used. It is based on a guide that the interviewer adjusts in real time according to the interviewee's answers; (ii) the unstructured interview is based on a subject, theme, or keyword but no precise questions are planned in advance. This type of interview constrains the interviewee's discourse the least. The choice of subject, theme, or keyword must be carefully considered during the interview design phase, as it is crucial for initiating and guiding the discussion; (iii) the structured interview is the third type of interview commonly described in the literature. Questions are asked in a predefined order. This type of interview requires precise answers from the interviewees and is similar to a questionnaire. However, contrary to questionnaires, which mainly use close-ended questions, structured interviews favor open-ended questions.

6 Other, more quantitative, methods can be used to analyze interviews. They rely on statistical lexical analysis. Such analysis may require the use of specific software (content analysis programs or text mining, for instance). Since this method of quantitative analysis is presented in Chapter 13, it is not discussed here.

7 The parent population is made up of all the people who are the subject of the survey. It corresponds to the population from which a sample has been obtained.

8 Other media such as maps, videos, or drawings are not presented in this chapter but could be used in social surveys.

9 Nutrient level is a factor that strongly governs the ecological functioning of floodplain lakes. Nutrients influence the composition of plant communities and therefore how the ecosystem functions. Floodplain lakes can be categorized, on the basis of this composition, into three functional classes, from oligotrophic (low nutrient availability) to eutrophic (high nutrient availability). Moreover, the nutrient level of floodplain lakes has visual impact: plant cover and the composition of plant communities are affected, as well as water transparency. This ecological parameter may therefore influence perception.

References

Åberg, E.U. and Tapsell, S. (2013). Revisiting the River Skerne: the long-term social benefits of river rehabilitation. *Landscape and Urban Planning* 113: 94–103.

Babbie, E.R. (2010). *The Practice of Social Research*. Belmont, CA: Wadsworth Cengage Learning.

Baker, S., Eckerberg, K., and Zachrisson, A. (2014). Political science and ecological restoration. *Environmental Politics* 23(3): 509–524.

Barraud, R., Germaine, M.-A., and Sneddon, C.S. (2017). Dam removal: social, cultural and political issues. *Water Alternatives* 10(3): 648–654.

Berelson, B. (1952). *Content Analysis in Communication Research*. New York: Free Press.

Bickman, L. and Rog, D.J. (eds) (2009). *The SAGE Handbook of Applied Social Research Methods*. Thousand Oaks, CA: SAGE Publications.

Buijs, A.E. (2009). Public support for river restoration: a mixed-method study into local residents' support for and framing of river management and ecological restoration in the Dutch floodplains. *Journal of Environmental Management* 90(8): 2680–2689.

Cottet, M., Piégay, H., and Bornette, G. (2013). Does human perception of wetland aesthetics and healthiness relate to ecological functioning? *Journal of Environmental Management* 128: 1012–1022.

Cuaz, M., Meuret, B., and Piégay, H. (1996). Social surveys of users and property owners: what interest for river management? *Revue de géographie de Lyon* 71(4): 353–362.

Dufour, S. and Piégay, H. (2009). From the myth of a lost paradise to targeted river restoration: forget natural references and focus on human benefits. *River Research and Applications* 25(5): 568–581.

Dunn, K. (2000). Interviewing. In: *Qualitative Research Methods in Human Geography* (ed. I. Hay), 50–82. Oxford: Oxford University Press.

Egan D., Hjerpe, E.E., and Abrams, J. (ed.) (2011). *Human Dimensions of Ecological Restoration: Integrating Science, Nature, and Culture*. Washington DC: Island Press.

Flaminio, S., Cottet, M., and Le Lay, Y.-F. (2015). À la recherche de l'Yzeron perdu: quelle place pour le paysage dans la restauration des rivières urbaines? *Norois* 237(4): 65–79.

Fowler, F.J. Jr. (2013). *Survey Research Methods (Applied Social Research Methods)*. Thousand Oaks, CA: SAGE Publications.

Fox, C.A., Magilligan, F.J., and Sneddon, C.S. (2016). "You kill the dam, you are killing a part of me": dam removal and the environmental politics of river restoration. *Geoforum* 70: 93–104.

Gideon, L. (2012). *Handbook of Survey Methodology for the Social Sciences*. New York: Springer.

Johnson, E.S., Bell, K.P., and Leahy, J.E. (2018). Disamenity to amenity: spatial and temporal patterns of social response to river restoration progress. *Landscape and Urban Planning* 169: 208–219.

Junker, B. and Buchecker, M. (2006). Social science contributions to the participatory planning of water systems: results from Swiss case studies. In: *Topics on System Analysis and Integrated Water Resources Management* (eds A. Castelletti and R. Soncini-Sessa), 243–255. Amsterdam: Elsevier.

Junker, B. and Buchecker, M. (2008). Aesthetic preferences versus ecological objectives in river restorations. *Landscape and Urban Planning* 85: 141–154.

Junker, B., Buchecker, M., and Müller-Böker, U. (2007). Objectives of public participation: which actors should be involved in the decision making for river restorations? *Water Resources Research* 43(10). https://doi.org/10.1029/2006WR005584.

Larrue C. (2014). *Analyser les politiques publiques d'environnement*. Paris: L'Harmattan.

Le Calvez, C. (2017). Les usagers confrontés à la restauration de la continuité écologique des cours d'eau: approche en région Bretagne. PhD thesis. Rennes 2 University.

Le Calvez, C. and Hellier, E. (2017). Expérimenter la continuité écologique sur une masse d'eau fortement modifiée ou la mise au jour des tensions entre les représentations de l'Aulne canalisée. In: *Démanteler les barrages pour restaurer les cours d'eau: controverses et représentations* (eds R. Barraud and M.-A. Germaine), 115–128. Versailles, France: Éditions Quae.

Le Lay, Y.F., Piégay, H., Gregory, K., et al. (2008). Variations in cross-cultural perception of riverscapes in relation to in-channel wood. *Transactions of the Institute of British Geographers* 33(2): 268–287.

Pratt, G. (2009). Interviews and interviewing. In: *The Dictionary of Human Geography* (eds D. Gregory, R. Johnston, G. Pratt, et al.), 393–394. Chichester: Wiley.

Tunstall, S.M., Penning-Rowsell, E.C., Tapsell, S.M., et al. (2000). River restoration: public attitudes and expectations. *Water and Environment Journal* 14(5): 363–370.

Tunstall, S.M., Tapsell, S.M., and Eden, S. (1999). How stable are public responses to changing local environments? A "before" and "after" case study of river restoration. *Journal of Environmental Planning and Management* 42(4): 527–547.

Westling, E.L., Surridge, B.W.J., Sharp, L., et al. (2014). Making sense of landscape change: long-term perceptions among local residents following river restoration. *Journal of Hydrology* 519: 2613–2614.

Woolsey, S., Capelli, F., Gonser, T., et al. (2007). A strategy to assess river restoration success. *Freshwater Biology* 52: 752–769.

Yamashita, S. (2002). Perception and evaluation of water in landscape: use of Photo-Projective Method to compare child and adult residents perceptions of a Japanese river environment. *Landscape and Urban Planning* 62(1): 3–17.

13

Documents on River Restoration: Temporal and Spatial Analyses of Written Discourses

Emeline Comby[1], Bertrand Morandi[1], Yves-François Le Lay[1], Silvia Flaminio[2], and Helena Zemp[3]

[1] Université de Lyon, CNRS, Environnement Ville Société, Lyon, France
[2] Institut de Géographie et Durabilité, Université de Lausanne, Géopolis, Lausanne, Suisse
[3] Haute École d'Ingénierie – FHNW, Windisch, Aargau, Suisse

13.1 Introduction

Researchers investigating restoration policies and practices (e.g. Bernhardt et al. 2007) or discourse on restoration projects (e.g. Tunstall et al. 2000) often conduct surveys (e.g. questionnaires, interviews) of the different stakeholders involved (e.g. practitioners, elected officials, scientists, local residents). The open-ended questions in the questionnaires and the transcribed interviews produced as part of the research take the form of written documents. Other written material includes administrative and engineering documents, newspaper articles, communication documents, and scientific literature. This chapter focuses on this latter type of material that predates the research. This material is especially worth investigating because it is more plentiful and increasingly accessible in the digital age. The methods we discuss in this chapter can be used for extracting evidence or information from documents on any subject.

Documents yield a wealth of valuable material for research into river restoration. Documentary approaches can be employed to situate the context of river restoration across a range of scales from local to international (Figure 13.1). They can be used for evaluating a river restoration policy on the scale of a particular project on the basis of local sources. Some understanding of the drivers and impediments involved may be achieved before restoration work by comparing the specific context of the project with widely available information about river restoration policies at both regional and national scales (e.g. Morandi et al. 2017). And documentary approaches enable us to examine potential commonalities and differences between different discourses about restored rivers and restoration practices internationally (e.g. Cottet et al. 2015).

Documentary material enables researchers and managers to understand river restoration projects not just spatially but also in terms of time. In describing a context, a documentary approach can question the main past and present issues. Analysis of documentary material

River Restoration: Political, Social, and Economic Perspectives, First Edition. Edited by Bertrand Morandi, Marylise Cottet, and Hervé Piégay.
© 2022 John Wiley & Sons Ltd. Published 2022 by John Wiley & Sons Ltd.

Spatial scale

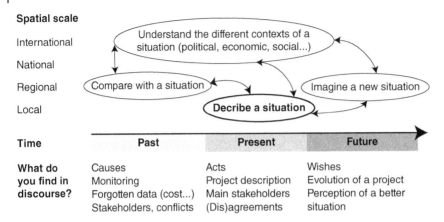

International

National

Regional

Local

Time	Past	Present	Future
What do you find in discourse?	Causes Monitoring Forgotten data (cost...) Stakeholders, conflicts	Acts Project description Main stakeholders (Dis)agreements	Wishes Evolution of a project Perception of a better situation

Figure 13.1 Why use a documentary approach to study restoration?

during the implementation of a river restoration project can help monitor public support for the project and identify any potential change in discourse on rivers. After implementation, such an approach can be integrated into the project evaluation process. New uses, perceptions, and pros and cons relating to river restoration work can be analyzed. This approach can therefore fulfill the growing need for feedback on restoration projects. It can contribute through an evidence-based approach to the understanding of successes, failures, and controversies. Such feedback about past actions enables river restoration stakeholders to adjust or improve their present and future projects.

A documentary approach providing descriptive information is a means to analyze different social processes (Fairclough 2010) such as the construction of perceptions and practices by residents, managers or scientists. Documents are not only sources of precise information but they also compose a discourse on the environment that can be analyzed as such (Dryzek 2012). According to Foucault (1966), discourse expresses a way of being, living, and speaking. Discourse is a practice with material consequences. That is why, as part of a political ecology approach to restoration, discourses are entangled with power relationships (Swyngedouw 2015). Studying discourses by means of documentary analysis provides an understanding of how a situation becomes a social problem. A social problem is to be understood as a putative condition defined as a problem in different public arenas (Hilgartner and Bosk 1988). Discourse analysis helps establish the present-day understanding of the topics and problems involved. It can offer a "genealogical" perspective that examines "how series of discourse are formed, . . . what were the specific norms for each, and what were their conditions of appearance, growth and variation" (Foucault 1971, pp. 62–63). Documentary approaches imply a focus on the various stakeholders and the interests they defend (Comby et al. 2019). It is interesting in the field of river restoration to understand when tensions or conflicts arise which stakeholders play a salient role in the debates and in implementing the restoration project, and why some people agree and others disagree with the project's objectives. Ecological river restoration "needs to be understood not only as a technical task but as deeply embedded in social and political processes" involving negotiations between different interest groups, choices among planning projects, and compliance with a regulatory framework planning and legal regulations (Baker et al. 2014, p. 518).

Based on examples from the scientific literature, this chapter provides methodological information about implementing documentary approaches by which to analyze, discuss, and support river restoration. In Section 13.2 we present the kind of documentary material that can be used to analyze discourse on river restoration, then in Section 13.3 we detail different qualitative and quantitative methods used to analyze such documentary material and the type of results they yield, and in Section 13.4 we discuss the pros and cons of documentary methods versus survey methods.

13.2   What kind of documentary material can be used for what purpose?

Many types of documents – legal provisions, administrative documents, technical memos, newspaper articles, websites, etc. – can be used to study stakeholders' discourses and to collect specific information about river restoration (Table 13.1). The development of computerization and the Internet have seen exponential growth in the number of documents potentially available to researchers. Faced with this mass of documents, researchers compiling corpora for documentary analysis must make choices based on scientific questions and hypotheses as well as on practical issues.

13.2.1   Building a corpus: the initial stage of documentary analyses

A corpus is a set of documents compiled by researchers specifically for analysis. To be defined as a corpus, a set of documents must meet five criteria (Bommier-Pincemin 1999). First, researchers put together a corpus to answer specific questions or hypotheses. The documents making up the corpus are chosen in preference to others because of the subject matter or the scope of the study. Some corpora pre-exist the study (e.g. legal documents), but a corpus needs to be compiled specifically for that analysis. Researchers have to select the relevant documents to feed their analysis. Second, a corpus must be coherent in terms of the topics covered or the data sources. For example, to ensure consistency, researchers often use only one type of document (e.g. newspaper articles, technical guides, legal texts) per corpus. Third, a corpus requires a sample of documents that is representative of a situation. When researchers can collect all of the documents pertaining to a situation, the corpus is "closed." In most cases, researchers cannot collect all documents exhaustively but have to select representative documents. Before putting together a corpus, researchers should study the situation as a whole so as to identify the main criteria behind the diversity. The corpus must be representative of the diversity of stakeholders, the genealogy of the situation, the variety of sites, and so on. Fourth, the corpus must be of sufficient volume in terms of document numbers and document size. If the corpus is too small, the questions that can be answered may be too specific. Nowadays, corpora are growing ever larger. Fifth, the corpus must be as complete as possible. It is often scarcity of time, access, and money that puts an end to the collection of documents.

Criticism of potential documentary materials is a necessary prerequisite when compiling any corpus. The researcher needs to identify (i) what information is to be found in documents and how it can be used to address the research questions, (ii) the origin of discourses,

Table 13.1 Different types of documents useful for river restoration.

Document type	Objectives	Analysis		Access From easy (+) to difficult (+++)
		Spatial scale	Temporal scale	
Press documents (newspaper articles, TV news, radio broadcasts, etc.)	Understand main issues of river restoration Analyze discourses of stakeholders (managers, elected officials, residents, etc.) Identify chronology of restoration	Understand river restoration issues at global scales (using international or national press documents) or at regional or local scales (using local press documents)	Characterize past evolutions of river restoration issues and stakeholders' viewpoints over the long term Picture a situation at a specific moment Identify wishes for future state of rivers	(++) Recent press documents are published online, generally with paid access. Most of the oldest documents are paper archived in public or private libraries. These documents are subject to copyright
Administrative documents (authorization, subsidy, application, etc.)	Analyze objectives, resources, and public policy methods Understand river restoration practices and policies Find factual information about project context and implementation	Understand river restoration policies at regional or national scale Characterize a situation at local scale (river restoration programs or projects)	Characterize medium-term evolutions of river restoration policy and practices Provide information about past, current, or upcoming projects	(+++) Administrative documents are hard to access because of sensitive information they may contain and lack of a centralized database. Support of administrative services producing the documents is essential
Technical documents (guidelines, notes, reports, etc.)	Understand river restoration practices Find factual information about project context and implementation Understand stakeholders' viewpoints of river restoration	Characterize a situation at local scale (river restoration programs or projects) (using reports) Understand policies and practices at regional or national scale (using guidelines or notes)	Characterize long-term evolution of river restoration practices Identify trends in technical recommendations and anticipate impact on future river restoration practices	(++) Most recent technical documents are online and some databases reference these documents. Support from institutions or stakeholders publishing the documents is usually needed to access the oldest of them

Document type			
Legislative and regulatory documents	Identify legislative and regulatory issues of river management Analyze objectives, resources, and more generally the role of law in orienting public policies	Characterize topical issues at global or regional scale Compare river restoration regulations in different countries	Analyze long-term evolutions of river restoration regulations Picture legal and reglementary context at a specific point in time	(+) Most legal texts, including the oldest, are officially published and retrievable online or in public offices
Communication documents (flyers, brochures, websites, etc.)	Find detailed information about a project context and its implementation Analyze objectives and positions of stakeholders publishing the documents Identify target public of the document and river restoration project	Characterize (laudatory or negative) viewpoints of different categories of stakeholder on broad scale (regarding river restoration policies in general) or local scale (regarding a specific project)	Understand stakeholder viewpoints at a specific point in time, before, during, or after project implementation Long-term analyses are difficult	(+) Current communication documents are easily accessible online but there is no central database. Most old documents, especially website contents, are not archived and so hard to access
Scientific documents (peer-reviewed articles, posters, oral communications, etc.)	Identify main research topics Analyze theories and concepts Understand researchers' practices and perceptions	Analyze scientific practices at international or national scales Compare research trends across countries or regions It is rare to have a significant corpus of publications at local scale	Analyze evolution of research dynamics and topical issues Picture research issues or researchers' perceptions at a specific point in time	(++) Scientific documents are published online with paid access. They are also accessible in libraries of universities or research institutions. Some documents such as posters or oral communication are more difficult to retrieve

(iii) the context of discourse production, (iv) the occurrence of author bias or subjectivity, (v) the purpose of producing such documents, and (vi) the audience for the documents. For example, newspapers are convenient sources of information for understanding how past reality was built, without any memory bias (Gregory and Rowlands 1990) and with a record over quite a long period (Comby et al. 2019). Despite their use of personalization, dramatization, and novelty (Boykoff and Boykoff 2007), the media play a major role in *mediating* between science, policy, and the public. By making allowance for these issues, researchers can determine what analyses can be conducted and what interpretations can be made. In some cases it is difficult to clearly identify the origin of discourses. For example, when considering technical or administrative documents concerning a restoration project, researchers often have access to final versions of these documents but not to the different working versions. They have little if any information about discussions during technical committee meetings or public consultations, or about contributions from the various stakeholders involved in the project. Such documents contain hybrid discourses. Sometimes it may prove a good compromise to collect several corpora and develop trans-corpora analyses to better highlight and discuss the limitations of each corpus.

Depending on the nature of the documents targeted, researchers employ various strategies to build a corpus. They may go on the Internet, go to public libraries, documentation centers, or archives of specific institutions (Figure 13.2). Sometimes documents (especially recent electronic documents) are referenced in digital databases (online or offline) which are valuable resources for researchers. For example, many digital subscription databases referencing and providing access to newspapers (e.g. Access World News, Europresse, Factiva, Lexis-Nexis) or to scientific publications (e.g. Scopus, ISI Web of Science) can be searched with keywords. In many countries, public administrations also have digital databases allowing access to legal, administrative, and technical documents.[1]

In some cases, electronic documents can be downloaded directly from these digital databases, but in other cases, only references or abstracts of documents are provided, with no direct access to the documents in full. The absence of digital archives is one of the major difficulties, especially for historical or local documents (e.g. municipal registers, local newspapers) (Driedger 2007). Moreover, some documents are not referenced in any database. Consequently, researchers have to use physical archives first to identify documents and then to access the information they contain. This is time-consuming, as each paper document has to be pre-analyzed, digitized, and then archived by the researchers themselves.

For corpora comprising electronic or paper documents, researchers have to create a database specifically for the analysis containing each document and the related metadata (e.g. document type, author(s), date, publication location, text length, number of pictures).

Based on case studies from the literature, Sections 13.2.2–4 give examples of uses of different documentary materials for river restoration research: scientific publications, newspaper articles, and administrative documents.

13.2.2 Scientific publications: identifying trends in river restoration research

Peer-reviewed literature provides some interesting documentary material for research focused on scientific practices (e.g. Smith et al. 2014). Considering sciences to be socially and politically embedded (Callon 1986), some researchers focus on how society and policy

Figure 13.2 Documentary analysis relies on corpora collected from different sources like the Internet, public libraries, documentation centers, or archives of specific institutions: (a) the Contemporary Documentation Centre of Lyon Institute of Political Studies; (b) the archives of Loire-Bretagne Water Agency. Compilation can be facilitated by use of tools (e.g. camera, scanner) to digitize documents; (c) in Nancy AgroParisTech Documentation Center. *Sources:* (a) E. Comby 2010; (b) E. Milcent 2012; (c) B. Morandi 2010.

influence scientific research, innovation, technology, and planning, and vice versa. They cast light on who it is that participates in scientific debates about technological changes and how different groups strengthen their positions while undermining others (e.g. Pritchard 2011). Bouleau (2014) shows how scientific categorization contributes to redefining what water and rivers are and how they should be managed.

In river management, an understanding of the context in which scientific knowledge is produced appears to be an issue, especially with respect to restoration where there are close connections between the way ecological restoration is defined and the various disciplines of environmental science. In a recent study of the temporal and spatial dynamics of river restoration sciences, Morandi and Piégay (2017) take peer-reviewed publications as their research material. Updating this study, they compile a corpus of 1783 international scientific papers published from 1925 to 2017. This updated corpus was put together from the *Web of Science* online commercial database using the query: "(restor* OR rehabilitat* OR revital* OR renat*) AND (stream* OR river*)" on document titles.

The map based on this analysis shows that scientific publications focus on restoration projects mainly in North American, Europe, and China and on large rivers (e.g. the Mississippi, Yellow, Kissimmee, Rhine, and Colorado rivers) (Figure 13.3). These results raise the question of the regionalization of scientific knowledge on river restoration and of the transfer of conclusions about restoration practices from one specific environmental and sociocultural context (mainly large rivers in developed countries) to another.

Scientific publications are valuable material for investigating research trends in the domain of river restoration. They may also be used to analyze various definitions of the same term and its proximity with other concepts, to discover the (dis)advantages of management practices, to analyze different feedback from restoration projects internationally, and to analyze scientists' discourse. These documents are often found in electronic form and are quite easy to access although subject to payment of a subscription to online databases (e.g. Web of Science, Scopus). However, the large number of documents of this kind calls for a careful choice when it comes to wording search queries.

13.2.3 Newspaper articles: highlighting social problems and controversies

Because controversies are often addressed by the media, newspapers may be an important resource for building support around a cause and influencing debate (Jørgensen and Renöfält 2013).

In considering the example of the decommissioning of a dam on the Sélune river, one of the most controversial restoration projects in France, Le Lay and Germaine (2017) analyze two newspapers: a regional daily newspaper (*Ouest-France*) and a local weekly one (*La Gazette de la Manche*). All articles relating to the hydraulic infrastructure and published from the beginning of 2001 to the end of 2015 were collected to identify the genealogy of this restoration project. A corpus of 515 articles was formed.

Toponyms were extracted from newspaper articles to identify the emblematic places of the dam removal controversy (Figure 13.4): the floodplain (in Saint-Hilaire, Ducey, and Poilley), the recreational areas (with a focus on La Mazure), the portions considered polluted, the angling areas, etc. Some places did not seem to feed the controversy in the newspaper, which does not mean that the residents did not have an opinion about the political

Figure 13.3 Mapping of river names mentioned in titles of 1783 scientific publications in the field of river restoration from 1925 to 2017 (Web of Science database source). *Source:* B. Morandi 2021.

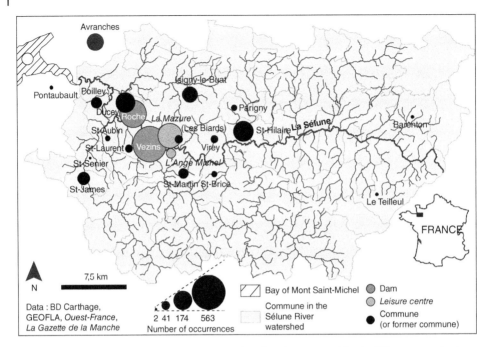

Figure 13.4 Mapping of the main toponyms related to dam removal in the Sélune watershed mentioned in 515 articles from regional (*Ouest-France*) and local (*La Gazette de la Manche*) newspapers. *Source:* Data from *Ouest-France* and *La Gazette de la Manche*.

decision to decommission the dam and restore the river. The distribution of the toponyms mentioned in the articles shows that there was very little mention of the different activities developed in the upstream part of the basin.

Newspaper articles are useful for understanding the geography and chronology of social problems, identifying the main topical issues, and examining stakeholders' discourses. Yet, the discourses of dominant stakeholders are overrepresented in newspapers. The press is of little avail when a situation entails no controversy and when researchers want to observe people who have no say. Consulting newspaper articles can be very time-consuming if they are in paper format and much easier when they are available through electronic databases.

13.2.4 Administrative documents: understanding river restoration practices and public policies

International, national, or local administrative agencies produce many documents each year to control, authorize, or fund river restoration programs or projects. It is worth studying these documents to understand the implementation of policies in the field and to identify potential divergences between policy orientations and actual situations of environmental management. Administrative documents are useful for gathering local and regional feedback. These documents are often full of practical details about river restoration design, implementation, or evaluation. For example, Morandi et al. (2016) use a documentary approach to address the question of the definition of river restoration by French practitioners. The authors study financial support documents to understand the uses of different concepts (e.g. restoration, rehabilitation, or renaturation) and related river management practices. The financial support documents are drafted jointly by river managers, who implement the projects, and French Water Agencies, which fund them.

These documents were collected from the archives of three Water Agencies in charge of funding restoration projects in different hydrographic districts in France. First, the digital reference database was requested, then a sample of the paper files was scanned. Search queries of their databases turned up 4089 documents for the period 1997–2011. Of those 4089 documents identified in databases, Morandi et al. (2016) scanned a random sample of 364 full-text documents.

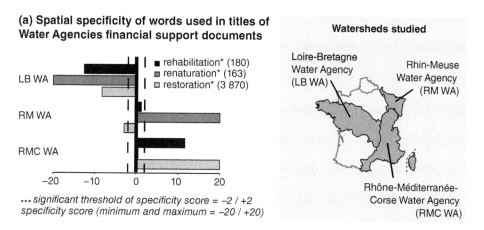

(a) Spatial specificity of words used in titles of Water Agencies financial support documents

- ■ rehabilitation* (180)
- ▨ renaturation* (163)
- ☐ restoration* (3 870)

LB WA

RM WA

RMC WA

-20   -10   0   10   20

··· *significant threshold of specificity score = –2 / +2*
specificity score (minimum and maximum = –20 / +20)

Watersheds studied

Loire-Bretagne Water Agency (LB WA)

Rhin-Meuse Water Agency (RM WA)

Rhône-Méditerranée-Corse Water Agency (RMC WA)

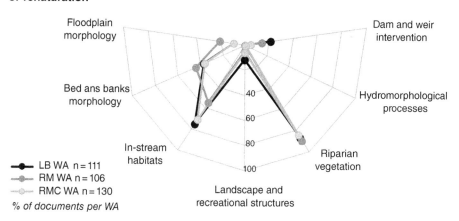

(b) Measures supported by Water Agencies under the terms restoration, rehabilitation or renaturation

Floodplain morphology

Dam and weir intervention

Bed ans banks morphology

Hydromorphological processes

In-stream habitats

Riparian vegetation

Landscape and recreational structures

40

60

80

100

━●━ LB WA n = 111
━◉━ RM WA n = 106
┄○┄ RMC WA n = 130

% of documents per WA

Figure 13.5 As applied to Water Agencies' administrative documents concerning financial support for river restoration: (a) textual analysis highlights spatial specificities (Lafon 1980) in the use of the concepts of restoration, rehabilitation, and renaturation; (b) whereas content analysis reveals the similarity of measures financed. The terms are different but the practices are the same. *Source:* for (a): Based on Lafon, P. (1980). Sur la variabilité de la fréquence des formes dans un corpus. *Mots* 1: 127–165.

Textual data analysis of the titles showed that the use of different concepts is related to regional institutional traditions rather than to fundamental differences in conceptual definitions. Analysis of the document contents revealed that, behind a variety of terms, different French Water Agencies and even different departments within the same Water Agency support quite similar restoration measures. There is no fundamental discrepancy in the conceptualization of restoration but different terms are used to express the same reality (Figure 13.5). Water technicians and engineers use the terms "restoration" and "rehabilitation" somewhat differently from the academic community.

Although their style may be somewhat cold and austere, administrative documents are crucial material for analyzing management practices and water policies. While generally informative, they may be short on information about stakeholders' viewpoints. Researchers may be denied access to such documents because of confidentiality issues and difficulties in cataloging them. Unless administrative documents are available in public archives, consulting them may prove complicated. The support of the agencies that produce these documents is often essential in identifying and accessing them.

13.3 What are the methods of documentary analysis and what results do they produce?

After creating a documentary corpus, different qualitative and quantitative methods can be used to analyze it (Table 13.2). We present four types of method commonly used by social researchers working on river restoration: (i) bibliometric analysis, (ii) content analysis, (iii) textual data analysis, and (iv) qualitative analysis. In this section, we illustrate each of these methods with examples of applications to river restoration in Switzerland, the USA, and France.

13.3.1 Methods of bibliometric analysis: counting the number of documents

The first method of documentary analysis focuses on document production and dissemination by using thorough quantitative methods. This type of analysis is known as "bibliometrics" (Mingers and Leydesdorff 2015). It originated in the research field of scientometrics and informetrics (Hood and Wilson 2001). Even if bibliometrics is often considered an informal and preliminary or complementary analysis, it is a foundational stage of any documentary analysis.

In environmental studies, bibliometrics is mainly used to illustrate the dynamics of research by way of the dynamics of publications. In river restoration, Smith et al. (2014, p. 251) highlight "the rise of restoration by plotting the increased references to river restoration in the literature." Although bibliometrics is traditionally applied to scientific output, it can be applied to other types of documentary material such as technical, communications, or press documents. Morandi et al. (2016) use the number of administrative documents related to river restoration to assess the relevance of this topic for French river management policies. In a study carried out in Switzerland, Zemp and Buchecker (2015) use bibliometric methods to analyze media coverage of river restoration (also called "renaturation" and "revitalization" in Switzerland) from 2000 to 2014. The media selected for

Table 13.2 Type of documentary analysis methods.

Method	Principles and objectives	Three main advantages	Three main limitations
Bibliometric analysis	Count the number of documents to ascertain trends	Relevant with the increasing number of documents Quick analysis of a large number of documents No need to read the documents	Very descriptive No explanation of the trends The first step of a further analysis
Content analysis	Create comprehensive, exclusive, "objective," relevant categories to sum up textual information	A picture of the text is sufficient (no need to use an OCR or to transcribe) Careful human reading of each text Possibility of conducting quantitative analyses	The time to code Inter-coder reliability and intra-coder reliability Impossible to add categories without reading again and highly dependent on the initial categories (researchers find what they are looking for)
Textual data analysis	Count the use of words in context to delay the use of interpretative categories	No need to read all documents Statistical analysis and also words in context The power of figures to convince	All text must be in digital format Statistical analyses are sometimes too complex Time to learn how to use the software
Qualitative analysis	Select extracts of texts to create a documented narrative	Description of specific and concrete situations Narratives can be easily re-used in communication documents Open to people with no say (not just majority or dominant stakeholders)	Hard to establish whether data are representative of the situation Strong emphasis on the researcher's subjectivity Need to trust the person interpreting the data

this study are three regional newspapers covering two separate cantons: Bern (*Berner Zeitung, Berner Oberländer*) and Valais (*Walliser Bote*). Online archives of the newspapers were used to initially identify articles about river restoration and revitalization as the main topic. All articles in which the key words or expressions (e.g. renaturation, river widening, natural waters) occurred in the headline or in the report were selected. The resulting corpus contained 683 articles.

Bibliometric analysis outlines the annual evolution of media coverage. More concretely, the research illustrates the salient features and dynamics of the information and communication flow, the framing of discourses, the stakeholders involved, and changes in risk perception. According to Zemp and Buchecker (2015), flooding in 1999 and 2005 might explain media attention to restoration projects which emerge as a new management

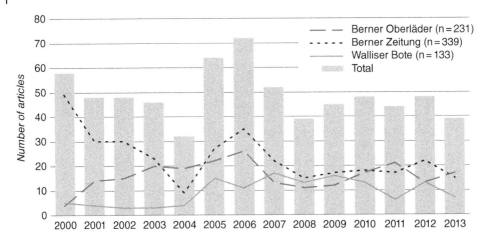

Figure 13.6 Change over time of media coverage evaluated through annual number of press articles about river restoration published for the period 2000–2013 in three regional Swiss newspapers: *Berner Zeitung*, *Berner Oberländer*, and *Walliser Bote*. *Source:* Modified from Zemp, H. and Buchecker, M. (2015). Die Renaturierung und Revitalisierung von FliessgewässernalsThema der Medien in eheralpinenländlichenRegionen der Schweiz (2000–2013). Eine MedienanalyseimRahmeneinerForschungskooperationmit dem Renaturierungsfonds des Kantons Bern (RenF). WSL Berichte, 36.

approach for mitigating floods in the Swiss context of highly channelized and impounded rivers. The authors also outline the role of management events such as conflict around the Belpau renaturation project (2000)[2] (especially in the *Berner Zeitung*) or the "Living Water" popular initiative (2005),[3] which can explain the specific media coverage (Figure 13.6). According to the authors, the relatively stable number of articles published between 2008 and 2013 "is an indication that the topic is considered to have a certain relevance" (Zemp and Buchecker 2015, p. 1).

13.3.2 Content analysis: categorizing documentary discourse

Content analysis is "a research technique for the objective, systematic, and quantitative description of manifest content of communication" (Berelson 1952, p. 18). This method explores qualitative data in quantitative form (Hayward and Osborne 1973). Before or after reading each text, categories are created in which to classify the information it contains (Boholm 2009). This standard method enables scientists to compare different texts (Mayring 2000).

Because the data are categorized, researchers can use different statistical analyses to describe the trends of all of the texts from a descriptive perspective and then explore the relationships between different variables from an explanatory perspective (Comby et al. 2014a). Content analysis is driven by hypotheses and may entail biases associated with coding and operators.

Content analysis can help understand the involvement and the viewpoints of different stakeholders in river restoration projects. Flaminio et al. (2015) represent the stakeholders who are either evoked (light gray) or interviewed (dark gray) in 177 newspaper articles from *Le Progrès* and published between 2003 and 2012 (Figure 13.7). The articles were all centered on the Yzeron river in Oullins (a municipality in the urban area of Lyon, in the Rhône river basin) where a restoration program was just beginning in 2012. While the project was being launched, many riverside residents were very concerned about the risk of flooding. The corpus of articles went back to the 2003 flood that caused the evacuation of 200 residents from their homes.

The content analysis showed that a great diversity of stakeholders was evoked by the press: environmental organizations (such as Cora, Frapna, and Naturama); local, regional, and national authorities, elected officials and institutions; residents (and in particular the riverside residents); and of course the "Sagyrc" in charge of the restoration project (the most mentioned and most interviewed stakeholder). This diversity also revealed how different scales and places are intertwined in the implementation of a restoration project. Focusing on the stakeholders interviewed, the content analysis also showed that the newspaper interviewed, first, elected officials and, second, riverside residents.

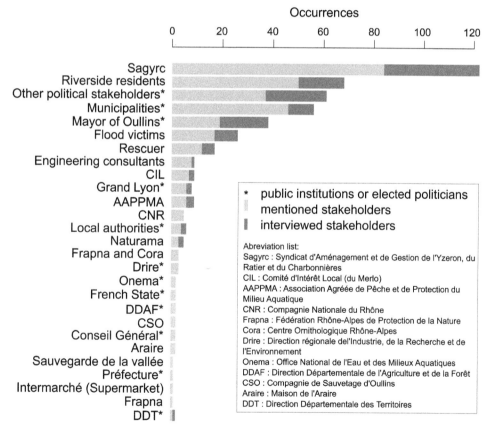

Figure 13.7 Stakeholders mentioned and interviewed in the daily regional press (*Le Progrès*, 2003–2012, n = 177) about the Yzeron River and its restoration project. *Source: Le Progrès,* Lyon, 2003.

13.3.3 Textual data analysis: analyzing the use of the words in context

To avoid the task of defining categories, as is the case with content analysis and to limit bias due to hasty interpretations by researchers, textual data analyses consider only written or spoken words (Lebart et al. 1998). Textual data analyses are based on the interpretation of statistical computations performed on texts (Comby et al. 2014b). Textometry, for example, combines factorial correspondence analyses with full-text search techniques (Heiden et al. 2010).

To study the restoration of the Sacramento River, the main data set addresses the description of the restoration project and its perception based on a regional daily newspaper, *The Sacramento Bee*. Comby (2015) collected newspaper articles from two digital databases: Access World News and Europresse. She constructed a query based on the words "Sacramento," "Central Valley," and "Delta" from 2005 to 2013. The research involved sorting through a huge number of articles. The final raw material included 1090 articles. The words "restoration(s)," "restore(d)," "restores," and "restoring" were quoted 576 times. She used co-occurrences to describe how words occurred together and to study the relationship between words. This task was facilitated by the use of TXM, an open-source software package (Heiden 2010), to select all words with a co-occurrence score higher than 8 (when the significance threshold is 2). This analysis helped understand the main aims ("habitat, ecosystem, imperiled, reliability, wildlife, improving, co-equal, species"), places ("Delta, acres, wetlands, tidal"), and policies ("plan, projects"). The results demonstrated that two political goals were co-equal in the corpus: restoring the Delta (California EcoRestore) and making California's water supply more reliable (California Water Fix). She focused on two fish species whose names were the most cited in the corpus to understand whether the discourse pattern changed over time (Figure 13.8).

The delta smelt is a threatened species. This fish was an icon of the so-called California water wars in the 1990s and is regarded as an ecological indicator of water and habitat quality. The numbers of delta smelt are at an all-time low, even though "CalFed" was in charge of an ecological restoration policy. In newspaper discourse, delta smelts were central until 2008, but afterward salmon became prominent. The salmon run verged on collapse in 2007–2008. Sport and commercial fishing seem to have been in jeopardy. Concerns have emerged regarding salmon migration upstream: to spawn salmon need to have suitable habitats and appropriate water flow. Anglers and fishermen interviewed by journalists considered that salmon were more necessary than delta smelt. These viewpoints can explain the increased newsworthiness of salmon. Furthermore, some stakeholders lost interest because the number of delta smelt was still very low despite the restoration projects. Although delta smelt were newsworthy for 20 years, they seem to have rallied fewer stakeholders to their cause. The understanding of temporal patterns of information enabled managers and scientists to propose strategies to different stakeholders to help them better respond to public interest at the right time.

13.3.4 Methods of qualitative analyses: creating documented narratives

A qualitative approach encompasses two main traditions: (i) a constructivist approach that considers different meanings of built environment in what is necessarily a subjective reality and (ii) a critical approach that examines power relationships, control, and ideology

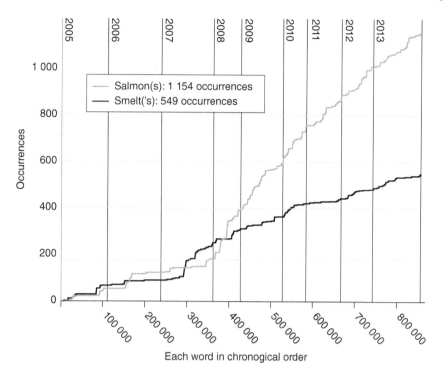

Figure 13.8 Cumulated frequency curves of the fish species evoked by media; their occurrences from 2005 to 2013 in *The Sacramento Bee. Source: The Sacramento Bee.*

(Hesse-Biber 2010). The qualitative methods may be criticized because of the major role of subjectivity and researcher interpretations.

For example, Comby (2015) compares media discourse in three regional newspapers along the Rhône river from 2002 to 2013: *Le Progrès* for the upstream stretch, *Le Dauphiné Libéré* for the middle stretch, and *La Provence* for the downstream stretch. Her qualitative approaches use excerpts of text to recreate narratives. She chooses significant or emblematic quotations, which can refer to dominant discourse or more minority views. This approach is relevant because the corpus is too small to create a large enough database to develop statistical analysis. Such an analysis sums up some spatial and temporal trends in discourse without being too time-consuming.

"It is high time to let the Rhône breathe" (*Le Progrès*, 10 August 2004). When journalists tackled restoration, in *Le Progrès* they dealt with water and habitat quality, whereas in *Le Dauphiné Libéré* they insisted on the quality issue but also on local development and floods. "Quality is a driver to achieve economic development, but we also need to protect ourselves from the river" *(Le Dauphiné Libéré,* 20 April 2002). During the 2000s, journalists evoked the restoration of oxbow lakes, whereas in the 2010s they tackled dike fields (known as the *Casiers Girardon*). This interest corresponds to the emergence of a new restoration policy and novel scientific developments. Unlike the other two newspapers, in *Le Dauphine Libéré*, journalists dealt with conflicts arising from the restoration policy. They gave broad media coverage to residents who wanted to reduce the risk of flooding and criticized the ecological policy of restoration. Finally, in *La Provence*, journalists barely mentioned the restoration projects.

13.4 When to use a documentary approach? The pros and cons compared to survey methods

A documentary analysis based on pre-existing documents has certain advantages over survey methods. Unlike most questionnaires or interviews, this kind of documentary analysis can provide a long-term perspective because discourses archived in written documents are often accessible. Some biographical and oral history interviews are also based on storytelling questions to track changes over time, but it can be quite difficult to find people who know the history of a river well, and many people forget what happened during river restoration projects. The temporal advantage of pre-existing documentary materials is valuable for monitoring restoration projects by comparing discourse written before and after the restoration work. While an interview often combines many different, and sometimes contradictory, perceptions and conceptions, a document provides a rather stable picture anchored in a specific period. Moreover, documentary data allow analysis to be carried out on large spatial scales with reduced investment in terms of time and money compared with survey methods. It is quite impossible to interview all river restoration specialists in the world, but it is possible to collect and analyze most of the scientific papers or technical documents published on river restoration.

However, documentary analysis with pre-existing documents also has its limits. First, from a temporal perspective, the researcher depends on historical collections. Legal or administrative documents and newspapers are often well catalogued in public archives. But many private documents such as personal correspondence of scientists and river managers, handouts from associations, or technical reports by consultants or private companies are not systematically collected, rigorously catalogued, or accessible. The risk for researchers is to address only dominant discourses by studying documents that can be readily consulted. The second limitation, compared with survey data, is that the documents consulted in archives have not been created and designed from an analytical perspective to test researchers' hypotheses. Consequently the documents may lack information. These documents may be published in paper format, meaning that, in some cases, it is time-consuming to collect and examine them.

Given all the pros and cons, documentary analyses are "often used in combination with other qualitative research methods as a means of triangulation" (Bowen 2009, p. 28). For example, Fox et al. (2016) use interviews, but also newspaper articles, transcripts of local and state hearings, and other textual sources to analyze controversies related to dam removal and river restoration. Documentary analysis can help researchers design their surveys, or check information collected by questionnaires or interviews. On the other hand, survey methods can also support interpretation and discussion of results from documentary analysis.

13.5 Conclusions

Documentary analysis is often used in a qualitative approach, while more quantitative approaches are still scarce in research today. However, recent work on river restoration and more broadly on river management has shown the value of these methods.

It has been demonstrated through different case studies that this approach can be used to analyze the objectives, resources, and strategies of public policies; to understand the main issues of river restoration; to identify the genealogy of restoration; to find factual information about a project's context and implementation; to analyze the discourses of the main stakeholders; to understand river restoration practices and stakeholders' viewpoints on river restoration; and to tackle controversy. If a restoration project becomes a social problem, a documentary analysis may help in ascertaining the interests and the concerns of the different stakeholders, identifying the main spaces and places where the project is topical, addressing the chronology of restoration, and studying conflicts and the different opinions or attitudes toward a problem or definitions of terms. A documentary analysis can provide a diagnosis of the pros and the cons of the project and may produce suggestions for making changes to the initial project. Therefore, a documentary analysis is a useful means of collecting feedback.

These methods open up interesting analytical perspectives. In particular, they make it possible to work on longer timescales and to study restoration as part of long-term territorial dynamics. It is useful to shed light on the discrepancy between a top-down policy (such as the European Union's Water Framework Directive or national water policy) and the actual situation. A documentary analysis is often useful at a local scale for understanding the context and events. Nevertheless, such analysis may also be conducted at an international scale when interviews are impossible.

Given the diversity of materials and methods, researchers have to make numerous decisions. These choices relate to the questions asked by researchers, but also to the practical constraints of collecting and analyzing documents. The main questions to be addressed concern the creation of a corpus and the methods of analyzing it.

Notes

1 In France, for example, the Documentary Portal Water and Biodiversity catalogs more than 50 000 online documents: https://www.documentation.eauetbiodiversite.fr/ (accessed 11 December 2020).
2 The Belpau project was aimed at both flood protection and floodplain ecological restoration along the river Aar.
3 The "Living Water" popular initiative is a democratic process by which Swiss citizens propose a project of river renaturation to the Swiss Federal Government.

References

Baker, S., Eckerberg, K., and Zachrisson, A. (2014). Political science and ecological restoration. *Environmental Politics* 23(3): 509–524.

Berelson, B. (1952). *Content Analysis in Communication Research*. New York: Free Press.

Bernhardt, E.S., Sudduth, E.B., Palmer, M.A., et al. (2007). Restoring rivers one reach at a time: results from a survey of US river restoration practitioners. *Restoration Ecology* 15(3): 482–493.

Boholm, M. (2009). Risk and causality in newspaper reporting. *Risk Analysis* 29(11): 1566–1577.

Bommier-Pincemin, B. (1999). Diffusion ciblée automatique d'informations: conception et mise en œuvre d'une linguistique textuelle par la caractérisation des destinataires des documents. PhD dissertation. University Paris IV – Sorbonne.

Bouleau, G. (2014). The co-production of science and waterscapes: the case of the Seine and the Rhône rivers, France. *Geoforum* 57: 248–257.

Bowen, G.A. (2009). Document analysis as a qualitative research method. *Qualitative Research Journal* 9(2): 27–40.

Boykoff, M.T. and Boykoff, J.M. (2007). Climate change and journalistic norms: a case-study of US mass-media coverage. *Geoforum* 38(6): 1190–1204.

Callon, M. (1986). Eléments pour une sociologie de la traduction: la domestication des coquilles Saint-Jacques et des marins-pêcheurs dans la baie de Saint-Brieuc. *L'Année Sociologique* 36: 169–208.

Comby, E. (2015). Pour qui l'eau? Les contrastes spatio-temporels des discours sur le Rhône (France) et le Sacramento (Etats-Unis). PhD dissertation, University of Lyon.

Comby, E., Le Lay, Y.-F., and Piégay, H. (2019). Power and changing riverscapes: the socio-ecological fix and newspaper discourse concerning the Rhône river (France) since 1945. *Annals of the American Association of Geographers* 109(6): 1671–1690.

Comby, E., Le Lay, Y.-F., and Piégay, H. (2014a). The achievement of decentralized water management through broad stakeholder participation: an example from the Drôme River catchment area in France (1981–2008). *Environmental Management* 54: 1074–1089.

Comby, E., Le Lay, Y.-F., and Piégay, H. (2014b). How chemical pollution becomes a social problem: risk communication and assessment through regional newspapers during the management of PCB pollutions of the Rhône river (France). *Science of the Total Environment* 482–483: 100–115.

Cottet, M., Piola, F., Le Lay, Y.-F., et al. (2015). How environmental managers perceive and approach the issue of invasive species: the case of Japanese knotweed s.l. (Rhône river, France). *Biological Invasions* 17: 3433–3453.

Driedger, S.M. (2007). Risk and the media: a comparison of print and televised news stories of a Canadian drinking water risk event. *Risk Analysis* 27(3): 775–786.

Dryzek, J.S. (2012). *The Politics of Earth: Environmental Discourses*. Oxford: Oxford University Press.

Fairclough, N. (2010). *Critical Discourse Analysis: The Critical Study of Language*. London: Longman Applied Linguistic.

Flaminio, S., Cottet, M., and Le Lay, Y.-F. (2015). A la recherche de l'Yzeron perdu: quelle place pour le paysage dans la restauration des rivières urbaines. *Norois* 237: 65–79.

Foucault, M. (1966). *Les mots et les choses: une archéologie des sciences humaines*. Paris: Gallimard.

Foucault, M. (1971). *L'ordre du discours: leçon inaugurale au Collège de France prononcée le 2 décembre 1970*. Paris: Gallimard.

Fox, C.A., Magilligan, F.J., and Sneddon, C.S. (2016). "You kill the dam, you are killing a part of me": dam removal and the environmental politics of river restoration. *Geoforum* 70: 93–104.

Gregory, K.J. and Rowlands, H. (1990). Have global hazards increased? *Geography Review* 4(2): 35–38.

Hayward, R. and Osborne, B.S. (1973). The British colonist and the immigration to Toronto of 1847: a content analysis approach to newspaper research in historical geography. *Canadian Geographer* 17(4): 391–402.

Heiden, S., Magué, J.-P., and Pincemin, B. (2010). TXM: une plateforme logicielle open-source pour la textométrie: conception et développement. *Statistical Analysis of Textual Data: Proceedings of 10th International Conference* 2: 1021–1032.

Hesse-Biber, S. (2010). Qualitative approaches to mixed methods practice. *Qualitative Inquiry* 16(6): 455–468.

Hilgartner, S. and Bosk, C.L. (1988). The rise and fall of social problems: a public arenas model. *American Journal of Sociology* 94(1): 53–78.

Hood, W.W. and Wilson, C.S. (2001). The literature of bibliometrics, scientometrics, *and informetrics. Scientometrics* 52(2): 291–314.

Jørgensen, D. and Renöfält, B.M. (2013). Damned if you do, dammed if you don't: debates on dam removal in the Swedish media. *Ecology and Society* 18(1): 18.

Lafon, P. (1980). Sur la variabilité de la fréquence des formes dans un corpus. *Mots* 1: 127–165.

Lebart, L., Salem, A., and Berry, L. (1998). *Exploring Textual Data*. Dordrecht, the Netherlands: Springer.

Le Lay, Y.-F. and Germaine, M.-A. (2017). Déconstruire? L'exemple des barrages de la Sélune (Manche). *Annales de Géographie* 715(3): 259–286.

Mayring, P. (2000). Qualitative content analysis. *Forum: Qualitative Social Research* 1(2): 10.

Mingers, J. and Leydesdorff, L. (2015). A review of theory and practice in scientometrics. *European Journal of Operational Research* 246: 1–19.

Morandi, B., Kail, J., Toedter, A., et al. (2017). Diverse approaches to implement and monitor river restoration: a comparative perspective in France and Germany. *Environmental Management* 60(5): 931–946.

Morandi, B. and Piégay, H. (2017). *River Restoration in France: Changes in Definitions and Techniques over Space and Time: Outlook for the Future*. Vincennes, France: French Biodiversity Agency.

Morandi, B., Piégay, H., Johnstone, K., et al. (2016). Les Agences de l'eau et la restauration: 50 ans de tension entre hydraulique et écologique. *Vertigo* 16(1). https://doi.org/10.4000/vertigo.17194.

Pritchard, S. (2011). *Confluence: The Nature of Technology and the Remaking of the Rhône*. Cambridge, MA: Harvard University Press.

Smith, B., Clifford, N.J., and Mant, J. (2014). The changing nature of river restoration. *Wiley Interdisciplinary Reviews: Water* 1(3): 249–261.

Swyngedouw, E. (2015). *Liquid Power: Contested Hydro-Modernities in Twentieth-Century Spain*. Cambridge, MA: MIT Press.

Tunstall, S.M., Penning-Rowsell, E.C., Tapsell, S.M., et al. (2000). River restoration: public attitudes and expectations. *Water and Environment Journal* 14(5): 363–370.

Zemp, H. and Buchecker, M. (2015). *Die Renaturierung und Revitalisierung von Fliessgewässern als Thema der Medien in eher alpinen ländlichen Regionen der Schweiz (2000–2013). Eine Medienanalyse im Rahmen einer Forschungskooperation mit dem Renaturierungsfonds des Kantons Bern (RenF)*. WSL Berichte, 36.

14

Participatory Approaches: Principles and Practices for River Restoration Projects

Alba Juárez-Bourke and Kirsty L. Blackstock

Social, Economic and Geographical Sciences Department, The James Hutton Institute, Aberdeen, Scotland

14.1 Introduction

Stakeholder participation is increasingly acknowledged as crucial for the successful management of social–ecological systems (Stringer et al. 2006). A stakeholder is defined as anyone who affects or is affected by an issue. In the field of water management, this trend is evidenced by the EU's Water Framework Directive, which emphasizes the role of stakeholder and public involvement in water management (European Commission 2000; Euler and Heldt 2018). More recently, the Organization for Economic Cooperation and Development published *12 Principles on Water Governance*, which includes the promotion of stakeholder engagement (OECD 2015a), and these insights have cascaded into the river restoration domain (INBO 2018). However, participation is far from being mainstreamed. Szałkiewicz et al. (2018) found that over half of the European river restoration projects they assessed did not involve stakeholders in their design or implementation. The traditional command-and-control paradigm in industrialized countries, which presumes water resources are predictable and controllable, is beginning to give way to a more adaptive approach, to which stakeholder participation is integral (Pahl-Wostl et al. 2007). For river restoration projects in particular, stakeholder participation has been shown to promote a convergence of views among stakeholders (Henze et al. 2018) and has been identified as vital for their success, as it can help resolve or mitigate conflicts between stakeholder interests, and can boost project acceptance (Woolsey et al. 2007).

Stakeholder participation can be understood in very different ways, from information provision, to consultation, to involvement in decision-making. The arguments put forward for adopting participatory approaches include increased (i) effectiveness, (ii) efficiency, (iii) legitimacy, and (iv) democracy. First of all, participatory processes can be more effective; because of their complexity, social–ecological systems cannot be understood by one single person or at a single scale. Involving different stakeholder groups, such as local users, in the process can add valuable insights for the interpretation of dynamic ecosystems (Folke et al. 2005). In addition, stakeholder participation can highlight the existence of conflicting perspectives, allowing for a better understanding of the problem, helping to better manage the system (Carmona et al. 2013). Second, involving different stakeholders can help to make decisions more transparent, increasing trust toward public institutions (Huitema et al. 2009). Stakeholders are therefore more likely to implement decisions, thus improving the efficiency of management projects (Stoll-Kleemann and O'Riordan 2002; Reed 2008). Third, it is argued that including those who are affected by decisions is a democratic right and can enhance the legitimacy of decisions made (Stoll-Kleemann and O'Riordan 2002). Finally, involving different stakeholder groups can strengthen democracy by promoting social learning and community cohesion, as well as by improving people's ability to participate in future processes (Blackstock 2017).

Nevertheless, stakeholder participation should not be considered a panacea for successful river restoration projects; it can be challenging or inappropriate in some cases. For instance, the time and resources required to carry out such processes can be restrictive, particularly in large-scale projects (Stringer et al. 2006; Maynard 2013). It can also result in the participation of only people with the most time and resources, to the disadvantage of others (Platteau and Abraham 2002). Poorly designed participatory projects may also result in "stakeholder fatigue," leading to unwillingness to participate in future activities. This can occur, for example, if the process is too time- or energy-consuming, if it is not transparent, or if participants feel that their input is ignored (OECD 2015b). Participatory processes might also be undermined if they are not adequately supported by their social and institutional context (Leitch et al. 2015).

The number of examples from academic literature on stakeholder participation in river restoration is increasing as the research on restoration matures, demonstrating how participatory approaches can contribute to some of the benefits outlined above. For instance, Junker et al. (2007) show how participation can promote the emergence of trust toward a project. In another study, Petts (2007) explains how the engagement of local citizens in a river restoration project enabled participants to define their priorities, reach an agreement on the actions needed, and recognize the constraints faced by the project, while Hostmann et al. (2005) describe how stakeholder participation can facilitate more consensus-oriented decisions. Participatory river restoration projects have also been shown to improve the understanding of the riparian system by drawing on people's experiential knowledge (Maynard 2013).

In this chapter we discuss two questions. What are good practice principles for participatory river restoration? What methods could be used for participatory river restoration?

14.2 What are good practice principles for participatory river restoration?

Participation can involve different levels of engagement. Arnstein's (1969, p. 216) "ladder of participation" (Figure 14.1) represents this range and illustrates a gradient of stakeholder engagement: from lower rungs of engagement, such as dissemination of information, to active engagement as the higher rungs. This metaphor implies higher engagement level to be preferable. However, depending on the goals and the context of a project, such

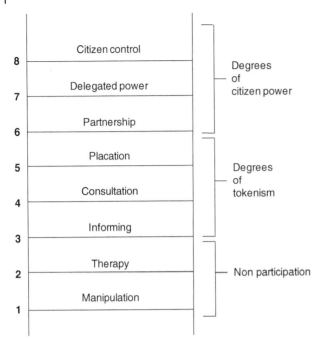

Figure 14.1 Arnstein's "ladder of participation." *Source:* Arnstein, S. R. (1969). A ladder of citizen participation. *Journal of the American Institute of Planners* 35(4): 216–224.

as if stakeholders do not have the capacity to influence the outcomes, this may not always be the case (Richards et al. 2004).

The following overlapping principles are synthesized from the literature on governance of water and other natural resources, with specific river restoration examples where appropriate. Observing these guidelines can contribute to selecting an appropriate approach and to implementing it in a fair, effective, and efficient manner.

14.2.1 Having clear goals

The organizers of a participatory restoration project must have clear goals as well as a clear understanding of the purpose and expected results of engaging stakeholders (e.g. participation to increase project acceptance or to improve management decisions). This will determine which stakeholder groups to involve and in what way to involve them (Reed 2008). For instance, a particular stakeholder group may have a positive effect on achieving some objectives but have a negative effect on other objectives. Similarly, a certain objective might be positively affected by the participation of one group but negatively affected by another group (Schultz et al. 2011). This principle can be difficult to implement in river restoration projects when there can be conflicts between the goals of local residents (e.g. recreational amenity, security from flooding) and goals held by other stakeholders (e.g. habitat protection, water quality) (Druschke and Hychka 2015; Chou 2016); this requires careful deliberative design to find common ground. Therefore, the stakeholders involved and the method of engagement chosen must be suited to the objectives of the participatory process.

14.2.2 Identifying the stakeholders, and ensuring inclusiveness and equity

Deciding whom to include in the project can be challenging, and to do so requires a careful and systematic stakeholder analysis, identifying those likely to affect the process and those who are affected by it, taking into consideration their core motivations and expectations as well as their interactions (OECD 2015b). This is important to avoid overlooking any people or groups of people who might be affected by the project (Reed et al. 2009). This step is essential. Heldt et al. (2016) suggest that up to three-quarters of previous river restoration projects have failed to achieve their outcomes, owing to a lack of stakeholder involvement. It is important to take into account the existing power imbalances, in order to avoid under- or overrepresentation, or the exclusion of certain groups (OECD 2015b). For instance, care should be taken to involve people with fewer resources, who are typically underrepresented in participatory processes (Huitema et al. 2009). The process should ensure that all groups have sufficient information and the language and medium used are accessible to all; training may be necessary (OECD 2015b). Specific arrangements may be needed to accommodate travel, accessibility, and childcare needs (e.g. Petts 2008). In addition, it is important to consider the knowledge systems of the different stakeholders (Henze et al. 2018). It can be a good practice to put in place several different ways in which people can participate (Metcalf et al. 2015). Lauer et al. (2018) draw attention to the importance of inclusive processes for overall stakeholder satisfaction in their Montana restoration case study, whilst Lave (2016) highlights the link between restoration and environmental justice outcomes.

14.2.3 Taking into account the context

When designing a participatory restoration project, it is important to take into consideration the institutional, political, geographical, and sociocultural aspects, as these will be decisive in the selection of an appropriate approach (Reed 2008). For example, whether a community has a strong culture of participation in decision-making processes is likely to determine people's ability and willingness to participate. Where there have been conflicts over water resources, trust between stakeholder groups may have been eroded, which could affect people's willingness to engage. For instance, in the Upper Hunter Valley, in Australia, the historical top-down approach to river works and lack of transparency has led to mistrust and skepticism regarding restoration processes (Spink et al. 2010; see also Fidelis and Carvalho 2015 for a Portuguese example). Sarvilinna et al. (2018) found that willingness to participate in restoration was influenced by factors particular to the context, illustrating how design must account for these factors. The process should also be flexible enough to respond to changing circumstances (OECD 2015b).

14.2.4 Promoting empowerment

The participatory process should ensure that participants have the capacity to influence the outcomes (OECD 2015b). Stakeholders are best engaged from the outset of a project, at the planning stages, when their involvement may have a meaningful impact, and this should continue throughout the entire process (Reed 2008; Metcalf et al. 2015). The importance of supporting and developing the energy of committed citizens and residents is repeated in many river restoration case studies (Barthélemy and Armani 2015). If the project cannot be influenced by stakeholders' input, a participatory approach beyond information provision is not appropriate (Richards et al. 2004). Some river restoration projects have sought to empower citizens through involving them as citizen scientists (Huddart et al. 2016; Edwards et al. 2018), although, for real empowerment, these citizens would need to collaborate or decide on future actions based on these data, rather than just provide information to experts.

14.2.5 Being transparent

The facilitators of an engagement process should openly and clearly communicate to participants the goals of a project, explain to them how their input will be used and taken into account, and inform them about the progress of the project. The process also requires transparency about limitations, costs, uncertainties, and delays in the project (Rowe and Frewer 2000; Metcalf et al. 2015). This is crucial for building stakeholders' trust and increasing their support for the project. Lack of transparency can lead to creating false expectations and consequently stakeholder dissatisfaction, which in turn can result in stakeholder fatigue, as Metcalf et al. (2015) found in their study of a river restoration in Montana.

14.2.6 Allocating sufficient resources

Participatory processes can be resource intensive, in terms of time, funding, and skills. Sufficient resources should be allocated specifically to the engagement process, ensuring that there is enough funding to cover logistical expenses as well as hiring staff and training them in communication and facilitation skills (Leitch et al. 2015). In addition, the process should allow enough time for trust to be built and for stakeholders to gain awareness about the project (OECD 2015b). The restoration of a tributary of the River Thames in London was designed to involve local people from the beginning of the project. However, the limited resources available meant that these efforts were ineffective (Eden and Tunstall 2006). Furthermore, it is essential to ensure there are sufficient resources to maintain engagement through all phases, including maintenance of restoration measures, as Fliervoet et al. (2017) found in their Dutch case study.

14.2.7 Evaluating the participatory process

Evaluation allows restoration projects to learn about what worked. Bedarkar et al. (2018) add to earlier calls for evaluation of river restoration processes to ensure future projects are effective and inclusive (see also Junker et al. 2007), and to avoid wasting resources on projects likely to fail (Heldt et al. 2016). Evaluation should be adequately resourced (Better Evaluation 2017) and it is more efficient if data collection is planned in advance. Methods can be simple quantitative measures, such as taking a register of attendance, collecting website visitor statistics, and having a distinct budget line in the accounts; but information on the quality of the process and changes in social processes, such as trust and empowerment, are equally important and could be collected using short feedback forms at all events.

Using developmental evaluation techniques (Patton 2010) allows any restoration project to stay aligned to original objectives, or to transparently adapt.

14.3 What methods could be used for participatory river restoration?

It is important to ensure that the methods are selected in light of the principles previously outlined and match the objectives and form of participation that is desired for the specific restoration project. Table 14.1 sets out methods we have classified under three distinct headings: information provision about the restoration project; consultation on options for restoration; and active involvement in deciding on restoration interventions, with the potential to manage these interventions. These map onto Figure 14.2: inform – information provision; consult – consultation; and active involvement – involve, collaborate, or delegate. Any project may involve all types at different times, and methods listed under one heading could be adapted to be used for another heading. The main message is that using any method requires adaption to ensure it is tailored to the specific needs of the participants as well as the project. In order to select the method for participation, the project manager needs to define: why (goals, expected results), when (what stage of the project),

Table 14.1 Methods for engaging stakeholders.

Type	Why	When	How	Who
Information provision	Provide information on the need for restoration and outcomes sought	At the start of a project Throughout the project After completion	Written (in print, online) Visual methods (videos, infographics) Face to face (meetings) Orally (podcast, radio, television)	Anyone with an interest in the project Those less engaged Indirect stakeholders
Consultation	Elicit opinions on proposals Help select solutions	Scoping stages Milestones throughout the project	Written (online or printed) Face-to-face (at meetings, focus groups, citizen juries, workshops) Social media or online polls Other methods (art, competitions, etc.)	Anyone with an interest in the project Specific interest groups Individuals with technical expertise
Active involvement	Develop solutions; make decisions; manage or implement measures	Throughout the project At mutually identified milestones	Face-to-face (meetings, discussion forums) Online meetings	Participants in a steering group or expert committee Representatives of formal organizations

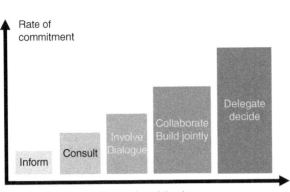

Figure 14.2 Types of participation and levels of required commitment. *Source:* INBO - International Network of Basin Organisations (2018). The handbook for the participation of stakeholders and the civil society in the basins of rivers, lakes and aquifers. INBO.

how (what medium to use), and who (who to involve), including at which scale one might work. It may be sufficient to use methods requiring less commitment or motivation in some cases. Heldt et al. (2016) found that attitudes toward their restoration case study were positive despite engagement being limited to information provision and consultation, although Lauer et al. (2018) found that satisfaction in their study in the United States was positively associated with the degree of control over decision-making.

14.3.1 Information provision regarding restoration projects

This category covers methods that are primarily used to provide information to anyone with an interest in a river restoration project. It refers to one-way provision from the project managers to interested parties and does not confer control over decision-making (Figure 14.3). Information provision is often used at the start of a project but is also useful throughout the project to keep stakeholders up to date with progress, and once a project is completed to provide a legacy of what was achieved and the lessons learned. Education about river restoration is a particular form of information provision that provides a long-term foundation for the success of any restoration project (Lehtoranta et al. 2017; Vian et al. 2018). Increased knowledge is often related to increased satisfaction of local residents, who are crucial to support the ongoing maintenance of the restoration measures (Sheng et al. 2019).

Information provision is generally written, in print or online, although increasingly visual methods such as videos and infographics are used. Information can also be provided in a face-to-face format, such as a presentation at a meeting, or orally through podcasts, radio, or television. Often, these are used in combination, such as giving a presentation at a local meeting and directing interested stakeholders to more information via other routes. All these methods can be highlighted through social media campaigns targeted to relevant communities of practice and place. In their study, Metcalf et al. (2015) illustrate the use of outreach information campaigns as a tactic to increase awareness of restoration activities in the US.

Figure 14.3 Information provided about a river restoration project in Scotland. *Source:* Marina Piper (Dee Catchment Partnership).

The language and concepts should be accessible – not necessarily simple but well explained (Margerum and Whitall 2004) and their relevance to the recipient clearly laid out. A message will be digested and acted upon if it is relevant, contextualized, and the recipient is clear about what they are supposed to do with the information and how they might be affected (Burton et al. 2007). Information provision is open to anyone, provided they are willing and able to access the methods used. It is often seen as low cost compared to more active methods, as online tools now mean that no specialist expertise is required to build an online presence. However, a well-designed, well-targeted, and well-maintained suite of information provision approaches can be time-consuming, and making the material accessible, attractive, and visible to the audience may require a great deal of resources.

14.3.2 Consultation on potential restoration decisions

Unlike information provision, consultation methods are explicitly designed to elicit information from participants, but unlike active involvement, there is no sharing of decision-making powers. Consultations are traditionally written pieces explaining the project, the issue being consulted on, and what the consultees are being asked to do and by when. These consultations are generally hosted online and/or in print form asking for written responses, but responses can also be elicited in meetings, focus groups, citizen juries, or workshops, or preferences generated through the use of online polls. This may be useful at the scoping stages of a project, where local knowledge can be very important to provide an understanding of the costs and benefits involved and to ensure there is a collective agreement on the problem, before starting to search for solutions (Wesselink et al. 2011). More commonly, consultation processes are used at milestones in projects to help select between different solutions. These later consultation processes are often criticized for their latent power dynamics (Lukes 1974), potentially skewing outcomes by offering constrained choices to a predefined problem, whereas stakeholders may have framed the restoration issues and solutions differently.

Consultation processes often require complementary information provision techniques to ensure consultees can make an informed response, and potential consultees are aware of the opportunity to respond (INBO 2018). It is good practice to provide a digest of consultation responses and a commentary on how these responses were used. Consultation methods are generally open to anyone willing and able to access the methods used, although specific interest groups or individuals with technical expertise may be targeted. Consultation approaches do not have to be particularly resource intensive beyond the time required to develop, disseminate, analyze, and evaluate the responses. The more successful the consultation in terms of responses, the more resources will be required to make use of the data.

14.3.3   Active involvement in decision-making and implementation

The main difference between active involvement and consultation is that the participants actively involved have more power to contribute to decisions throughout the restoration project, from discussing issues during the planning phase to influencing how measures are implemented and managed (Figure 14.4). Participants tend to be more informed about the specifics of the project and to be able to give more valuable support. Many commentators believe that the complex, long-term, and integrated nature of river restoration requires active involvement to ensure the objectives are achieved (Metcalf et al. 2015; Bedarkar et al. 2018).

Figure 14.4   Workshop participants engaged in discussions over a river restoration project of River Dee in Aberdeenshire, Scotland. *Source:* Rachel Creaney (James Hutton Institute).

Active involvement requires these participants to dedicate time to the project (see Figure 14.2). These processes tend to be face-to-face through regular meetings, although it is possible to have virtual meetings or input via discussion forums (Blackstock 2017). Generally, actively involved participants are pre-selected by the project organizers to represent formal organizations' perspectives. This formalization of representation to those with mandates to act on behalf of wider stakeholders can help to ensure support for implementation and management of restoration measures (see Blackstock et al. 2014). More rarely, projects have involved self-selecting interested parties (Lane et al. 2011). However, owing to the time and energy required for intense deliberation over time, numbers tend to be restricted.

Active involvement is often considered the best as it is higher on Arnstein's (1969) ladder of participation. However, the tendency to focus on formal representatives with a mandate for a recognized community of interest or place can exclude the more diffuse wishes of the "unorganized publics", destabilizing legitimacy (Behagel and Turnhout 2011). Second, this exclusion can reinforce a perception that river restoration is a technical issue, when the sustainability of river restoration perhaps lies in a greater public appreciation of its importance (Vollmer et al. 2015; Vian et al. 2018). Finally, even active involvement participatory processes often remain advisory (Antunes et al. 2009), owing to path dependencies set by funding constraints, statutory requirements, or negotiations with individual riparian owners.

One way to allow active decision-making by anyone with an interest in the project is to hold a binding referendum to make decisions. This differs from a consultation approach in that the outcome is the decision. However, as the literature on referenda makes clear, such an approach requires extremely careful question design and a strong complementary program of information provision. The approach is more democratic in allowing anyone with an interest to exercise their right to vote, but could still be an exercise in latent power through the closing down discipline of voting on a single option (Stirling 2008) and, like consultation, could be subject to capture by interest groups who vote tactically to ensure a particular outcome (Singleton 2000).

Active involvement requires significant resources in terms of meeting facilities or professional online forum design, whilst a referendum will probably also require considerable resources to ensure the response rate is seen as high enough to be legitimate. As with consultation processes, actively involved participants may be "free" for the restoration project but reimbursement for travel costs may be required to allow a wider selection of participants to attend and ensure the process conforms to principles around equity and empowerment.

14.4 Conclusions

Stakeholder participation is increasingly expected, if not required, in natural resource management. It is therefore important to understand how best to do this. For river restoration, it can be useful to build on the experiences and ideas about participation in other areas of natural resource management and water management; this chapter draws on these wider debates, as well as from river restoration experiences.

Although much of what can be learned about stakeholder engagement in river restoration is shared with other areas of environmental management, there are challenges that particularly affect participatory processes in river restoration; one of these is the value that different stakeholders place on environmental, social, and economic outcomes. Though this issue is not exclusive to rivers, fluvial areas are often in high demand, and their uses in conflict with each other, requiring a careful design that is clear about the project's purpose of engaging stakeholders. Another challenge is the uncertainty and long timeframes of river restoration projects, which require a high commitment from participants and a willingness to work collaboratively; it can be challenging to break away from the traditional engineering approaches that many stakeholders expect. Finally, rivers are often simultaneously subject to other management processes and policies, such as flood management, urban development, and transport, adding to the complexity of a restoration project.

Project managers should not expect a straightforward process; it is likely to be challenging and resource intensive. If carried out without following the principles outlined in this chapter, it could jeopardize the success of the project and future projects; if participants felt that their participation had no effect on the outcome, or if the results did not match their expectations, they might not be willing to participate in the future and even oppose future projects. However, participation that is carefully planned with clear objectives and sufficient resources, and involves stakeholders in a fair, equitable, and transparent way, is likely to lead to a more effective, efficient, legitimate, and democratic restoration project.

References

Antunes, P., Kallis, G., Videira, N., et al. (2009). Participation and evaluation for sustainable river basin governance. *Ecological Economics* 68: 931–939.

Arnstein, S.R. (1969). A ladder of citizen participation. *Journal of the American Institute of Planners* 35: 216–224.

Barthélémy, C. and Armani, G. (2015). A comparison of social processes at three sites of the French Rhône river subjected to ecological restoration. *Freshwater Biology* 60: 1208–1220.

Bedarkar, M., Dhingra, U.G., Mishra, M., et al. (2018). Analysis of global river restoration experiences: learnings and policy measures in the Indian context. *Asian Journal of Water, Environment and Pollution* 15: 203–215.

Behagel, J. and Turnhout, E. (2011). Democratic legitimacy in the implementation of the Water Framework Directive in the Netherlands: towards participatory and deliberative norms? *Journal of Environmental Policy & Planning* 13: 297–316.

Better Evaluation (2017). Evaluation costing. http://www.betterevaluation.org/en/evaluation-options/calculate_evaluation_costs (accessed 14 August 2017).

Blackstock, K. (2017). Participation in the context of ecological economics. In: *Routledge Handbook of Ecological Economics* (ed. C.L. Splash), 341–350. London: Routledge.

Blackstock, K., Waylen, K., Marshall, K., et al. (2014). Hybridity of representation: insights from river basin management planning in Scotland. *Environment and Planning C: Government and Policy* 32: 549–566.

Burton, R., Dwyer, J., Blackstock, K., et al. (2007). *Influencing Positive Environmental Behaviour Among Farmers and Landowners: A Literature Review: Report to DEFRA*. Socio-Economic Research Group, The Macaulay Land Use Research Institute.

Carmona, G., Varela-Ortega, C., and Bromley, J. (2013). Participatory modelling to support decision making in water management under uncertainty: two comparative case studies in the Guadiana river basin, Spain. *Journal of Environmental Management* 128: 400–412.

Chou, R.-J. (2016). Achieving successful river restoration in dense urban areas: lessons from Taiwan. *Sustainability* 8: 1159.

Druschke, C.G. and Hychka, K.C. (2015). Manager perspectives on communication and public engagement in ecological restoration project success. *Ecology and Society* 20. https://doi.org/10.5751/ES-07451-200158.

Eden, S. and Tunstall, S. (2006). Ecological versus social restoration? How urban river restoration challenges but also fails to challenge the science: policy nexus in the United Kingdom. *Environment and Planning C: Government and Policy* 24: 661–680.

Edwards, P.M., Shaloum, G., and Bedell, D. (2018). A unique role for citizen science in ecological restoration: a case study in streams. *Restoration Ecology* 26: 29–35.

Euler, J. and Heldt, S. (2018). From information to participation and self-organization: visions for European river basin management. *Science of the Total Environment* 621: 905–914.

European Commission (2000). Directive 2000/60/EC of the European Parliament and of the Council establishing a framework for community action in the field of water policy. *Official Journal of the European Communities L327*: 1–73.

Fidelis, T. and Carvalho, T. (2015). Estuary planning and management: the case of Vouga Estuary (Ria de Aveiro), Portugal. *Journal of Environmental Planning and Management* 58: 1173–1195.

Fliervoet, J.M., van den Born, R.J., and Meijerink, S.V. (2017). A stakeholder's evaluation of collaborative processes for maintaining multi-functional floodplains: a Dutch case study. *International Journal of River Basin Management* 15: 175–186.

Folke, C., Hahn, T., Olsson, P., et al. (2005). Adaptive governance of social-ecological systems. *Annual Review of Environment and Resources* 30: 441–473.

Heldt, S., Budryte, P., Ingensiep, H.W., et al. (2016). Social pitfalls for river restoration: how public participation uncovers problems with public acceptance. *Environmental Earth Sciences* 75: 1053.

Henze, J., Schröter, B., and Albert, C. (2018). Knowing me, knowing you: capturing different knowledge systems for river landscape planning and governance. *Water* 10: 934.

Hostmann, M., Bernauer, T., Mosler, H.J., et al. (2005). Multi-attribute value theory as a framework for conflict resolution in river rehabilitation. *Journal of Multi-Criteria Decision Analysis* 13: 91–102.

Huddart, J.E., Thompson, M.S., Woodward, G., et al. (2016). Citizen science: from detecting pollution to evaluating ecological restoration. *Wiley Interdisciplinary Reviews: Water* 3: 287–300.

Huitema, D., Mostert, E., Egas, W., et al. (2009). Adaptive water governance: assessing the institutional prescriptions of adaptive (co-)management from a governance perspective and defining a research agenda. *Ecology and Society* 14: 26.

INBO (International Network of Basin Organisations) (2018). *The Handbook for the Participation of Stakeholders and the Civil Society in the Basins of Rivers, Lakes and Aquifers*. INBO.

Junker, B., Buchecker, M., and Müller-Böker, U. (2007). Objectives of public participation: which actors should be involved in the decision making for river restorations? *Water Resources Research* 43. https://doi.org/10.1029/2006WR005584.

Lane, S.N., Odoni, N., Landström, C., et al. (2011). Doing flood risk science differently: an experiment in radical scientific method. *Transactions of the Institute of British Geographers* 36: 15–36.

Lauer, F.I., Metcalf, A.L., Metcalf, E.C., et al. (2018). Public engagement in social-ecological systems management: an application of social justice theory. *Society & Natural Resources* 31: 4–20.

Lave, R. (2016). Stream restoration and the surprisingly social dynamics of science. *Wiley Interdisciplinary Reviews: Water* 3: 75–81.

Lehtoranta, V., Sarvilinna, A., Väisänen, S., et al. (2017). Public values and preference certainty for stream restoration in forested watersheds in Finland. *Water Resources and Economics* 17: 56–66.

Leitch, A.M., Cundill, G., Schultz, L., et al. (2015). Principle 6: broaden participation. In: *Principles for Building Resilience: Sustaining Ecosystem Services in Social-Ecological Systems* (eds R. Biggs, M. Schlüter, and M.L. Schoon), 201–225. Cambridge: Cambridge University Press.

Lukes, S. (1974). *Power: A Radical View*. London: Macmillan.

Margerum, R.D. and Whitall, D. (2004). The challenges and implications of collaborative management on a river basin scale. *Journal of Environmental Planning and Management* 47: 409–429.

Maynard, C.M. (2013). How public participation in river management improvements is affected by scale. *Area* 45. https://doi.org/10.1111/area.12015.

Metcalf, E.C., Mohr, J.J., Yung, L., et al. (2015). The role of trust in restoration success: public engagement and temporal and spatial scale in a complex social-ecological system. *Restoration Ecology* 23: 315–324.

OECD (Organization for Economic Cooperation and Development) (2015a). OECD Principles on Water Governance. http://www.oecd.org/governance/oecd-principles-on-water-governance.htm (accessed 14 August 2017).

OECD (Organization for Economic Cooperation and Development) (2015b). *Stakeholder Engagement for Inclusive Water Governance*. Paris: OECD Publishing.

Pahl-Wostl, C., Sendzimir, J., Jeffrey, P., et al. (2007). Managing change toward adaptive water management through social learning. *Ecology and Society* 12. https://doi.org/10.5751/ES-02147-120230.

Patton, M.Q. (2010). *Developmental Evaluation: Applying Complexity Concepts to Enhance Innovation and Use*. New York: Guilford Press.

Petts, J. (2007). Learning about learning: lessons from public engagement and deliberation on urban river restoration. *The Geographical Journal* 173: 300–311.

Petts, J. (2008). Public engagement to build trust: false hopes? *Journal of Risk Research* 11: 821–835.

Platteau, J.-P. and Abraham, A. (2002). Participatory development in the presence of endogenous community imperfections. *Journal of Development Studies* 39: 104–136.

Reed, M., Graves, A., Dandy, N., et al. (2009). Who's in and why? A typology of stakeholder analysis methods for natural resource management. *Journal of Environmental Management* 90: 1933–1949.

Reed, M.S. (2008). Stakeholder participation for environmental management: a literature review. *Biological Conservation* 141: 2417–2431.

Richards, C., Blackstock, K., and Carter, C. (2004). *Practical Approaches to Participation*. Aberdeen: Macaulay Institute.

Rowe, G. and Frewer, L.J. (2000). Public participation methods: a framework for evaluation. *Science, Technology, & Human Values* 25: 3–29.

Sarvilinna, A., Lehtoranta, V., and Hjerppe, T. (2018). Willingness to participate in the restoration of waters in an urban–rural setting: local drivers and motivations behind environmental behavior. *Environmental Science & Policy* 85: 11–18.

Schultz, L., Duit, A., and Folke, C. (2011). Participation, adaptive co-management, and management performance in the world network of biosphere reserves. *World Development* 39: 662–671.

Sheng, W., Zhen, L., Xiao, Y., et al. (2019). Ecological and socioeconomic effects of ecological restoration in China's Three Rivers Source Region. *Science of the Total Environment* 650: 2307–2313.

Singleton, S. (2000). Co-operation or capture? The paradox of co-management and community participation in natural resource management and environmental policy-making. *Environmental Politics* 9: 1–21.

Spink, A., Hillman, M., Fryirs, K., et al. (2010). Has river rehabilitation begun? Social perspectives from the Upper Hunter catchment, New South Wales, Australia. *Geoforum* 41: 399–409.

Stirling, A. (2008). "Opening up" and "closing down": power, participation, and pluralism in the social appraisal of technology. *Science, Technology, & Human Values* 33: 262–294.

Stoll-Kleemann, S. and O'Riordan, T. (2002). From participation to partnership in biodiversity protection: experience from Germany and South Africa. *Society & Natural Resources* 15: 161–177.

Stringer, L., Dougill, A., Fraser, E., et al. (2006). Unpacking "participation" in the adaptive management of social–ecological systems: a critical review. *Ecology and Society* 11: 39.

Szałkiewicz, E., Jusik, S., and Grygoruk, M. (2018). Status of and perspectives on river restoration in Europe: 310 000 euros per hectare of restored river. *Sustainability* 10: 129.

Vian, F.D., Martínez, M.S., and Izquierdo, J.J.P. (2018). Citizen participation as a social shift tool in projects of urban fluvial space recovery: a case study in Spain. *Urban Forestry & Urban Greening* 31: 252–260.

Vollmer, D., Prescott, M.F., Padawangi, R., et al. (2015). Understanding the value of urban riparian corridors: considerations in planning for cultural services along an Indonesian river. *Landscape and Urban Planning* 138: 144–154.

Wesselink, A., Paavola, J., Fritsch, O., et al. (2011). Rationales for public participation in environmental policy and governance: practitioners' perspectives. *Environment and Planning A* 43: 2688–2704.

Woolsey, S., Capelli, F., Gonser, T., et al. (2007). A strategy to assess river restoration success. *Freshwater Biology* 52: 752–769.

15

Economic Benefits: Operationalizing their Valuation in River Restoration Projects

Sylvie Morardet

UMR G-EAU, Université de Montpellier, AgroParisTech, BRGM, CIRAD, INRAE, Institut Agro, IRD, Montpellier, France

15.1 Introduction

The increasing acknowledgment of the usefulness of ecosystems for society and the need to protect them has given rise to the concept of ecosystem services (ES), mainstreamed by the Millennium Ecosystem Assessment (MEA). At the interface between socioeconomic and ecological perspectives, ES have been defined as "benefits humans derive from nature" (MEA 2005) or "the direct and indirect contributions of ecosystems to human well-being" (de Groot et al. 2010). The diversity and importance of services provided by river ecosystems is nowadays widely acknowledged (MEA 2003; Finlayson et al. 2005; Grizzetti et al. 2016) and the ES concept is increasingly incorporated into river management policies. Applying an ES approach in river management allows a more holistic and systemic analysis of all present and potential direct river uses, as well as benefits provided by river functioning. It enables the consideration of temporal dynamics of ES and the complementarities or competitions between river uses. It also helps communicating the intentions and objectives of river management policies. Although the ES approach is not reducible to ES economic valuation, it provides a useful framework to structure the economic valuation of the benefits of river protection or restoration.

From an economic perspective, rivers can be considered assets that provide flows of diverse goods and services to people. On the other hand, actions undertaken by humans to satisfy their evolving needs have impacted the rivers' state and functioning and their capacity to sustain these service flows over time (Finlayson et al. 2005).

Valuation can be defined "as the process of assessing the contribution of a particular object or action to meeting a particular goal" (Liu et al. 2010). The neoclassical economic theory, which supports most ES economic valuation exercises, is rooted in an anthropocentric and utilitarian approach. In this approach, the value of goods and services, including ES, derives from the utility they provide to humans, either directly or indirectly, and depends on individual preferences (MEA 2003; Pearce et al. 2006). When deciding about a course of action, individuals are supposed to be rational: they are able to order all the

River Restoration: Political, Social, and Economic Perspectives, First Edition. Edited by Bertrand Morandi, Marylise Cottet, and Hervé Piégay.

possible states of nature according to their preferences and choose among them the state that maximizes their utility (Vatn 2004; Chevassus-au-Louis et al. 2009). The approach also assumes that all goods and services, including ES, can be substituted to each other in terms of the utility they provide. Consequently, values attributed by individuals to goods and services can be measured, in monetary terms, by their willingness to pay (WTP) to benefit from a good, a service, or their willingness to accept (WTA) a compensation for a decrease in their utility. Preferences are assumed to be strictly individual, and collective preferences are conceptualized as the sum of all the individual preferences. According to these principles, the economic value of ES provided by a river can be derived from the observation of the choices made by individuals about their use of river services, and from the costs endured to benefit from these services.

The approach to economic valuation of the benefits of river restoration presented in this chapter is based on the neoclassical approach, as it is the most commonly used. Since the 1990s, ecological economists have challenged utility as the sole principle of individual and collective choices, and therefore proposed other valuation approaches, funded on a broader and less anthropocentric conception of values. For example, in the case of river restoration, an alternative approach would be to assess the costs of maintaining a certain state of river functioning, regardless of its contribution to human well-being (Levrel et al. 2012). These approaches are discussed in Section 15.2.3.1.

In the neoclassical approach, economic valuation of the benefits of river restoration projects aims at providing a common metric to compare flows of goods and services provided by rivers with and without restoration. It consists of (i) defining the values attached to rivers by individuals and the society as a whole, (ii) assessing how these values are modified by restoration projects, and (iii) choosing and implementing the most appropriate methods to measure these value changes. It should be noted that, although economic values are frequently expressed in monetary terms, they can also be measured using other units (e.g. quantities of water used, number of users, level of satisfaction with the state of the river, etc.).

The reasons to undertake such an evaluation are diverse (MEA 2003; Amigues and Chevassus-au-Louis 2011; Gómez-Baggethun and Barton 2013; Kumar 2010). Laurans et al. (2013) have broadly categorized them into three types:

- to inform and raise awareness of the general public, stakeholders, and decision-makers about the importance of services provided by rivers and justify river conservation or restoration measures;
- to support decision-makers in choosing the most appropriate restoration options to implement in specific contexts (e.g. selection of restoration measures, river stretches to be restored); and
- to design incentives to guide the behavior of river users or instruments to secure financial streams for river restoration, or to calculate monetary compensations for damages incurred by river users due to certain developments (e.g. flood damages, water pollution).

While the economic valuation of benefits can contribute to economic analyses for decision-support purpose (the second type of use, e.g. cost–benefit analysis), it should not be confused with them (Hérivaux and Gauthey 2018).

If the costs of river restoration projects are usually easy to calculate based on accounting measures of expenses of past projects, assessing and valuing the benefits from river restoration is a much more difficult task (Brouwer and Sheremet 2017). In the neoclassical framework, the economic valuation of benefits associated with river restoration presents four main difficulties, linked to the characteristics of the goods and services provided by rivers. First, as most ES are not traded on markets (e.g. bathing), their value (i.e. the WTP/WTA of individuals) cannot be measured from market prices. Economists have developed monetary valuation methods, described in Section 15.2.3.1, to overcome this difficulty. Second, people are not necessarily aware of some of the services provided by rivers (e.g. water purification, flood protection). It is therefore impossible for them to assign a value to it. Third, impacts of restoration projects on ES may occur at different times and extend throughout a long period after the restoration intervention. Valuation of their benefits thus implies choosing the relevant time horizon to consider, and the appropriate discount rate[1] to compare benefits at different times, two issues highly debated by economists (Pearce et al. 2006; Arrow et al. 2014). Finally, goods and services provided by rivers result from complex interactions between ecosystems and socioeconomic systems, hence the need to adopt a multidisciplinary and participatory approach that combines a diversity of knowledge to characterize the object of the valuation process.

In the present chapter, we describe and discuss the main steps to assess the economic value of the benefits associated with river restoration projects, using the neoclassical economic framework.

15.2   Main phases of a valuation study and points of attention

We distinguish four main phases in the economic valuation of changes in river ES associated with river restoration, each of them being composed of several steps (Figure 15.1).

15.2.1   Phase 1: Defining the objectives of the valuation study

The definition of valuation objectives (among the three types of use described in the introduction) is essential. Indeed these objectives will influence the elements of the rivers considered in the analysis (e.g. value of all the ES supplied by rivers in a given watershed, or value of a specific function of the river ecosystem), the population concerned by the river changes, and the choice of valuation methods (Bouscasse et al. 2011), as well as the needed degree of accuracy of value estimates (Pearce et al. 2006).

Objectives and type of information provided by economic analysis vary with the geographical scale of the project (river basin, network, or reach) and the phase of the river management cycle, at which it occurs (Table 15.1).

As river restoration project affects many river users, it is of particular importance to involve them in the definition of, or at least to make clear, the objectives of the valuation exercise. After setting the general objectives of the valuation exercise based on its timing within the decision process, objectives can be refined taking into account the analysis of the river system.

Table 15.1 Objectives and information provided by the economic valuation of river restoration benefits in relation to the phases of river management process (Amigues and Chevassus-au-Louis 2011; Salvetti 2013; de Groot et al. 2006; Loubier 2015). *Source:* Sylvie Morardet.

River management process phase	Objectives of the economic analysis	Information provided by economic valuation of benefits	Spatial and time scales
Problem definition and objectives	Raising awareness of the general public about environmental benefits of restored ecosystems	List and quantity of present goods and services provided by rivers, number and characteristics of beneficiaries Identification of potential goods and services permitted by river restoration	All scales
Identification of appropriate measures	Assessing the contribution of present and restored rivers to population well-being and local/regional economy Legal requirements (WFD)	Economic value of direct and indirect uses of the river in the present situation and under restoration	Large river basin, medium to small catchment
Ex ante assessment of the plan	Convincing decision-makers to adopt the plan Protecting some categories of river users, helping resolving conflicts among river users and stakeholders Identifying additional measures to ease plan implementation	Detailed expected benefits of the proposed restoration plan: e.g. cost of artificial infrastructures to prevent or mitigate natural hazards related to degraded rivers; benefits of restored rivers (Spatial) distribution of benefits across stakeholder groups	Catchment, restoration sites
River management plan implementation	Justifying and securing public funds for restoration project Designing economic incentives Informing litigation processes	WTP beneficiaries for restored river ecosystem services (e.g. improved water quality, avoided soil erosion, and flood, enhanced cultural amenities) WTA compensation for foregone benefits of economic agents adversely affected by river restoration (e.g. landowners, farmers) Influence of river governance aspects on WTP/WTA Defining compensation amount	All scales
Monitoring and ex post evaluation	Assessing the economic impacts of the plan/project Comparing multiple projects	Real benefits at implementation and in the long run	Mid-term, long-term

WFD: Water Framework Directive.

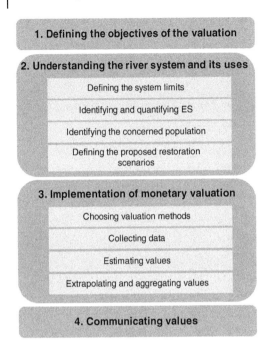

Figure 15.1 Main phases of the economic valuation process (Bouscasse et al. 2011; Brouwer et al. 2009; Grizzetti et al. 2016). *Source:* Sylvie Morardet.

15.2.2 Phase 2: Understanding the river system and its uses

Assessing the economic value of river ES with and without restoration requires a thorough understanding of the river functioning and its uses by the society (Laurans and Cattan 2000; Bouscasse et al. 2011; Grizzetti et al. 2016), which calls for the collaboration of economists with biophysical scientists. This phase of particular importance for the relevance, accuracy, and reliability of economic valuation results comprises four steps (Figure 15.1):

15.2.2.1 Defining the river system limits and identifying the ecosystem services provided without restoration

The very first step of phase 2 of Figure 15.1 consists of specifying the limits and the scope of the ecosystem under study, which may range from a limited stretch of the river, a network, a whole catchment, or even several catchments in a larger basin. Once the ecosystem is delineated, the next step is to understand its biophysical functioning and to identify the ES it supplies.

A popular conceptual framework, known as "the ecosystem services cascade" has been proposed under The Economics of Ecosystems and Biodiversity (TEEB) initiative, which links the biophysical structures and processes of ecosystems to their functions, the ES, the benefits provided to society, and finally their values (de Groot et al. 2010; Haines-Young and Potschin 2010). Before being able to value the costs and benefits of river restoration programs, one first needs to consider how aquatic ecosystems are functioning and what services they provide to society in a particular setting, as well as how aquatic ecosystems respond to the various pressures of human activities (e.g. alterations of water quantity and quality, changes in the physical morphology of the river and its biological components) (Grizzetti et al. 2016).

Based on the classification of ES proposed by the MEA or further refined by the Common International Classification of Ecosystem Services (CICES), several authors have tailored a list of ES adapted to economic valuation of freshwater ecosystems (e.g. Finlayson et al. 2005; Brander et al. 2006; Brouwer et al. 2009; Amigues and Chevassus-au-Louis 2011; Grizzetti et al. 2016; Bergstrom and Loomis 2017) (Table 15.2).

From an economic perspective, several issues have to be considered in the identification of ES:

- The distinction between the final ES (directly useful for human populations) and the intermediate services (incorporated into ecological processes contributing to other services): to avoid double counting in the valuation process, only the final services should be considered (Fisher et al. 2009).
- The spatial relationships between the area where ES are generated by ecosystems and the area where people benefit from them (Brouwer et al. 2009; Fisher et al. 2009).
- The variations of ES provision over time: from an economic perspective, the reference situation is not the present state of the river but its likely state without restoration, at a given time horizon. This is because both ecosystem functioning and social preferences for ES evolve over time.
- The synergies and trade-offs between ES: for example, increasing water supply for irrigation above a certain level may have a detrimental effect on the water purification service. Conversely, improving river flows can have a positive impact on both water purification (a regulating service) and recreational fishing (a cultural service).
- Services are produced through complex interactions between biophysical and socioeconomic systems, and human-made capital is sometimes necessary to benefit from the service (e.g. infrastructures to access a recreational site along a river). It is therefore sometimes difficult to precisely identify the contribution of the ecosystem to the benefit perceived by people.
- The distinction between the capacity of an ecosystem to provide ES and the real use of ES by people (Schröter et al. 2014; Villamagna et al. 2013): for example, the magnitude of the flood protection service of a wetland not only depends on its capacity to store river flows but also on the size of the population living downstream and being protected.
- River ecosystems are interconnected with terrestrial ecosystems and it might be important to consider these interactions for the provision of water-related services (e.g. the role of forest in hydrological processes at catchment level) (Grizzetti et al. 2016).
- To assist in understanding the functioning of the river systems, Grizzetti et al. (2016) propose a framework linking human interventions and river uses (drivers of changes) to aquatic ecosystems' states and finally to ES provision. They also list several approaches and tools that can help the biophysical assessment of rivers (e.g. Burkhard et al. 2009; Bagstad et al. 2013). To ease further steps of economic valuation these authors also suggest using ES indicators for water-related ES. Because of limitations of the scientific knowledge regarding river functioning, it is not always possible to identify and quantify all the ES provided by rivers, nor is it possible to account for all interactions between ecosystem components, ES, and human uses, which results in uncertainties regarding economic values.

Table 15.2   List of ecosystem services provided by rivers (Finlayson et al. 2005; Maresca et al. 2011; Kumar 2010; Amigues and Chevassus-au-Louis 2011; Grizzetti et al. 2016). *Source:* Sylvie Morardet.

Provisioning
Aquaculture and professional fishing
Extraction/exploitation of mineral products
Fibers and other materials (wicker, rush, peat)
Water withdrawal for domestic use
Water withdrawals for agricultural use
Water withdrawals for industrial use
Raw (biotic) materials
Water use for energy production (abiotic)
River transport (abiotic)
Regulating
Floods and flood prevention
Drought mitigation
Prevention of geomorphological disorders
Water purification and waste treatment
Regulation of parasites and pathogens
Regulation of harmful and invasive species
Local climate regulation
Maintaining population and habitats
Air quality regulation
Social and cultural
Recreational fishing and hunting
Freshwater sports
Other freshwater recreation and tourism
Landscape (aesthetic appreciation)
Art (source of inspiration)
Subject matter for research and education
Spiritual and symbolic appreciation

15.2.2.2   Identifying the population concerned by river restoration

One of the most difficult steps of a valuation study is to characterize the population concerned by the valuation exercise (Grizzetti et al. 2016; Katossky and Marical 2011) in terms of size, location and sociodemographic features. There is no robust and widely recognized method to do it. The geographical entity from which the concerned population is originating is sometimes referred to as the "economic jurisdiction" (e.g. Bateman et al. 2006). Theoretically, it should include all the individuals whose well-being is impacted, positively or negatively, by the restoration. In practice, number and location of beneficiaries of river services may vary with the ES to be valued.

Each individual may benefit from the river in several ways (e.g. as a consumer of drinking water, as a practitioner of recreational activities, or as an inhabitant protected from flood), and preferences between these services may vary across individuals. Indeed, economists usually disaggregate the total economic value of ecosystems into direct use, indirect use, and nonuse values. Direct use values derive from direct interactions with the ecosystems and their services (e.g. fish consumption, recreational activities). Indirect use values are the benefits supplied by regulating services (e.g. the self-purification capacity of rivers). Nonuse values include existence value (satisfaction of knowing that a good or service exists, regardless of how and by whom it is used) and heritage value (satisfaction of knowing that future generations will be able to benefit from it) (MEA 2003; Pascual and Muradian 2010). These values reflect the altruistic and ethical nature of individual preferences. Typically, provisioning, regulating services, and some cultural ES are associated with use values, while other cultural services are related to nonuse values (Chevassus-au-Louis et al. 2009).

It is therefore important to consider both users and nonusers of the river ES, as the preferences of these two groups over restoration scenarios may vary considerably (Bateman et al. 2006; Jørgensen et al. 2013; Hanley et al. 2003). This clearly raises the question of how values related to various ES can be aggregated over different subpopulations.

From an operational point of view, it is recommended to identify the relevant population for each ES separately. This is all the more important when the proposed restoration differentially affects the bundle of services supplied by the river, thus favoring certain beneficiaries over others.

Identifying the relevant population is of particular importance when valuation methods are based on survey (see Section 15.2.3.1) as it will influence the sample size ("How many people should be surveyed?"), the sampling strategy ("Where should the survey be undertaken?" "How to stratify the sample?"), and the rules for aggregating survey results (see Section 15.2.3.4).

For most ES, the population living in the river watershed is probably a reasonable basis for determining the reference population. However, for some specific ES, the reference population can be larger (e.g. river features attracting recreationists from outside the watershed) or smaller (e.g. share of the population protected against floods by a restored river stretch). For most ES, the distance from the river is an important variable to consider. Economists (Hanley et al. 2003; Bateman et al. 2006) have used the notion of distance decay to refer to the negative influence of distance on WTP to benefit from an environmental good or service, especially in the case of recreational ES.

Available sources of information regarding the concerned population include: data from population census (number of inhabitants, population density, sociodemographic characteristics), geographic information on the location of habitats and their distance from watercourses, database on water withdrawals per sector compiled by water management institutions, information from water supply utilities on their water sources and users, database of enterprises by sector, survey of attendance at riverside recreation sites, etc. Spatial analysis tools provided by geographic information systems can be very useful to manage and analyze these various sources of information.

A general rule is to be transparent about the hypotheses adopted for identifying the population concerned by the restoration. When possible, sensitivity analysis is recommended to assess the impact of uncertainty regarding the benefiting population on valuation estimates. Finally, the analyst will tailor hypotheses about the relevant population, depending on the objective of the valuation exercise, and thus the expected degree of accuracy of value estimates.

15.2.2.3 Defining the proposed restoration scenarios and the subsequent changes in ecosystem services provision

Once the present functioning of the river ecosystem is described, it is necessary to define the proposed restoration scenarios (length and localization of the river stretch, type of restoration, time and duration of restoration works, etc.) and their goals (restoration of habitats, water quality improvements, etc.).

- What will be the changes in rivers following the restoration? And when will they occur?
- What will be the effects on ES and benefits for river users? And when?

This step is particularly challenging, as a change in one feature of the river may affect positively or negatively several ES, and one specific ES may depend on several river ecological features, as demonstrated by Keeler et al. (2012) for services related to water quality. Therefore, the analysis requires a close collaboration between scientists and practitioners from biophysical and social sciences to identify the linkages between ecosystem state and ES delivery, considering synergies and trade-offs among ES.

The changes in ecosystems expected from restoration need to be defined in terms of quantity, quality, localization, and timing (see Brauman et al. 2007 for services related to hydrological functioning). Both spatial and temporal scales are important to consider because the effects of ecological processes vary with space and time, but also because river beneficiaries are sensitive to these dimensions. In England and Wales, Metcalfe et al. (2012) show that residents value more local improvements in river quality than the same changes at national level. Using the same valuation framework, Glenk et al. (2011) found that residents in Scotland preferred short-term improvements in river and lake quality than long-term ones.

To value the impacts on ES of biophysical changes expected from restoration, it is necessary to quantify these changes, using metrics that are relevant with respect to the various river uses (e.g. population size of fish species, length of river with public access for recreational uses, length of river with water quality suitable for various uses). Ringold et al. (2013) used a comprehensive list of river beneficiaries. Focusing on the ways each category of beneficiary interacts with the river, they developed ES indicators that make sense both from the point of view of ecological functioning and from the perspective of beneficiaries (Table 15.3).

The assessment of benefits requires also additional information on the abundance or scarcity of the services provided by the river to be restored (Ringold et al. 2013), or the existence, in the vicinity, of substitutes for the various uses (Jørgensen et al. 2013).

In addition to biophysical characteristics, it is also important to specify some governance and financing aspects of the restoration scenarios: what are the property rights regarding the river uses? Which organization will be in charge of implementing the project? How will it be funded and what are the possible channels for collecting river users' financial contributions? Indeed, the influence of these elements on river beneficiaries' WTP has been demonstrated in some cases (e.g. Chaikaew et al. 2017, on the role of implementing agency).

Table 15.3 Example of biophysical characteristics of interest and ES indicators for two types of river beneficiaries (Ringold et al. 2013). *Source:* Sylvie Morardet.

Beneficiaries	Ecosystem services	Relevant biophysical characteristics of the river	Complementary goods (necessary to enjoy the ES) and information (to assess the benefits)	ES indicators
Catch-and-release anglers	Recreational fishing (cultural service)	Size and abundance of recreational fishes Aesthetic features of the fishing sites Usability (including accessibility of the fishing sites; characteristics of the streambed) Water safety (e.g. chemicals, pathogens, and parasites)	Infrastructures to access the site Distance to fishing sites Local abundance of recreational fishing sites Fishing regulations and crowding at the sites	Proportion of stream length with a given fishing quality and road access
Crop irrigators	Withdrawals of water for irrigation (provisioning service)	Quantity and temporal variability of water: daily average flow and standard deviation for each day during the irrigation season Probability of flooding and flood levels during the cropping season Quality of water and suitability for irrigation (salinity, sediments)	Presence of farmland at a given distance of the river site	Stream length in salinity safety classes according to stream size and presence of farmland

It is not always possible to quantify all the ES impacted by restoration project because some services, such as regulating and cultural services, are difficult to capture, not measurable, or subject to many uncertainties (Feuillette et al. 2015). Furthermore, the existing budgets for valuation studies do not always allow for the quantification and valuation of all ES, in particular when the size of the restoration project is small or very large (Bergstrom and Loomis 2017). This often leads to include in the valuation exercise only the ES that ES beneficiaries find most important.

15.2.3 Phase 3: Implementation of monetary valuation methods

15.2.3.1 Diversity of valuation methods

A wide range of methods is available to value the benefits derived from ecosystem changes due to river restoration in monetary terms. These methods fall in three broad categories:

The first category gathers all the methods used for "primary" valuation exercises. It includes four types of methods:

- Price-based methods: when real transactions exist in direct relation with ES provision, values of ES changes can be derived from individuals' behavior on these markets.
- Cost-based methods use costs incurred in the absence of the ES as a proxy of ES values. Three situations are considered: replacing the ES by an artificial infrastructure (replacement costs), protecting people against damages due to the loss of ES (preventive expenditure or mitigation costs), or suffering the damages associated with changes in ES provision (avoided damage costs).
- Revealed preferences methods are based on the observation of individual choices in actual transactions of a marketable good, which includes environmental attributes. Beneficiaries' preferences relative to these attributes are "revealed" by their behavior on this associated market.
- Stated preferences methods simulate a market for ES by means of surveys on hypothetical changes in the provision of ES. Scenarios are proposed to the survey respondents that describe the proposed ecosystem changes (including time and location of provision), the institutional setting in which the changes would occur, and the way these changes would be financed. The respondents are then asked to explicitly or implicitly state their willingness to pay for a positive change or their willingness to accept a compensation for a negative change. These methods assume that people answer as if they were in a real market.

More details can be found on this first category of methods in the literature (Pearce et al. 2006; Pascual and Muradian 2010). Table 15.4 gives an overview of specific methods pertaining to these four types, summarizing their main advantages and limits and specifying the type of ES that can be valued.

The second category of methods, generally referred to as "benefit transfer," is used when time and budget do not allow undertaking primary ecological and economic studies. It consists of using ES values estimated from previous studies (study sites) to value services provided by ecosystems in a new site (policy site). Care should be taken to choose study sites sufficiently similar to the policy site under consideration in terms of ecological and socioeconomic characteristics and to adjust values to reflect important differences between sites and beneficiary populations. Several methods of transfer are available. The most sophisticated method, the meta-analysis, consists of using multiple primary valuation studies to build a statistical function relating the estimated values to the characteristics of the study sites, and applying this function to the characteristics of the policy site (Brouwer and Sheremet 2017). Pascual and Muradian (2010) reference several databases of existing valuation studies that can be used to undertake benefit transfer.

The third category, the mixed valuation methods, has emerged more recently following criticisms addressed to the valuation methods described above, based on the neoclassical economic framework (Raymond et al. 2014). Three main assumptions of this framework are targeted: (i) all values are commensurable using a same monetary metric; (ii) societal preferences can be adequately represented by aggregating values obtained from isolated individuals; and (iii) individuals have predefined static preferences for all existing goods (Martín-López et al. 2009). These assumptions and the associated valuation methods fail to take into account the multiple dimensions of values that people held for the environment

Table 15.4 Valuation methods for ecosystems services (de Groot et al. 2006; Grizzetti et al. 2015; Grizzetti et al. 2016; Chevassus-au-Louis et al. 2009; Pascual and Muradian 2010). *Source:* Sylvie Morardet.

	Valuation method	Description	Advantages	Limits	Type of ES
Direct market valuation (can only be used when market exist for goods related to ES)	Market price	Uses the existing prices for goods and services traded on markets	Data on prices and quantities of environmental resources used are relatively easy to obtain Prices reflect the actual preferences of individuals for goods and services that are traded	Market prices may not reflect the economic value of ES for society as a whole in case of market imperfections (e.g. subsidies) The approach is only valid if the activity using the environmental resource is sustainable	Provisioning and some cultural services (e.g. commercial fisheries)
	Factor income or production function	Estimates the value of an ES from its effects on the output of an economic activity	Widely used to estimate the impacts of ecosystems on productive activities such as farming or fishing Reliable statistical and economic techniques	Important amount of data needed Requires the modeling of the cause–effect relationships between the environmental resource and the economic output More difficult in the case of multiple-use systems Risk of double counting because of interdependencies between ES	Provisioning ES some Regulating ES (e.g. water use for irrigation; impact of water quality improvements on commercial fishery catches)

(Continued)

Table 15.4 (Continued)

	Valuation method	Description	Advantages	Limits	Type of ES
Cost-based valuation (assumption that the cost related to the maintenance of the environmental benefit is a reasonable estimate of its value)	Avoided damage cost	Costs that would have been incurred in the absence of the ES	Application of the precautionary principle	Requires the construction of damage functions Data intensive	Regulating ES (e.g. flood protection)
	Replacement cost	Costs of replacing the ES by human-made systems	Useful when ecological data are not available Less time consuming than valuing the benefits	Difficult to identically replace the ecosystem function with an artificial substitute (risk of underestimation of benefits) Do not consider individual and social preferences for natural/artificial way of supplying the service Cost of substitute is only an approximated measure of the benefit	Regulating ES (e.g. groundwater recharge: costs of obtaining water from another source)
	Mitigation or restoration cost	Cost of moderating effects of lost functions (or of their restoration)	Useful when preventive or mitigation technologies exist	Not always applicable, because of diminishing returns and difficulty of restoring previous ecosystem conditions	Regulating ES (e.g. flood protection: costs of preventive expenditures)

Revealed preferences (based on observation of individual choices in existing markets related to ES)				
Travel cost	Survey-based technique that derives WTP for environmental benefits at a specific site from the cost and time of traveling (and associated costs) to this site	Widely used to estimate the value of recreational sites in developed countries Results can be easily interpreted Rely on observed behavior	Restrictive assumptions about consumer behavior Results sensitive to statistical methods used ES values may be overestimated if visitors also traveled for other reasons Travel cost is only a lower bound of the economic value (e.g. for local visitors) Large data sets needed, complex statistical analysis Expensive and time consuming	Cultural ES (e.g. recreation use of rivers: swimming, boating, nature viewing, etc.)
Hedonic pricing	Value of an environmental amenity is obtained from property (labor) market prices	Allow to estimate separately the value of several nonmarket attributes (e.g. distance from the ecosystem, quality of the ecosystem) Rely on observed behavior	Environmental benefits should be of common knowledge to be reflected in property prices Information about environmental conditions of properties is limited Property prices may be explained by factors subject to bias (e.g. taxes, interest rates) Large datasets needed, complex statistical analysis Expensive and time consuming	Cultural ES (e.g. impact of the presence of water on property values)

(Continued)

Table 15.4 (Continued)

	Valuation method	Description	Advantages	Limits	Type of ES
Stated preferences (based on individual choices in hypothetical markets related to ES)	Contingent valuation	Survey-based technique that asks people how much they would be willing to pay (or accept a monetary compensation) for a change in ES provision	Allow to measure value of nonmarketed services Allow to measure nonuse values Could reveal potential conflicts among stakeholders	Potential differences between stated behavior and real behavior Time consuming and expensive Application to complex and unfamiliar ES questionable (important role of information provided to respondents) Results sensitive to numerous sources of bias in survey design and implementation Limited number of proposed scenarios	All ES (e.g. recreational uses of rivers, water quality improvements, biodiversity, flood protection, consumptive uses of water, aesthetic service)
	Choice experiment	Survey-based technique in which people are asked to choose among different hypothetical alternatives (each alternative consists of a combination of attributes of an ecosystem and a price associated with this combination)	Allow to measure value of nonmarketed services Allow to measure nonuse values Could reveal potential conflicts among stakeholders Provides value estimates for changes in specific attributes of an environmental resource Less biases than contingent valuation Several scenarios of change can be proposed	Potential differences between stated behavior and real behavior Time consuming and expensive Application to complex and unfamiliar ES questionable (important role of information provided to respondents) Requires specific expertise for questionnaire design and statistical analysis Cognitive complexity of choices (only a limited number of attributes can be considered)	All ES (e.g. recreational uses of rivers, water quality improvements, biodiversity, flood protection, consumptive uses of water, aesthetic service)

(Spash 2008; Wegner and Pascual 2011). To overcome such limitations, the concepts of shared or social values have been proposed to designate nonutilitarian dimensions of ecosystem values that are articulated in the frame of social processes (Kenter et al. 2015). Mixed valuation methods of environmental goods combine qualitative and quantitative methods. These methods consider "not just how much ES are worth but also what they mean to people" (Kenter 2016). In particular, assuming that preferences over environmental goods and services may be formed through a process of social exchange, rather than be predefined (Vatn 2004), deliberative valuation methods are proposed (Kenter et al. 2014; Spash 2007). During the process, participants, involved as a formal group or citizen juries, are invited to argue in an opening way about their perceptions of ES, and express the social WTP to maintain benefits delivered by the ecosystems to society as a whole.

15.2.3.2 Criteria for choosing a valuation method

River managers and decision-makers may be confused by this diversity of valuation methods. Most of time, the choice of the relevant method depends on several criteria.

The first one is the type of ES to be valued. "Some valuation methods are more appropriate for valuing particular ES than others. Regulation services have been mainly valued through avoided cost, replacement and restoration costs, or contingent valuation; cultural services through travel cost (recreation, tourism or science), hedonic pricing (aesthetic information), or contingent valuation (spiritual benefits –i.e. existence value-); and provisioning services through methods based on the production approach" (Martín-López et al. 2009). Table 15.4 provides examples of ES that have been valued using different valuation methods.

A second important criterion is the type of values provided by the rivers, as only stated preference methods can be used to estimate nonuse values.

The types of data required to value the benefits derived from changes in ES differ with the valuation method (Table 15.5). Therefore, economists in charge of evaluating a river restoration project will select the method(s) to be used depending on the availability of data on the study site. Direct market valuation and cost-based methods are usually less data intensive than the other methods.

Similarly, some methods, such as revealed and stated preference methods, require a specific economic expertise to design the survey and undertake the statistical analysis. This expertise may not be readily available in the organizations in charge of river restoration projects, which leads them to commission external experts. Consequently, stated preference methods are usually more time consuming and expensive than other valuation methods.

When the scope of the valuation exercise is quite large (i.e. large basin, country), it is often not realistic to undertake primary valuation studies for all ES in every location. In this case, and more generally when time and budget constraints are high, benefit transfer might be an appropriate method (Grizzetti et al. 2016).

The choice of a valuation method also depends on the required reliability and accuracy of value estimates. These are higher when the valuation study aims at estimating compensatory damages in a litigation procedure than when it is used to inform actual policy decisions, or only to scope the value of some policy options (Pearce et al. 2006). Stated preference methods and benefit transfer are generally considered the least reliable methods because of the uncertainties associated with the estimation techniques (Chevassus-au-Louis et al. 2009).

Table 15.5 Data requirements for various valuation methods. *Source:* Adapted from Grizzetti, B., Lanzanova, D., Liquete, C., et al. (2015). *Cook-Book for Water Ecosystem Service Assessment and Valuation.* Luxembourg: European Commission, Joint Research Centre.

Valuation method	Examples of ES provided by rivers	Data requirements
Market price	Fish supply	Price of the environmental good (fish) on the wholesale market
		Demand for the good (can be approximated by current good extractions/used amounts)
		Production cost of the industry extracting the good (fisheries)
Production function	Use of water for irrigation	Quantity and cost of production factors (including water)
		Level of production in relation with the use of the environmental good
		Market price of the produced good
Replacement costs	Water purification	Quantity of water purified by the river ecosystem
		Population benefitting from the clean water
		Cost of providing clean water with an alternative built infrastructure or from an alternative water source
Contingent valuation / Choice modeling	Recreational uses (bathing, boating, etc.)	Physical and ecological characteristics of the river ecosystem
		Scenario of change of the river ecosystem (e.g. change in the water quality or ecological status)
		(Declared) individual willingness to pay for the service (contingent valuation) or choices made by the participants during the survey (choice modeling)
		Socioeconomic characteristics of the respondents
		Socioeconomic characteristic of the population benefitting from the river
Travel costs	Recreational uses (bathing, boating, etc.)	Visitors' travel costs (including the value of time spent traveling)
		Other travel expenses (e.g. accommodation, specific equipment)
		Visitors' socioeconomic characteristics
		Distance from visitors' hometown to the visited river site
		Other locations visited during the trip
		Distance of the site from substitute rivers
		Biophysical and ecological characteristics of the river ecosystem
Hedonic pricing	River related amenities (e.g. views, microclimate, etc.)	Data on property sales (price, property characteristics, including location)
		Data on the river ecosystem (size, quality, ecological status)
		Size of the beneficiary population

15.2.3.3   Collecting data

Obviously, data collection methods are closely linked to the chosen valuation methods (Table 15.5). For methods based on surveys (travel costs, stated preferences methods), questionnaire development is usually a critical step, given the different biases that can result from the way questions are formulated, and therefore requires a specific expertise. The determination of the size and characteristics of the sample, as well as the targeted population, derives from the identification of the population benefiting from the restoration project (see Section 15.2.2.2). The questionnaire may be administered face-to-face, by mail, by telephone, or more recently by Internet (Fleming and Bowden 2009). The budget and time allocated to the survey, the number of people to be interviewed and their spatial distribution, and the complexity of the scenarios subjected to valuation must be considered when choosing the survey administration mode (Bouscasse et al. 2011). Brouwer and Sheremet (2017) note significant differences in WTP estimates depending on survey mode.

For methods based on secondary data (costs-based methods, hedonic pricing, benefit transfer), the main difficulty is to identify and access to the relevant studies. Fortunately, several initiatives to develop databases for environmental valuation studies have been launched such as Environmental Valuation Reference Inventory (EVRI), EnValue, or the Review of Externality Data of the European Commission.

15.2.3.4   Estimating, extrapolating, and aggregating values

In the case of valuation methods using individual data (revealed or stated preferences), the statistical (or econometric) analysis results in the estimation of the value of a change in the provision of a given ES for an average individual in the beneficiary population or, in some cases, for different groups within that population. It also allows studying how this value varies according to certain characteristics of the river or of its beneficiaries.

To be able to assess the value of the total benefits associated with an improvement of ES provision by a restored river, two operations are necessary (Figure 15.2):

- to extrapolate individual values for a given ES to the share of the population benefiting from it;
- to aggregate the values of changes in several ES for the whole beneficiary population.

The extrapolation entails several difficulties (Morrison 2000): (i) as underlined in section 15.2.2.2, defining the extent of the beneficiary population is not an easy task, (ii) preferences among the population might be heterogeneous, and (iii) some discrepancies may occur between the preferences of the interviewed sample and those of the whole population.

Similarly, the aggregation of several ES values to assess total value of changes due to river restoration also requires special precautions. First, it is important to identify the functional linkages between the different river ES and the uses they allow to avoid double counting (the same ecological function can participate in the provision of several ES; conversely, the provision of a given ES may require several functions). Second, it might be necessary to use different methods, each with its own pros and cons, to value the same ES, because they do not capture exactly the same value components. Finally, the same individual may benefit from more than one ES provided by a restored river. Therefore, the values estimated for each category of beneficiaries cannot be simply added without due consideration. Bouscasse et al. (2011) propose a process to aggregate ES values, estimated through various methods.

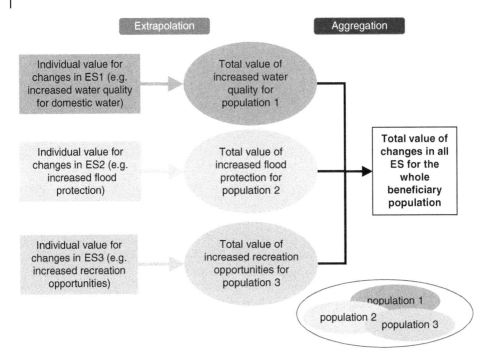

Figure 15.2 Extrapolation and aggregation of ES individual values. *Source:* Sylvie Morardet.

15.2.4 Phase 4: Communicating values to the public

Communicating the results of valuation studies to the public, stakeholders, and decision-makers is a crucial step of the analysis (ACTeon 2013), but is often neglected. According to Amigues and Chevassus-au-Louis (2011), more than the lack of reliability of valuation methods, the main reason environmental valuations are underused in the decision-making process is the difficulty to translate their results into operational conclusions for river managers.

Beyond their operational use by river managers, information provided by valuation studies also contributes to the appropriation of the ecological stakes of environmental protection by elected officials and citizens (Amigues and Chevassus-au-Louis 2011). This requires limiting the use of technical vocabulary, adapting the messages to the different targets, while making explicit the assumptions underlying the analysis, and the uncertainties affecting the results.

As underlined by many academics and professionals, economic valuation of ES should be seen as part of a general process of deliberation about programs of actions and policies. In this respect, it must be transparent about the economic issues underlying the river restoration projects, in particular about the distribution of their costs and benefits among the present and future river users and beneficiaries (Laurans and Cattan 2000).

While communicating about uncertainties is important, Feuillette et al. (2015) caution against the strategic use of economic valuation results that could be made by some stakeholders to influence the decision. Referring to the experience of valuation studies carried out for the implementation of the European Union's Water Framework Directive, these authors cite the example of industrialists, impacted by projects to restore the good ecological status of water bodies on which they are located, who have requested a valuation study in order to prove that the benefits of the planned restoration were not commensurate with their costs.

15.3 Conclusions

In a context of high costs of river restoration measures, economic valuation of their benefits aims to raise awareness of the public, stakeholders, and decision-makers, to support decision-makers in their policy and management choices, or to provide information to design policy instruments. In this chapter, we proposed to adopt an approach based on ES to value the benefits associated with river restoration projects, in particular to take into account the nonmarketed values of rivers. The most challenging steps of the valuation process consist in (i) identifying ES provided by restored and unrestored rivers, (ii) identifying the population concerned by the restoration project, (iii) characterizing changes in ES provision with metrics relevant both from the point of view of ecosystem functioning and of river beneficiaries, (iv) choosing the most appropriate valuation methods in relation to the context of the river restoration project, and (v) communicating the results of valuation exercises to the public and stakeholders and discussing them.

Several types of limitations affect the valuation process, leading to uncertainties regarding the values of ES provided by restored rivers. The first type of limitation pertains to technical limits associated with the neoclassical framework (e.g. difficulty to identify the concerned population and its spatial distribution, or multiple biases associated with the various valuation methods). Further research is thus needed to improve the robustness of valuation methods and reliability of results. A second category of limits originates from the complexity and connectivity of river ecosystems and the lack of scientific knowledge about it. A multidisciplinary scientific effort should be undertaken to better understand the relationships between the ecological processes underlying the provision of ES and the benefits derived by individuals and society as a whole.

The third group of limitations concerns the integration of the results of economic valuation of ES associated with river restoration into the decision-making process: because of their uncertainties, these results should be considered as only one element among others to inform the decision-making process about river restoration. Similarly, caution should be exercised against the strategic use of evaluations by some stakeholders.

The last, but not the least, set of limitations stems from ethical and philosophical concerns about the neoclassical framework (e.g. anthropocentric and utilitarian conceptions of value) focus on individual self-regarding agents, independent of the social context (Vatn 2004; Laurans and Maris 2017; Kallis et al. 2013; Parks and Gowdy 2013). In particular some academics and practitioners argue for the elaboration of transdisciplinary evaluation frameworks that explicitly state the objectives and uncertainties of the evaluation, identify the different types of values attached to rivers (e.g. shared values, social values, cultural values), and incorporate them into the public debate.

Note

1 The discount rate is the rate at which an individual is ready to exchange a quantity of goods or services at present with the same quantity in the future.

References

ACTeon (2013). Guide pratique pour la mise en œuvre d'analyses socio-économiques en appui à l'élaboration de SAGE et contrats de rivière. Agences de l'Eau, Ministère de l'Ecologie, du Développement durable et de l'Energie.

Amigues, J.-P. and Chevassus-au-Louis, B. (2011). *Evaluer les services écologiques des milieux aquatiques: enjeux scientifiques, politiques et opérationnels*. Vincennes, France: ONEMA.

Arrow, K.J., Cropper, M.L., Gollier, C., et al. (2014). Should governments use a declining discount rate in project analysis? *Review of Environmental Economics and Policy* 8(2):145–163.

Bagstad, K.J., Semmens, D.J., Waage, S., et al. (2013). A comparative assessment of decision-support tools for ecosystem services quantification and valuation. *Ecosystem Services* 5: 27–39.

Bateman, I.J., Day, B.H., Georgiou, S., et al. (2006). The aggregation of environmental benefit values: welfare measures, distance decay and total WTP. *Ecological Economics* 60(2): 450–460.

Bergstrom, J.C. and Loomis, J.B. (2017). Economic valuation of river restoration: an analysis of the valuation literature and its uses in decision-making. *Water Resources and Economics* 17: 9–19.

Bouscasse, H., Defrance, P., Duprez, C., et al. (2011). Evaluation économique des services rendus par les zones humides: enseignements méthodologiques de monétarisation. Commissariat Général du Développement Durable, Ministère de l'Ecologie, de l'Energie, du Développement durable et de la Mer.

Brander, L.M., Florax, R.J.G.M., and Vermaat, J.E. (2006). The empirics of wetland valuation: a comprehensive summary and a meta-analysis of the literature. *Environmental and Resource Economics* 33(2): 223–250.

Brauman, K.A., Daily, G.C., Duarte, T.K., et al. (2007). The nature and value of ecosystem services: an overview highlighting hydrologic services. *Annual Review of Environment and Resources* 32: 67–98.

Brouwer, R., Barton, D., Bateman, I., et al. (2009). Economic valuation of environmental and resource costs and benefits in the Water Framework Directive: technical guidelines for practitioners. Aquamoney European Research Project Report, Institute for Environmental Studies, VU University Amsterdam, The Netherlands.

Brouwer, R. and Sheremet, O. (2017). The economic value of river restoration. *Water Resources and Economics* 17(C): 1–8.

Burkhard, B., Kroll, F., Müller, F., et al. (2009). Landscapes' capacities to provide ecosystem services: a concept for land-cover based assessments. *Landscape Online* 15: 1–22.

Chaikaew, P., Hodges, A.W., and Grunwald, S. (2017). Estimating the value of ecosystem services in a mixed-use watershed: a choice experiment approach. *Ecosystem Services* 23: 228–237.

Chevassus-au-Louis, B., Salles, J.-M., Bielsa, S., et al. (2009). Approche économique de la biodiversité et des services liés aux écosystèmes: mission présidée par Bernard Chevassus-au Louis. Centre d'Analyse Stratégique, Rapports et Documents No. 18.

de Groot, R., Stuip, M., Finlayson, M., et al. (2006). Valuing wetlands: guidance for valuing the benefits derived from wetland ecosystem services. Ramsar Technical Report No. 3. *CBD Technical Series No. 27.*

de Groot, R., Fisher, B., Christie, M., et al. (2010). Integrating the ecological and economic dimensions in biodiversity and ecosystem service valuation. In: *The Economics of Ecosystems and Biodiversity: The Ecological and Economic Foundations* (ed. P. Kumar), 9–40. London: Earthscan.

Feuillette, S., Levrel, H., Blanquart, S., et al. (2015). Évaluation monétaire des services écosystémiques: un exemple d'usage dans la mise en place d'une politique de l'eau en France. *Natures Sciences Sociétés* 23(1): 14–27.

Finlayson, C.M., Cruz, R.D., Davidson, N., et al. (2005). *Millennium Ecosystem Assessment: Ecosystems and Human Well-Being: Wetlands and Water Synthesis.* Washington DC: World Resources Institute.

Fisher, B., Turner, R.K., and Morling, P. (2009). Defining and classifying ecosystem services for decision making. *Ecological Economics* 68(3): 643–653.

Fleming, C.M. and Bowden, M. (2009). Web-based surveys as an alternative to traditional mail methods. *Journal of Environmental Management* 90(1): 284–292.

Glenk, K., Lago, M., and Moran, D. (2011). Public preferences for water quality improvements: implications for the implementation of the EC Water Framework Directive in Scotland. *Water Policy* 13(5): 645–662.

Gómez-Baggethun, E. and Barton, D.N. (2013). Classifying and valuing ecosystem services for urban planning. *Ecological Economics* 86: 235–245.

Grizzetti, B., Lanzanova, D., Liquete, C., et al. (2015). *Cook-Book for Water Ecosystem Service Assessment and Valuation.* Luxembourg: European Commission – Joint Research Centre.

Grizzetti, B., Lanzanova, D., Liquete, C., et al. (2016). Assessing water ecosystem services for water resource management. *Environmental Science and Policy* 61: 194–203.

Haines-Young, R., and Potschin, M. (2010). The links between biodiversity, ecosystem services and human well-being. In: *Ecosystem Ecology: A New Synthesis* (eds D.G. Raffaelli and C.L. Frid), 110–139. Cambridge: Cambridge University Press.

Hanley, N.D., Schläpfer, F., and Spurgeon, J. (2003). Aggregating the benefits of environmental improvements: distance-decay functions for use and non-use values. *Journal of Environmental Management* 68(3): 297–304.

Hérivaux, C. and Gauthey, J. (2018). *Les bénéfices liés à la protection des eaux souterraines: pourquoi et comment leur donner une valeur monétaire?* Vincennes, France: Agence Française pour la Biodiversité.

Jørgensen, S.L., Olsen, S.B., Ladenburg, J., et al. (2013). Spatially induced disparities in users' and non-users' WTP for water quality improvements: testing the effect of multiple substitutes and distance decay. *Ecological Economics* 92: 58–66.

Kallis, G., Gómez-Baggethun, E., and Zografos, C. (2013). To value or not to value? That is not the question. *Ecological Economics* 94: 97–105.

Katossky, A. and Marical, F. (2011). Evaluation économique des services rendus par les zones humides: complémentarité des méthodes de monétarisation. Commissariat Général du Développement Durable, Ministère de l'Ecologie, de l'Energie, du Développement durable et de la Mer.

Keeler, B.L., Polasky, S., Brauman, K.A., et al. (2012). Linking water quality and well-being for improved assessment and valuation of ecosystem services. *Proceedings of the National Academy of Sciences of the United States of America* 109(45): 18619–18624.

Kenter, J.O. (2016). Deliberative and non-monetary valuation. In: *Routledge Handbook of Ecosystem Services* (eds M. Potschin, R. Haines-Young, R. Fish, et al.), 271–288. London: Routledge.

Kenter, J.O., O'Brien, L., Hockley, N., et al. (2015). What are shared and social values of ecosystems? *Ecological Economics* 111: 86–99.

Kenter, J., Reed, M.S., Irvine, K., et al. (2014). *UK National Ecosystem Assessment Follow-on. Work Package 6: Shared, Plural and Cultural Values of Ecosystems.* UNEP-WCMC LWEC, UK.

Kumar, P. (ed.) (2010). *The Economics of Ecosystems and Biodiversity: Ecological and Economic Foundations.* London: Earthscan.

Laurans, Y. and Cattan, A. (2000). Une économie au service du débat: l'évaluation économique des services rendus par les zones humides. In: *Fonctions et valeurs des zones humides* (eds E. Fustec and J.C. Lefeuvre), 295–309. Paris, France: Dunod.

Laurans, Y. and Maris, V. (2017). Radiologues plutôt qu'opticiens: la biodiversité bouscule le rôle des économistes. *Humanité et Biodiversité* 3 (Regard critique N°9).

Laurans, Y., Rankovic, A., Billé, R., et al. (2013). Use of ecosystem services economic valuation for decision making: questioning a literature blindspot. *Journal of Environmental Management* 119: 208–219.

Levrel, H., Hay, J. Bas, A., et al. (2012). Coût d'opportunité versus coût du maintien des potentialités écologiques: deux indicateurs économiques pour mesurer les coûts de l'érosion de la biodiversité. *Natures Sciences Sociétés* 20(1): 16–29.

Liu, S., Costanza, R., Farber, S., et al. (2010). Valuing ecosystem services: theory, practice, and the need for a transdisciplinary synthesis. *Annals of the New York Academy of Sciences* 1185: 54–78.

Loubier, S. (2015). Appui méthodologique pour l'évaluation économique des projets territoriaux. Rapport d'étude. Agence de l'Eau Rhône-Méditerranée Corse, Irstea, UMR G-eau, Montpellier, France.

Maresca, B., Mordret, X., Ughetto, A.L., and Blancher, P. (2011). Évaluation des services rendus par les écosystèmes en France. *Développement durable et territoires* 2(3).

Martín-López, B., Gómez-Baggethun, E., González, J.A., et al. (2009). The assessment of ecosystem services provided by biodiversity: re-thinking concepts and research needs. In: *Handbook of Nature Conservation* (ed. J.B. Aronoff), 261–282. Hauppauge, NY: Nova Science Publisher.

MEA (Millennium Ecosystem Assessment) (2003). *Ecosystems and Human Well-Being: A Framework for Assessment.* Washington DC: World Resources Institute.

MEA (Millennium Ecosystem Assessment) (2005). *Ecosystems and Human Well-Being: Synthesis.* Washington DC: World Resources Institute.

Metcalfe, P.J., Baker, W., Andrews, K., et al. (2012). An assessment of the nonmarket benefits of the Water Framework Directive for households in England and Wales. *Water Resources Research* 48(3): W03526.

Morrison, M. (2000). Aggregation biases in stated preference studies. *Australian Economic Papers* 39(2): 215–230.

Parks, S. and Gowdy, J. (2013). What have economists learned about valuing nature? A review essay. *Ecosystem Services* 3: e1–e10.

Pascual, U. and Muradian, R.C. (2010). The economics of valuing ecosystem services and biodiversity. In: *The Economics of Ecosystems and Biodiversity: The Ecological and Economic Foundations* (ed. P. Kumar), 183–256. London: Earthscan.

Pearce, D., Atkinson, G., and Mourato, S. (2006). *Cost–Benefit Analysis and the Environment: Recent Developments*. Paris: OECD Publishing.

Raymond, C.M., Kenter, J.O., Plieninger, T., et al. (2014). Comparing instrumental and deliberative paradigms underpinning the assessment of social values for cultural ecosystem services. *Ecological Economics* 107: 145–156.

Ringold, P.L., Boyd, J., Landers, D., et al. (2013). What data should we collect? A framework for identifying indicators of ecosystem contributions to human well-being. *Frontiers in Ecology and the Environment* 11(2): 98–105.

Salvetti, M. (2013). *Les évaluations économiques en appui à la gestion de l'eau et des milieux aquatiques*. Vincennes, France: ONEMA.

Schröter, M., Barton, D.N., Remme, R.P., et al. (2014). Accounting for capacity and flow of ecosystem services: a conceptual model and a case study for Telemark, Norway. *Ecological Indicators* 36: 539–551.

Spash, C.L. (2007). Deliberative monetary valuation (DMV): issues in combining economic and political processes to value environmental change. *Ecological Economics* 63(4): 690–699.

Spash, C.L. (2008). Deliberative monetary valuation and the evidence for a new value theory. *Land Economics* 84(3): 469–488.

Vatn, A. (2004). Environmental valuation and rationality. *Land Economics* 80(1): 1–18.

Villamagna, A.M., Angermeier, P.L., and Bennett, E.M. (2013). Capacity, pressure, demand, and flow: a conceptual framework for analyzing ecosystem service provision and delivery. *Ecological Complexity* 15: 114–121.

Wegner, G. and Pascual, U. (2011). Cost–benefit analysis in the context of ecosystem services for human well-being: a multidisciplinary critique. *Global Environmental Change* 21(2): 492–504.

Part VI

Conclusions

16

Social, Economic, and Political Stakes of River Restoration: A Dynamic Research Field Facing Several Challenges to Strengthen Links with Practitioners

Marylise Cottet, Bertrand Morandi, and Hervé Piégay

Université de Lyon, CNRS, ENS de Lyon, Environnement Ville Société, Lyon, France

16.1 Humanities and social sciences now fully engaged within the field of restoration

The place of humanities and social sciences (HSS) within the restoration sciences no longer has to be justified. Approaches are increasingly interdisciplinary, and calls to develop work on the societal issues of restoration have been widely relayed and often heard. In 2000, in the conclusion to their book *Restoring Nature*, Gobster and Hull write that "Contributions from the humanities and social sciences are needed to help decide restoration goals, to justify them in a competitive social context, and ultimately to plan, implement, and maintain desired states of nature" (Gobster and Hull 2000, p. 299). Twenty years after the publication of this reference work, HSS research in the field of restoration has multiplied and diversified, particularly with respect to rivers. If this growing involvement is always to be placed into perspective with regard to the overall scientific production in this field, it is important to highlight the work carried out and the results achieved. It is in this spirit that the present book was conceived. The various contributions show how HSS researchers, through their analytical and critical work, contribute to the strengthening, and sometimes to the reorientation, but always to the improvement of river restoration practices and policies. These inputs thus fully support the adaptive approaches advocated for river restoration. They provide a space for research and reflection in which the social, political, and economic questions raised by river restoration are linked to those relating to ecology, biology, hydrology, and geomorphology.

This chapter provides a synopsis of the contributions structuring the different parts of the book. It also highlights the operational perspectives opened up by this work. These operational perspectives are supported by several original reports written by managers of restoration initiatives in Malaysia and the United States. The issues covered are also compared with the methodological approaches described in Part 5 of this book. We therefore propose to address four main points in this conclusion:

We will see, on the basis of the chapters in Part 2 of the book, how new thinking in environmental ethics makes it possible to think about, or even rethink, the links between societies and rivers in the context of restoration. The application of political ecology approaches to restoration highlights the political stakes of these links and the socio-economic power relationships that underlie them. 2) We then return to the perspectives opened up by the chapters in Part 3, in terms of the analyses of socio-political processes for improving governance approaches. Critical considerations of participation are particularly represented. 3) From the contributions in Part 4, we will focus on the progress made in assessing the effects – positive or negative – of restoration projects. This research is based on rich conceptual frameworks for the valuation of ecosystem services, of the "total economic value" or the measurement of "sense of place". 4) Finally, we will see how interdisciplinary approaches and the links between HSS and restoration practitioners can be strengthened with a view to improving river restoration practices and policies.

16.2 Analysis of people–river relationships: from ethics to politics

The importance that discussions on the definition of river restoration objectives have assumed in recent decades can be seen as a signal of the integration – gradual but real – of the debate on human and social values in the conception of restoration approaches. The definition proposed by Davis and Slobodkin – "Ecological Restoration is the process of restoring one or more valued processes or attributes of a landscape" (2004, p. 2) – places environmental valuation processes at the center of the remediation process. Within the sphere of river restoration, schematic concepts evolve between positions oriented by the question "What rivers did we have?" and more anthropocentric positions guided by the question "What rivers do we want?" Whatever the position defended within this debate, it is certain that the concept of restoration, which implies a modification of the environment as much as of the relationship to the environment, makes considerations of environmental ethics particularly fertile and concrete. According to Palmer et al. (2014, p. 420), "Environmental ethics is the study of ethical questions raised by human relationships with the nonhuman environment," and specifically, "Ethical questions are those about what we ought to do." These latter points have been actively debated in the field of restoration from the 1980s to the present day, particularly because of the growing importance of the concept of restoration in environmental management policies.

The reinterpretations of these ethical debates proposed by Henry Dicks (Chapter 2) and Dan Hikuroa et al. (Chapter 3), particularly demonstrate how the idea of separation between humans and nature, long discussed by the social sciences, has long structured discussions in the field of restoration. Henry Dicks reminds us that the passionate discussions between "conservationist" and "restorationist" have opposed, in the North American context, the notion of artificiality to a naturality rooted in the myth of wilderness. However, as Jamie Linton reminds us (Chapter 4), there is not a single way of thinking about restoration, but diversified paths that are based on different world visions, "intellectual presuppositions," "cultural dispositions," and multiple "material interests." These are some of the alternative paths that the authors of this part open up to us, particularly by their incorporation of non-Western philosophies in their thinking. The work proposed by Dan Hikuroa et al. highlights the perspectives opened up by Maori cosmology, which considers humans as an integral part of nature in a kin-based relationship with all the constituent elements of the universe. This more organic vision of the world is also encouraged by Henry Dicks, who mobilizes the concept of biomimicry to conceive a restoration ethic based on the imitation of nature, and not on its domination, thus approaching certain non-Western philosophies such as Daoism. As the concept of restoration is becoming increasingly globalized, these new perspectives are particularly exciting. They re-establish in the geographical context, and particularly in the sociocultural context, a central place for consideration of the values of river restoration. In addition to the diversity of these values, which Dan Hikuroa et al. highlight with the concept of river pluralism, it is also their evolution over time that needs to be considered. Consideration of the fact that environmental ethics are plural and evolving appears to be a prerequisite for improving the implementation of river restoration.

If the ideas proposed in this part 2 allow us to nurture the debates on environmental ethics, the different authors also give us an insight into the political implications of these debates. For Dan Hikuroa et al., the ethical question "To what ideals should [a] river be restored" is immediately followed by the question "Who should make these decisions?" which is very political. The historical perspective proposed by Dan Hikuroa et al. shows how management policies integrated the Maori concepts of the relationship with nature, the most symbolic example being the recognition of the rights of the Whanganui River. If "the ethical claims are prescriptive rather than descriptive or predictive" (Palmer et al. 2014, p. 420), the effectiveness of the ethical regulation is particularly dependent on its political implementation. Therefore, when the diversity of restoration values is taken into account, power and dominance struggles, and consequently tensions and even conflicts, will arise. The analysis and highlighting of power relationships linked to differences in values are, as Jamie Linton shows very well through the example of the controversies linked to the restoration of the ecological continuity of rivers, central to the work of political ecology. The posing of environmental ethics questions requires thought about how companies and individuals organize themselves to define environmental problems, and the restoration measures to be implemented to solve them. In the different chapters, the authors clearly show that the considerations of environmental ethics in the field of river restoration open up questions of political philosophy whose scales of discussion far exceed those of the environmental restoration project.

By placing themselves at a higher scale than that of the projects, are these discussions disconnected from the action? What is their operational scope? Above all, the ethical considerations contribute to a state of open-mindedness. Placing river restoration into perspective with other systems of thought – whether they are geographically, historically, or culturally anchored, or purely theoretical – contributes to a step back from the daily practices in which practitioners are engaged. By showing that elsewhere, in other contexts, people think differently about the relationships between societies and rivers and attribute other significance to rivers and their restoration, ethical considerations show that alternatives exist. They reveal that river restoration, as it is defined and implemented in a given time and place, is one action among others, underpinned by values.

This off-beat view, while inviting us to take a step back from restoration practices, raises questions about the transferability of such ideas. Can the forms of restoration inspired by non-Western philosophies, as described by Henry Dicks and Dan Hikuroa et al., be transferred to other contexts and serve as a model for restoration policies such as those pursued in the West? Considering the strong cultural anchoring of public policies, nothing is less certain, at least in the short and medium term. However, these discussions provoke the imagination; they are a great asset for activating innovation, to imagine a different tomorrow. Thus, ethical considerations are truly operational if they penetrate the spheres of politicians and stakeholders who convey a certain number of ideas to the political authorities (economic actors, NGOs, etc.). At the time of writing this conclusion (October 2020), a Swiss public interest association (id-eau, imagination durable pour l'eau douce – sustainable thinking for freshwater) is publishing, at 21 000 km from New Zealand and the Whanganui River, "*L'Appel du Rhône*" (the call of the Rhône) for the recognition of the legal personality of the Rhône river. It is, by their own description, "a collective grassroots mobilization, based on what other countries in the world have already done." Ethical considerations and their ensuing policies therefore stimulate debate, even if they are not directly transferred as they stand.

Of course, it is sometimes difficult for the personnel involved in the restoration sector to take ownership of these ideas, because they may have to "make do" with the current policy and respond to its demands within a limited timeframe. It is difficult because the ideas are not always the most accessible: they are based on a language and a theoretical background that is sometimes complex and requires time to assimilate. Further difficulty also comes from the critical positioning that underlies these ideas. Thinking differently about restoration inevitably leads to questioning what has been achieved in favor of river quality, often in small steps, often with difficulty, thanks to a great deal of commitment and determination on the part of those involved in restoration. Such ethical considerations may not represent the consensus thought. They lead us to take a critical look at the actions undertaken, and can expose or even destabilize certain parties. The political game described by Jamie Linton shows the extent to which environmental action is based on precarious balances of power struggles and compromises. It is understandable that some actors involved in the implementation of river restoration policy are reluctant to engage in thinking that would call into question the work so far accomplished. And even if it is always enriching for managers to take a step back from their own practices in order to better understand them – and subsequently accept or adapt them – we believe that these ethical considerations are also of particular benefit to political players, who, through their debates and decision-making power, are creating tomorrow's restoration policy.

16.3   Understanding of governance and power relationships between stakeholders

The major surveys carried out in the field of restoration in the early 2000s show that many countries have undertaken ambitious policies to restore good ecological quality to rivers. At an international level, projects – and consequently feedback – have multiplied. Most of the time, this feedback is produced by researchers in the natural sciences and responds to scientific and technical issues. They are particularly concerned with the restoration measures carried out and their effects on the ecology or hydromorphology of rivers; fewer works question the social and political issues raised by the implementation of projects. As well as questioning the effects of restoration, it is also interesting to question the way in which the actions that produce these effects are implemented. The chapters that make up Part 3 are all based on the premise that "ecological restoration also needs to be understood not only as a technical task but as deeply embedded in social and political processes," to quote the words of Baker et al. (2014, p. 518). The analysis of these processes today constitutes a research space occupied by social and political science researchers, with a particular focus on identifying ways to improve restoration policies.

The various contributions collected in Part 3 first allow us to contextualize river restoration practices and to identify the sociopolitical factors that can influence their implementation. The notion of interdependencies proposed by Carter et al. (Chapter 5) is particularly fertile, as it enables the restoration process to be linked to political, territorial, and cognitive processes that influence it, but also go beyond it. The interaction of restoration policies with other public policies, the inclusion of the project in the more integrated development of the territory concerned by the project, and the introduction of local knowledge in the definition of objectives are all processes whose analysis allows lessons to be drawn for the implementation of restoration. Catherine Carré et al. (Chapter 6) also emphasize the importance of the national context of projects by showing how restoration practices have evolved over time and differ between countries. The authors identify not only the regulatory framework but also public funding and modes of governance as influential factors in the definition of projects that ultimately determine the sociopolitical success or failure of restoration approaches. Within governance, particular importance is given by the authors to the local development of action plans and the search for consensus before the project is carried out. The management feedback provided by WWF-Malaysia for a river restoration project on the island of Borneo provides a good illustration of the value of co-constructing projects with local stakeholders (Box 16.1).

In all chapters, governance – even if not always explicitly named – is a central issue. According to a general definition, governance is "the range of political, institutional, and administrative rules, practices and processes (formal and informal) through which decisions are taken and implemented, stakeholders can articulate their interests and have their concerns considered, and decision-makers are held accountable for water management" (OECD 2015, in Woodhouse and Muller 2017, p. 226). The contributions forming this book pay particular attention to the stakeholders involved in the governance of restoration. Who participates in defining the problems and defining the solutions? Who ultimately makes the decision? Caitriona Carter et al. insist on the fact that, in addition to the diversity of interests and society-river relationships, tensions around restoration projects are often linked to the distribution of decision-making powers among the stakeholders. The various contributions clearly show that participation – and in particular public participation – is today an issue structuring thinking in the field of restoration, and more broadly concerning the environment in general. On the basis of a solid analysis of dam removal operations in France and the USA, Marie-Anne Germaine et al. (Chapter 7) show how the involvement

---

Box 16.1   Social conditions and influences on riverbank restoration in the Long Semadoh Highland areas of the Heart of Borneo, Malaysia.

The Heart of Borneo (HoB) is an area of importance for conservation and sustainable development. It hosts the largest remaining connected forests in Borneo and serves as a "water tower" with many of the headwaters of the island's main rivers located within the HoB.

WWF-Malaysia[1] works with the Lun Bawang communities in Long Semadoh, a village cluster located in the Upper Trusan catchment, to promote sustainable agricultural practices for forest and freshwater conservation within the HoB. Almost all of the traditional economic activities of the Lun Bawang here are related to rice plantation. Income generated is related to the amount of land available for paddy cultivation. Loss of any paddy field areas could significantly affect the Lun Bawang's livelihoods. Land ownership and its arrangements for use to support livelihoods therefore play a significant role in their strategies to increase income or reduce loss.

The decline in riverbank integrity has partly be due to the community's income strategies. In some areas, riparian vegetation had been cleared to increase the paddy field areas. The removal of the riparian vegetation reduced the structural integrity of the riverbank and made it more susceptible to erosion. A strategy had been to move rocks and sediments in the river to divert flows away from the paddy fields. This changed the river flows temporarily but affected stretches further down the river due to the changed flow shifting its erosion and deposition sites to get back to its equilibrium state. This had hastened the worsening of the riverbank erosion conditions along the Upper Trusan river stretch.

The communities, through the Alliance of the Indigenous People of the Highlands of Borneo (FORMADAT)[2] had approached WWF-Malaysia in 2017 for assistance to address erosion which is causing them to lose valuable land for paddy cultivation. WWF partnered with the University of Nottingham Malaysia (UNM)[3], who had the hydromorphological expertise, to assist in a study to determine the causes of erosion and to recommend ways to restore affected riverbanks. The aim is to enable riverbanks to be restored and strengthened over time through bioengineering measures, reduce further losses of valuable paddy fields to erosion, enhance knowledge and capacity of the local communities to understand erosion causes, and apply greener options for flow and erosion management, and to improve overall catchment management for better long-term resilience of both the land and the livelihoods of the people.

The sense of urgency to reduce land loss had influence on the community's motivation and willingness to cooperate with any party that can assist them in any way they can. Taking a more natural or bioengineering approach to address their erosion issues was not a high priority of theirs, compared to just stopping the erosion any way they could. It took many consultations, and the sharing of study results and examples of practices in other parts of the world, before communities in the upper sections of the river agreed to give it a try.

Throughout the river restoration project, from the inception of the hydrological study in 2017 to the implementation of recommended measures in 2019, many consultations had been done with the communities and installations had been jointly done. This included a community perception survey conducted during the hydrological study phase to ascertain the perception of the communities toward causes of erosion in their area. Many had the perception that erosion was mainly the cause of logging activities upstream, which cause the river flow changes. The sharing of the hydrological study results with the communities took this perception into account, being careful to also demonstrate through the science that on-site activities, such as riparian clearing and buffalo trampling at the degraded riverbanks, contributed to the problem and to present solutions that took all responsible parties into account, along with the message that there was a need to work together to resolve the problems.

Following the study, four sites along the affected riverbanks were restored with bioengineering measures. The installations themselves were again influenced by the consultations and the livelihood activities of the communities apart from being weather dependent. Installations of the structures had to revolve around the time the communities were available, usually after their paddy planting in July and before harvesting time in January. This had a direct impact on the effectiveness of the installations. Each installation needed to be complemented with the planting of vegetation and building fences to manage buffaloes' movements into the riparian area. At many of the sites, only the bioengineered structures were able to be completed in time and this resulted in failure of the structures to withstand high flow events when it occurred. This had been a lesson learned for all parties. Designs and installation processes are now being improved to accommodate this important time constraint.

One of the measures to reduce erosion risks is to reduce further clearing of vegetated areas and limit the expansion of roads that contribute to clearing of areas. In July 2019, a new road was cleared to link paddy fields at the uppermost parts of the river to the village. Anecdotal sharing from the community in the village indicated that the road enabled them to transfer heavy bags of harvested paddy from their paddy fields to their home easily with trucks. In the past, they had to walk through the paddy field tracks to bring their harvest home. This had discouraged some owners of land in the upper areas to develop their paddy fields, but with the improved accessibility, there are plans now to develop the abandoned paddy fields. So, while the road may not be environmentally desirable, it is important to the life and livelihoods of the community.

Through the various community dialogues and sharing of experience and knowledge by WWF and UNM, the local communities are eager to learn options to tackle the erosion problem. There are continued interests by the local communities to work together to improve the designs and process, although priorities are still centered on their livelihood generation activities and the preference for a quick solution that requires less time and effort. Implementing river restoration effectively here requires a good understanding of the community's sense of priority and their sense of urgency about the loss of their lands and their livelihoods and the ability to adapt restoration measures to these needs.

Belinda Lip[a], McKenzie Augustine Martin[a], Christopher Gibbins[b] ([a]WWF-Malaysia; [b]School of Environmental and Geographical Sciences, University of Nottingham Malaysia) of often-neglected categories of stakeholders has enriched and improved restoration projects. The question of the inclusion or exclusion of certain categories of stakeholders from participation in processes is also at the heart of the thinking of Nora Buletti et al. (Chapter 8).

The contribution of these different works is first of all reflective. They allow us to permanently re-emphasize the political dimension of the action by always returning to the objectives of the restoration and the actors who define them. Today, participation constitutes an important field of research and often appears as a way to re-politicize environmental action, including the issue of knowledge production, as shown by Carter et al. However, Buletti et al. warn of the risk of the technicalization of participation itself. Acceptance of the project and the permanent search for consensus are nowadays questioned as action-guiding principles. The authors, notably Nora Buletti et al. and Caitriona Carter et al., insist on the importance that divergent points of view, tensions, and even controversies can bring to the renewal of environmental action, both in its form and in the solutions it proposes.

The multiplication of case studies in HSS also makes it possible to cross-reference and compare different ways of organizing governance, and more broadly sociopolitical processes in different contexts. This comparative approach is at the heart of the work of Catherine Carré et al., who are interested in projects conducted in Colombia, Spain, and Canada. Case studies also allow us to recontextualize and question actions. It is, for example, interesting for a manager who engages in participatory approaches in France, a country with a representative democracy and a centralizing tradition, to witness the critical thinking on participation in Switzerland, a federal country with a more direct implication of citizens with public decisions through the federal voting procedure. The work on interdependencies proposed by Caitriona Carter et al. also makes it possible to position current governance trends in the field of restoration within a growing movement of questioning of expertise and public policies, particularly in Western societies.

Demonstrating the importance of interdependencies also allows questioning of the scales of intervention. Restoration projects benefit from other approaches undertaken at the level of the territories in which they are implemented, such as the raising of awareness of local populations and training for management stakeholders. In the same way that the ecological stakes of restoration are considered according to the natural sciences within the theoretical framework of the river system and the watershed, sociopolitical stakes must be interpreted according to what happens within the territories in which the restoration is implemented. As perfectly explained by Caitriona Carter et al., restoration projects are often implemented at the river reach or watershed scale but are linked to other territorial policies conducted at other spatial scales. Reflections on the issue of dam removal without considering the issues of energy transition or tourism development policies may thus seem counterproductive. Beyond the project, policy analysis helps to rethink the coordination and interlinking of governance processes. The governance of the restoration project is not independent of other decision-making bodies at the territorial level.

Finally, the contribution of research lies not only in the results it produces but also in the conduct of the research itself. Indeed, the analysis of governance processes is generally based on interview campaigns that make it possible to question the stakeholders who took part in the action, consequently leading these stakeholders to take a critical look at the action, and also at their role in it. Studies in the HSS create time for reflection and introspection, which operators who are engaged full-time in the action may not otherwise easily find.

16.4   Evaluation of socioeconomic effects of river restoration projects

Work conducted in HSS helps to emphasize the fact that societies are influenced by restoration projects, just as ecosystems are. A restoration project tends to act on the biophysical qualities of a river. As such, it modifies its appearance, its functionality, and the availability of the services that it produces and that are likely to be exploited by society. A restoration project thus results in a modification of the links, established over time, between an ecosystem and the society that depends on it: modification of the living environment, but also of the resources on which it relies to perpetuate itself, modifications that can be perceived as a benefit or a harm. This strong dependence makes it necessary to consider the socioeconomic impacts of restoration projects. The aim is to ensure that projects are socioeconomically viable, and that societies are sufficiently resilient to adapt to the changes brought about. It is also a question of verifying that certain sections of society – by virtue of their practices or geographical location – are not unjustly harmed by the changes brought about. This attention to the common good and justice is what governs public action in democratic systems. From an ethical point of view, it is difficult to accept that ecological restoration, under the guise of pro-environmental action, can deviate from this rule. This is all the truer as it is society that bears the financial cost. In the current context of budgetary restraint, the public is in fact entitled to demand accountability for the budgetary effectiveness of public action, as the overall benefit – from an ecological, but also a social, point of view – must exceed the costs of the action. The experience of the managers of California's Central Valley provides a very good illustration of the environmental, but also the social and political benefits, of considering restoration projects from the perspective of "multiple benefit" projects (Box 16.2).

The contributions in Part 4 are consistent with this perspective. They aim to shed light on the evaluation of the societal benefits of restoration and propose approaches for carrying out such evaluation. All the contributions anchor their thinking at the project level, and consider that the evaluation of social benefits makes sense both before and after project implementation. Evaluation of social benefits before implementation should accompany the decision on whether to implement an action according to its social utility, and can assist in arbitrating between different restoration scenarios. As shown by the work of Bergstrom and Loomis (Chapter 9), pre-project evaluation approaches can be used as a preliminary step in cost-benefit analyses in specific cases where they mobilize economic evaluation methods. Evaluations can also take place after the work is completed, in a follow-up approach such as that developed by Garcia et al. (Chapter 10) or van den Born et al. (Chapter 11). The challenge is then to estimate the social benefit derived from a restoration action and to produce feedback for future actions.

The approaches proposed in Part 4 for valuing the social benefits of restoration can be divided into two, depending on whether they are based on economic or noneconomic valuation methods. The first approach is proposed by Bergstrom et al., who develop a holistic approach to measure the total economic value (TEV) of a restoration project, i.e. "the theoretically appropriate measure of the total benefits of river restoration policy and projects." They suggest the use of a conceptual framework to qualify all the interactions between people and rivers that give

---

Box 16.2   Multiple Benefits of Riverine Ecosystem Restoration in California's Central Valley.

Only 150 years ago, the Central Valley of California was a vast floodplain wetland that supported 525 000 ha of seasonal wetlands, 400 000 ha of riparian forests, and large populations of fish and wildlife. Ninety-seven percent of these wetlands were destroyed by agricultural land conversions and a system of levees and dams constructed in the early to mid-twentieth century to prevent floods and irrigate agricultural fields. Today more than 2000 kilometers of low gradient rivers and waterways are bordered by levees and managed as "regulatory floodways."

Starting in the 1980s, environmental organizations began advocating for the restoration of the region's wetlands and forests. Efforts to restore wetlands on agricultural fields outside of the regulatory floodways for migratory birds, particularly waterfowl, have been very successful, but efforts to restore habitat in the regulatory floodways were strongly resisted for decades by the farmers and engineers charged with maintaining the system who argued that any habitat restoration would imperil public safety and the agricultural economy. A century of laws and policies prioritizing flood protection over all other purposes thwarted all attempts at habitat restoration in the regulatory floodway.

Hurricane Katrina along with large floods in 1997, 2004, and 2006 prompted California voters to approve a $4 billion bond measure to upgrade the levee system and compelled the state legislature to pass the Central Valley Flood Protection Act of 2008 mandating several reforms including a new flood plan for the region. Environmental nongovernmental organizations (NGOs) organized themselves to advocate for habitat restoration for endangered species as part of the new flood plan and associated investments. Initially, flood engineers and farmers framed the issue as public safety vs. habitat for endangered species. Both sides viewed losing the debate as an existential threat to their interests whether they be agriculture, endangered species, or public safety. To break this stalemate, the NGOs worked with media and communication experts, informed by decades of public opinion polls about environmental values, to reframe the issue in a manner that would galvanize public support. At public meetings regarding the new plan they all delivered a consistent message:

"Public safety is, and must be, the number one priority for river management. And the best way to keep communities safe from flooding is to give rivers more room to accommodate the large floods that climate change will bring. Giving rivers more room will provide other benefits, including clean water, riverside parks and trails, and habitat for fish and wildlife."

Although it was a major adjustment for the environmental NGOs to subordinate their emphasis on endangered species protection, the new message was simply an acknowledgment of reality: no politician or government official, and very few people, would openly choose habitat restoration over public safety. But many didn't want to choose between the two, and the solution of expanding floodways gave decision-makers the opportunity to choose both.

The message was effective. The Central Valley Flood Protection Board adopted a new plan that called for expanding floodways, advancing habitat restoration objectives, and prioritizing multibenefit flood management projects.

Although the concept of multibenefit projects was immediately compelling, many of the more serious restoration and flood management practitioners were left wondering how to design and finance the relatively vague concept of a "multibenefit project." Engineers and flood managers knew how to design and pay for levees that could withstand a 100-year flood, but they did not understand what a multibenefit flood management project was, let alone how to design and pay for it.

An interdisciplinary team of engineers, social scientists, and ecologist developed a values based framework around four societal goals, for defining and quantifying multibenefit projects: (i) public health and safety, (ii) ecosystem function, (iii) economic stability, and (iv) enriching experiences to embody all of the quality of life benefit that accrue from nature-based experiences. Under each goal, the interdisciplinary team developed a list of metrics to suggest design criteria for multibenefit projects including ecosystem restoration metrics designed to recover endangered species. Recognizing that the process of planning and adaptive management is more important than the actual plan, the framework leaves open the possibility for practitioners to revise or add metrics that better describe how to advance the four societal goals.

Rather than arguing over a limited pot of funds and geographic space, engineers and restoration practitioners are now collaborating to advance a bigger vision that more taxpayers are interested in financing. The four societal goals enabled both restoration practitioners and flood managers to better explain how the investments in the flood system will yield dividends for everybody, while the metrics helped practitioners and engineers design projects with a broader base of funding and public support. Expanding floodways and restoring ecosystem function is far more expensive than strengthening existing levees, but it yields a healthier, more resilient outcome in the same way that a nutritious and delicious meal delivers better outcomes than cheap, fast food.

Although local flood management agencies were willing to advance multibenefit projects if the state would help fund them, they were very resistant to any unfunded mandates. Lessons from political science and community-based resource management informed design of the governance strategy to garner local support and increase the probability that the plan would actually be implemented. The president of the Flood Board, who had served for decades as a city manager, pushed the state to direct planning funds to regional planning efforts led by local flood management agencies. The state used the plan and associated metrics to describe the type of projects they would pay regions to implement, and let local decision-makers design projects that both satisfied local preference while attracting state and federal funds to advance public goods like carbon sequestration, groundwater recharge, and wildlife habitat.

Since the Central Valley Flood Protection Plan was first adopted in 2012, several multibenefit projects have been implemented or planned for the regulatory floodways of the Central Valley. The once arcane subject of floodways and multibenefit flood management is now a cornerstone of California's water resilience portfolio. Developed to adapt to a changing climate, California executive order N-10-19 calls for utilizing "natural infrastructure such as forests and floodplains" and "prioritizing multi-benefit approaches that meet multiple needs at once." Across the Central Valley, flood

> management engineers and restoration ecologists are now working together to plan and build multibenefit flood management projects along the region's rivers and regulatory floodways.
>
> *John Cain (River Partners)*

rise to economic value and propose methods to measure these values, all based on economic evaluation (standard market valuation techniques or "nonmarket valuation techniques," such as revealed preference methods and stated preference methods). With the Yarqon as a case study, Garcia et al. also use an economic approach and assess the feasibility of a restoration project by comparing its costs and benefits with respect to the provision of cultural services associated with the river. The second approach proposed in these chapters is based on noneconomic valuation methods. It is then a question of qualifying the benefit of a restoration project on the basis of the perception that society (i.e. the individuals who make up the community) has of it. The Caldes stream case described by Garcia et al. and the Wigger case described by van den Born et al. focus on the benefits perceived after the project is completed, while the Waal case, proposed by van den Born et al., puts into perspective the benefits expected before the project and those actually perceived after the project. For this work, the authors use both quantitative and qualitative data produced by questionnaire surveys.

Despite the diversity of approaches, the chapters agree on several points: first of all, the importance given to spatial and temporal issues in the evaluation of the social benefits of restoration. The chapters invite reflection on the definition of an appropriate territorial scale for the evaluation of these benefits. The contours of the territory to be taken into account when assessing the benefits of restoration vary according to the stakeholder being considered, particularly in terms of their relationship to the river and the territory they cross. The various contributions in Part 4 explore discontinuities within the territories concerned by restoration projects, as well as potential sources of inequalities in the benefits derived from the work. Many are the questions that deserve to be asked when discussing a restoration project, including "Which territories benefit from it?", "Which suffer from it?", and "Are the territories that fund these projects the same as those that benefit from them?" With this in mind, the chapters propose methods for setting restoration projects in a more territorialized manner, for example through the care given to sampling in survey protocols, or through the exploration of participatory mapping. These also invite us to think about timeframes when designing the evaluation of restoration projects. Indeed, they underline the existence of different timeframes between the time of the project, the time of the ecological benefits, and the time when these benefits are perceived by the stakeholders; much time may lapse between the implementation of the river restoration project and its evaluation. In addition, there are memory issues that are likely to affect the evaluation process, as pointed out by van den Born et al. For this type of project, it is difficult to define the optimal moment to evaluate the benefits between the time when the benefit is not yet perceived and the time when the benefit is forgotten or faded from memory because too much time has elapsed since the project was carried out.

Finally, all the authors who have dealt with evaluation of the social benefits of restoration underline the difficulty in evaluating the intangible benefits that populations derive from rivers: aesthetic inspiration, cultural identity, a sense of belonging, and spiritual experience, as well as leisure activities linked to rivers. Some of these intangible benefits have already been well studied, while the analysis of others still poses methodological problems. As a result, the restoration projects performed to date have tended to focus on aesthetic or recreational benefits, which are easier to assess, while neglecting other benefits. The chapters propose perspectives for making progress in this evaluation of intangible benefits, either through conceptual tools (sense of place and environmental values in the chapter proposed by van den Born et al.; ecosystem services in the chapters proposed by Bergstrom and Loomis and Garcia et al.) or through survey methods and participatory mapping. Although all the authors agree on the importance of evaluating intangible benefits, this issue seems all the more important for them when the restoration project concerns urban territories, where access to nature is often less convenient for the population. It can thus be noted that almost all of the case studies presented in the chapters of Part 4 relate to urban areas.

16.5   Strengthening collaborations between HSS and restoration stakeholders

While the interest in addressing HSS themes is now recognized in the field of restoration, the operational scope of this research may raise questions. Indeed, although research relating to the social, economic, and political stakes of restoration is most often thought of in relation to action, its conception (and in particular the definition of research objectives) is very often carried out independently of the restoration stakeholders. This can lead to a disconnection between the scientific questions and the questions posed by those involved in restoration. Ultimately, it may be difficult for these same stakeholders to take ownership of the research results. Fundamental research in HSS has shown real utility in terms of science and innovation, and it is not a question here of saying that all research relating to restoration should respond to purely operational objectives. However, it seems important to consider in parallel the modalities of interaction between the stakeholders in the restoration and HSS researchers, in order to co-construct projects. The aim is to establish spaces for exchange to clearly identify the needs raised by the restoration project in terms of knowledge production, and to define which stakeholders – scientists, consulting firms and consultants, managers, and even the public (in the case of participatory science approaches) – are best able to meet them and how. The difficulties encountered (by the authors of this conclusion) in proposing feedback on the experience of effective collaboration between HSS researchers and stakeholders in the restoration sector suggests that such experiences of co-construction of studies are still too rare, and deserve to be developed.

Moreover, it is not always easy for project leaders to mobilize scientists, particularly given the small community of HSS researchers engaged in these issues and the limited financial resources available for such studies. The commitment of researchers also assumes that the restoration project can raise scientific questions that go beyond the simple case study and operational issues, which is not the case in all intervention situations. How can actions carried out in the field of river restoration benefit from advances in research if the scientific communities do not have sufficient resources to become directly involved in projects? HSS researchers often point out the difficulties involved in transferring the results of their research from one restoration project to another. They encourage the specificities of the territories and the societies that live in them to be taken into account, and advocate a systematic adaptation of actions to their specific challenges. In this sense, it seems difficult for managers involved in the definition and implementation of river restoration projects to directly reuse research findings and apply them to new projects being implemented. As the scientific community cannot always invest directly in restoration projects, a question arises as to the transferability of its methods and protocols. The methods for transferring skills, and more broadly for the training of managers involved in the implementation of river restoration projects, are then questioned. Is it conceivable that river restoration practitioners themselves will adopt the protocols proposed by HSS researchers? As far as biophysical monitoring is concerned, there has been such a transfer of competence between scientists and managers in recent years. It is not uncommon today to see managers use indicator sets developed by scientists in ecology or hydromorphology and carry out field monitoring themselves. This appropriation seems more delicate in the case of HSS methods, on the one hand because this transfer of expertise can be poorly compatible with the scientific practices of these disciplines, which often refuse to establish standardized research protocols, and, on the other, because their use sometimes presupposes specific knowledge that is important for the collection, analysis, and interpretation of data, which are often qualitative.

The attention that HSS pay to the specificity of territories and societies leads them to construct questions that are specific to them, and therefore to develop methods that are adapted to each context. In this respect, the more methodological chapters of this book are rich in advice for defining the fields of questioning and the follow-up modalities for projects. Also note the keys proposed by Caroline le Calvez et al. (Chapter 12) for choosing between an interview-based or questionnaire-based survey, depending on the objectives of the study and the state of knowledge of the intervention area. In Chapter 13, Emeline Comby et al. present literature review methods that are rarely used outside of research work, and highlight their interest and complementarity to survey methods in the context of operational considerations of restoration. Chapter 14, proposed by Alba Juárez-Bourke et al., provides methodological solutions to encourage debate during the implementation of restoration approaches. Finally, let us note the various points of attention proposed by Sylvie Morardet (Chapter 15) to define the territory to be taken into account for an economic evaluation of the benefits of a project. This territory can be very different from the section on which the work is carried out. The points of vigilance and the scientific thinking proposed in the chapters can, for example, help managers to define the specifications of studies entrusted to design offices, so that nothing is overlooked during project management.

Even if managers have everything to gain by being trained in the issues and methods of participation (to define more integrated and socially just projects), specialized design or planning offices represent an indispensable link for considering the benefits of river restoration. Endowed with real expertise in the field of human and social sciences, they are at the interface between research and operational action. They are the local interlocutors able to conduct a regionalized analysis at the project level, and able to respond to the questions and operational expectations of managers. In this perspective, scientific work can provide design or planning offices with guidance and inspiration for the construction of study protocols, indicate points of vigilance, and help innovate in the production of regionalized data. Conversely, such offices can be carriers of innovation and provide feedback on the implementation of certain protocols. If such feedback is well documented and made available, research can then take it on board and perform meta-analyses that will allow us to take advantage of the experiences and consider possible generalizations. The co-construction of projects involving managers, design offices, and researchers would make it possible to strengthen public–private links in the production of knowledge. This evolution can be observed at the biophysical level, and would also make sense in the socioeconomic field.

Notes

1 WWF-Malaysia is a national conservation trust affiliated to the WWF Global network. It is a nonprofit organization focused on environmental conservation in Malaysia. There are offices based in Sabah and Sarawak states of Malaysia on the island of Borneo and its headquarters is in the state of Selangor, in the Peninsular Malaysia. The Sarawak Conservation Programme focuses on forests, freshwater, and species conservation, including conservation initiatives within the HoB.
2 FORMADAT is a transboundary, grassroots initiative that aims to increase awareness and understanding of the communities of the Highlands, maintain cultural traditions, build local capacity, and encourage sustainable development in the HoB without risking the degradation of the quality of the social and natural environment.
3 UNM is a university based in Semenyih, Kuala Lumpur, Malaysia. The Project is conducted in collaboration with expertise from Dr. Christopher Gibbins from the schools of Environmental and Geographical Sciences, Dr. Teo Fang Yen from the School of Civil Engineering, and a PhD student, Leon Lip.

References

Baker, S., Eckerberg, K., and Zachrisson, A. (2014). Political science and ecological restoration. *Environmental Politics* 23(3): 509–524.

Davis, M.A. and Slobodkin, L.B. (2004). The science and values of restoration ecology. *Restoration Ecology* 12(1): 1–3.

Gobster, P.H. and Hull, R.B. (eds) (2000). *Restoring Nature: Perspectives from the Social Sciences and Humanities*. Washington DC: Island Press.

Palmer, C., McShane, K., and Sandler, R. (2014). Environmental ethics. *Annual Review of Environment and Resources* 39: 419–442.

Woodhouse, P. and Muller, M. (2017). Water governance: an historical perspective on current debates. *World Development* 92: 225–241.

Index

Please note that page references to Figures will be followed by the letter 'f', to Tables by the letter 't', while references to Notes will contain the letter 'n' following the note number

River Restoration: Political, Social, and Economic Perspectives, First Edition. Edited by Bertrand Morandi,
Marylise Cottet, and Hervé Piégay.
© 2022 John Wiley & Sons Ltd. Published 2022 by John Wiley & Sons Ltd.